Introduction to Probability and Statistics

FOURTH EDITION

B. W. Lindgren
G. W. McElrath
D. A. Berry

School of Statistics / University of Minnesota

INTRODUCTION TO *Probability and Statistics*

Macmillan Publishing Co., Inc.

New York

Collier Macmillan Publishers

London

Printed in the United States of America

Macmillan Publishing Co., Inc.
866 Third Avenue, New York, New York 10022

Collier Macmillan Canada, Ltd.

Library of Congress Cataloging in Publication Data

Lindgren, Bernard William, (date)
Introduction to probability and statistics.

Includes bibliographical references and index.
1. Mathematical statistics. 2. Probabilities.
I. McElrath, G. W., joint author. II. Berry,
D. A., joint author. III. Title.
QA276.L54 1978 519 76-30497
ISBN 0-02-370900-6 (Hardbound)
ISBN 0-02-979430-7 (International Edition)
PRINTING 11 YEAR 789

ISBN 0-02-370900-6

Preface

This book is intended as a text for a course of 30 to 45 hours that presents the basic notions of probability models, important ideas of statistical inference, and some useful statistical methods. Although written for those who do not intend to become professional statisticians, it is not slanted to a particular area of application, for we feel that the essentials taught in such a first course are common to all areas and best learned as part of a general discipline. The material presented provides the student with a good basis for taking up the particular quantitative methods of his major field.

Our aim has been to give, along with formulas and methods, some understanding of their power and proper use. They need not seem like hocus-pocus, when it is understood that they simply systematize and quantify what people already know and do intuitively.

As to mathematical level, we have used the fundamental concepts of integral and derivative in the discussion of continuous models. However, calculus is not used thereafter except in the optional material of Section 7.3 and Appendix A; any reader who grasps the idea of probability as represented by area under a curve, and who is not baffled by mathematical symbolism—functions, formulas, inequalities, absolute values, and the like—will have no difficulty following the development and using the methods.

The material covered is about the same as in earlier editions. However, the text has been reorganized and almost completely rewritten, and the examples and problems are mostly new. A new feature is the use of tables of "random numbers" in sampling experiments, illustrated in examples and called for in problems.

Instructors should find the book flexible and adaptable to courses of various lengths and complexions. Some problems, sections, and chapters are "starred" to alert the instructor to material that is more difficult and/or easily omitted. Most of the later chapters (say 13 through 19) might be considered optional, according to the available time, the needs of the students, and the fancy of the instructor.

It is expected that the student will attempt most of the problems, for it is from his *own* efforts in solving these that he will learn the most. Answers for selected problems are given at the back of the book. Many of the problems are clearly textbook exercises; many involve real data; and still others employ artificial data for real situations. A number of the problems and examples stem from uses of statistics in the student's everyday life as presented to him in the various mass communication media.

Try as we will to prevent it, we may uncover misprints in the early printings of the book. We would hope that anyone about to teach from it would write (to the first author, at the School of Statistics, University of Minnesota, Minneapolis 55455) to obtain an up-to-date list of errata.

B. W. L.
G. W. McE.
D. A. B.

Contents

1 PROBABILITY MODELS

2 COUNTING

3 DISCRETE RANDOM VARIABLES

INDEPENDENCE

5 BERNOULLI POPULATIONS

6 CONTINUOUS RANDOM VARIABLES

★7 THE POISSON PROCESS

8 NORMAL DISTRIBUTIONS

9 SAMPLES

ESTIMATION

★14 SEQUENTIAL TESTING

15 TESTS FOR CATEGORICAL POPULATIONS

16 CORRELATION AND REGRESSION

17 ANALYSIS OF VARIANCE

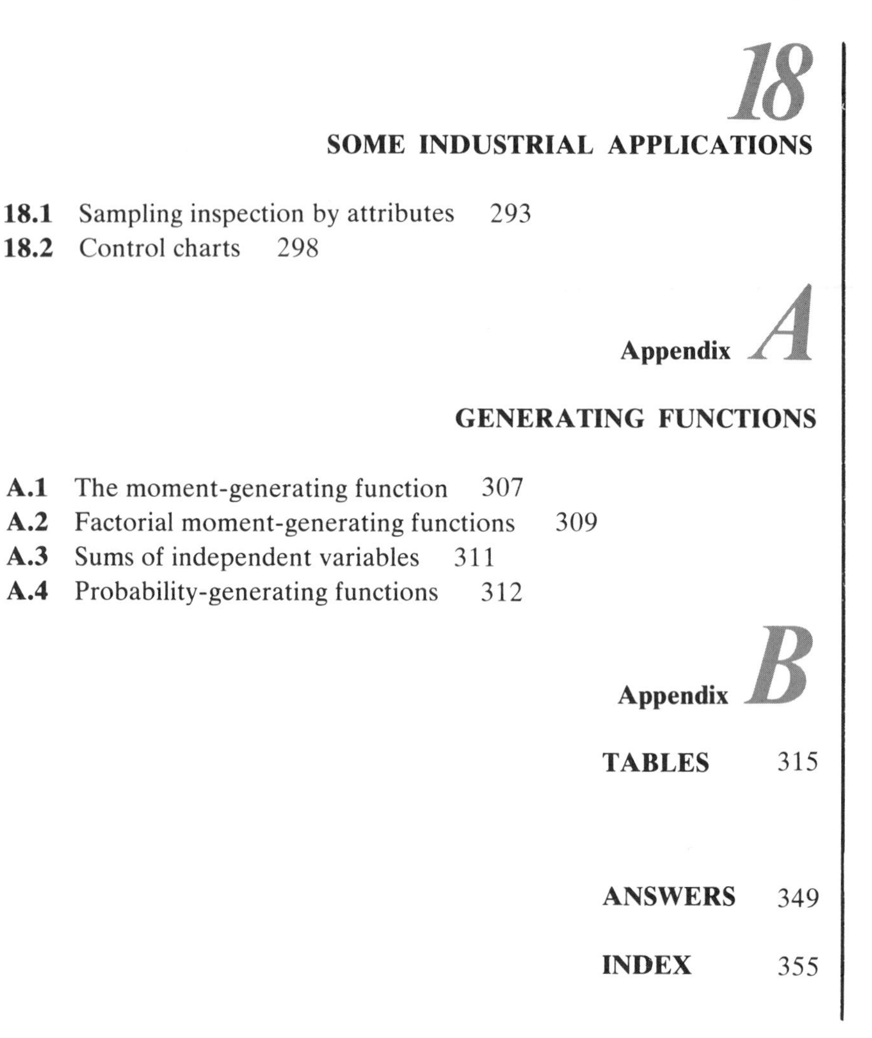

18 SOME INDUSTRIAL APPLICATIONS

Appendix A GENERATING FUNCTIONS

Appendix B

1

Probability Models

In making measurements of various kinds and in sampling populations of people or things, it is usually found that results vary from one observation to another. Because they cannot be perfectly predicted, they are thought to be subject to "chance." Such a measurement or sampling process is said to be an *experiment of chance*.

Even though the outcome of a single trial of an experiment of chance cannot be predicted, experience with sequences of trials has revealed a useful kind of regularity. A "model" or mathematical structure that describes this regularity is of value, and probability theory has evolved to provide models for experiments of chance.

The subject of *statistics* deals with the making of inferences about the correct or true probability model for a given experiment using the results of actual experimentation—carrying out the experiment one or more times to obtain data. Ordinarily, one does not and cannot know the true model in every detail, and this is why he obtains data and uses them as the basis for drawing conclusions about unknown aspects of the true model. Such conclusions may be close enough to the truth for practical purposes.

Thus, our study begins with probability models, proceeds to the gathering and description of data, and then takes up inference concerning the models.

1.1 Mathematical Models

In applying mathematics to any "real" situation, it is essential to construct a model for it. Consider a familiar example. Newton's second law of motion states that an object subject to a force moves with an acceleration proportional to the force: $F = ma$. The model is defined by this equation, which relates the force F, the acceleration a, and the mass m of the object. The *law* states that this mathematical model accurately describes the motion of an object. The law is not provable in the sense of logic, but it can be tested experimentally. Moreover, the basic equation leads mathematically to various consequences that can be checked. For instance, the distance a dropped object falls in a given time can be derived from the basic equation; this consequence in the model can then be checked by dropping objects and measuring their times of fall.

Similarly, geometry postulates mathematical points, lines, circles, and so on; these, together with certain axioms and rules of logic, constitute a model. Theorems of geometry are theorems about this model, *not* theorems about physical objects. However, people study geometry because the model is found to represent rather faithfully what is observed about the physical world. (Indeed, mathematicians even study models whose usefulness may not yet be apparent.)

Many of the models of classical physics and other sciences are *deterministic*. That is, given some initial configuration of elements, the outcome of an experiment is completely and precisely determined therefrom by the equations of the model. But there are many phenomena in which the cause mechanisms are so complex, and the initial configurations are so hard to measure, that any attempt to apply a deterministic model is rather futile. Still, some kind of mathematical description of these phenomena is desirable.

Consider, for example, the familiar experiment of tossing an ordinary die. When tossed onto a flat surface, the die will come to rest with one of its six faces on top. The determinist would say that the outcome of the toss is completely determined by the initial position and velocity and by Newton's laws. But actually to carry out such a determination is quite impractical, if not impossible.

People have been tossing coins and dice for centuries. They have found that when such experiments are repeated many times, any given outcome tends to occur about the same proportion of the time in every sequence of trials. More precisely, in any given sequence of trials, the proportion of trials in which a given outcome occurs tends toward a limiting value as the number of trials is increased without limit. The student would find it very instructive to toss something 100 times (or 500, if he or she has the patience), and after each trial to compute the proportion of trials, up to that point, in which a given outcome is observed. (Tossing a thumbtack would be especially educational, since trials in which it falls with point up and in which it falls with point down are not expected to be equally divided.) A tendency to a limiting value will be noticed. This limiting value will be found to be about the same in one long sequence of tosses as in another, and so it seems reasonable to postulate that there will always exist a limiting proportion that is a property of the object tossed and the way it is tossed—a proportion that could only be found exactly after infinitely made trials. This proportion is called the *probability* of the outcome.

Example 1-a An experiment of chance can be simulated on a computer. A computer was programmed to generate a sequence of 2000 "random numbers," numbers selected "at random" from the interval 0 to 1. A number greater than .5 was called "heads" and a number less than .5, "tails." Thus the experiment (like the toss of a coin) had two possible outcomes at each trial. After each trial the proportion of heads up to that point was computed and plotted, with results as shown in Figure 1-1. Observe that there does appear to be a tendency toward a limiting proportion somewhere near 1/2. ◄

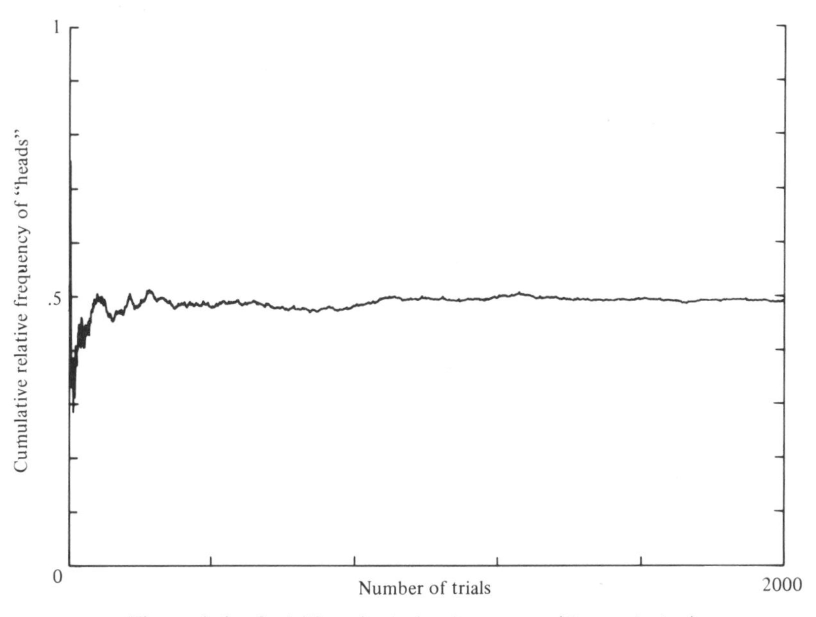

Figure 1-1 Stability of relative frequency (Example 1-a).

Suppose that the experiment is that of tossing a thumbtack, with possible outcomes "point up" and "point down." About all there is about this experiment that can be modeled is that there are the two possible outcomes, which occur, in a long sequence of trials, with frequencies whose ratio tends to limiting values. Similarly, the model for the toss of a coin will consist of the two outcomes *heads* and *tails*, together with a limiting proportion or probability for each. Most people will say that these probabilities should each be 1/2, and that if they are *not* 1/2, the coin is not "fair." And it is interesting to observe that even though the proportion 1/2 can only be realized in a long sequence of trials, people do bet even money on heads and tails in a *single* toss of the coin.

More generally, a probability model, or stochastic model, intended to represent an experiment of chance, is defined by a set or list or space of possible outcomes, together with probabilities of these outcomes. It is a mathematical structure that may or may not represent the experiment faithfully, as is the case with any mathematical model. Ultimately, its worth is determined by experience—by experimentation.

Although we have introduced the notion of a probability model for certain experiments that can be repeated indefinitely, there are experiments whose repetition is difficult to imagine. Thus, when horse A runs in race B, people say that he has a certain chance of winning, even though he cannot ever run this particular race again under the same conditions. And people may be willing to bet that Senator C will be elected President in the next election, given acceptable odds. And again, the probability of rain tomorrow is commonly given in weather reports, although tomorrow comes only once. Perhaps one might imagine that the world could be started over again, indefinitely often, and brought to the same point each time to observe in what proportion of worlds the horse wins or the Senator is elected or it rains tomorrow; but this gets a bit elusive.

To the extent that these experiments—the race, the election, and the weather—are not repeatable, the chances or odds or probabilities that are quoted or used for betting are simply measures of belief about an outcome, and beliefs are a personal matter. Even in the case of a coin, the probability 1/2 for heads could be thought of as a measure of belief; but many claim that there is an impersonal or objective quantity, a property of the coin, that should take precedence over any personal opinions if it is known. However, whether probability is conceived of as personal and subjective, or as impersonal and objective (both views have their uses), the mathematics of probability will be the same.

1.2 Sample Spaces and Events

The first ingredient of a mathematical model for an experiment of chance is the list or collection of possible outcomes: the things that can happen.

> DEFINITION. The *sample space* of an experiment of chance is the set of possible outcomes of the experiment.

This set can be finite, countably infinite, or uncountably infinite.† And being part of the model, it can be formulated according to the needs or whims of the model builder.

Example 1-b When two coins are tossed, one could say there are *two* possible outcomes: either they match or they do not match. The sample space of these two outcomes is adequate when the experiment is used to select one of two alternatives (such as who gets the choice of receiving at the beginning of a football game). But one might also list *three* possible outcomes, according to the number of heads showing, namely, 0, 1, or 2. Or, if the coins are distinguishable, one can list four outcomes: both heads, both tails, heads on one and tails on the other, or tails on one and heads on the other. Of these three formulations, the last—the most detailed—would have the widest applicability, for if one knew which of these four outcomes occurred, he could determine whether or not the coins match, as well as the number of heads. ◄

Example 1-c Calls come in to a service center in an uncoordinated, haphazard fashion. Observing the number of calls in a 10-minute period, say, is an experiment of chance. The outcome of the experiment, the number of incoming calls, is a nonnegative integer, with no obvious limit as to size. The sample space can be taken to be

$$\{0, 1, 2, 3, \ldots\}. \quad ◄$$

Example 1-d The students at a certain college are listed on a numbered list. One name is selected from the list in a manner thought of commonly as "at random." The sample space would be the set of all names.

However, it may be that only the *class* of the student whose name is drawn is of interest. Then the appropriate sample space would consist of the four outcomes freshman, sophomore, junior, and senior. ◄

† A set is *countable* if it is possible to count its elements—to place them into one-to-one correspondence with the "counting numbers" 1, 2, 3, ... This means that the elements can be *listed* in a sequence, possibly unlimited in length. A set is *uncountably* infinite if it is infinite but not countable. (The set of real numbers is uncountably infinite.)

Example 1-e The weight of a box of packaged cereal is variable and unpredictable, best thought of as the result of an experiment of chance. Even though the weight can be measured only to the nearest tenth of an ounce, say, the actual weight in the box can be any one of a continuum of values—a positive real number. The set of positive real numbers can be taken as the sample space. ◄

The points of a sample space represent the individual possible outcomes of an experiment, enumerated according to a scheme that is sufficiently detailed for the purposes at hand. However, having specified a sample space, one might be interested in some outcome or result that is not so precisely defined as a point in the sample space.

Example 1-f In counting arrivals at a service counter, as in Example 1-c, it may be of interest to note whether that number (in a 10-minute interval), exceeds 3, say. The condition of exceeding 3 is true of any of the numbers in this list: 4, 5, 6, ... Or perhaps a condition of interest is that there are not more than 4 arrivals. This condition defines the set of outcomes {0, 1, 2, 3, 4}, each one of which is not more than 4. ◄

> DEFINITION. An *event* is a set of outcomes in a sample space.

An event can be specified (1) by giving a list of the outcomes that make it up, or (2) by giving a descriptive phrase or condition that serves to characterize those outcomes, a condition that is satisfied by each outcome in the event and not satisfied by outcomes not in the event. An event is said to have happened or occurred when one of its outcomes is the result of a trial of the experiment.

Example 1-g A card is selected from a shuffled deck of playing cards. The sample space for this experiment can be taken to be the list of 52 cards in the deck, any one of which might be selected. Some events of interest might include

E: the card is a spade,
F: the card is a face card,
G: the card is black,
H: the card is an ace.

Condition E defines the subset of the sample space consisting of the 13 cards that are spades, and so on. ◄

It is convenient to have on hand definitions of certain operations with events or sets of outcomes, operations that correspond to natural ways of expressing conditions in terms of other conditions. First, conditions that define precisely the same set of outcomes are said to be equivalent or equal:

DEFINITION. Events E and F are *equivalent*, written $E = F$, if and only if the outcomes in E are precisely the same as the outcomes in F.

Equality of E and F means that the condition defining E implies and is implied by the condition defining F.

The condition that an event E has *not* occurred defines the set of outcomes that are not in E.

DEFINITION. The *complement* of an event E in a sample space Ω is the event consisting of those outcomes of Ω that are not in E. It is denoted by E^c.

The sample space Ω is itself an event; and since there are no elements in Ω that are not in Ω, there is no complement. Rather, the complement of Ω is a condition satisfied by no outcomes, and such a condition is said to define an *empty* set. Indeed, *any* condition satisfied by no outcomes defines this empty set, and the notation $\varnothing$ is often used for it. Clearly,

$$\Omega^c = \varnothing \quad \text{and} \quad \varnothing^c = \Omega.$$

Events that have outcomes in common are said to *intersect*, and the following operation is defined:

DEFINITION. The *intersection* of events E and F is the event whose outcomes are those that are contained in *both* E and F. It is written either EF or $E \cap F$.

The notion of intersection is also defined for any number of events, as the event each of whose outcomes lies in every one of them. Events that do not intersect are said to be *mutually exclusive* or *disjoint*:

DEFINITION. Events E and F are *disjoint* if and only if their intersection is empty: $EF = \varnothing$.

An event defined by saying that its outcomes satisfy condition E or condition F consists of all outcomes contained in either of the events E and F (or possibly in both), and is called their *union*:

> DEFINITION. The *union* of events E and F is the event whose outcomes are those contained in either E or in F or in both. It is written either $E + F$ or $E \cup F$.

This operation is also easily defined for any number of events, as the event consisting of all outcomes that lie in one or more of them.

And finally, a relation of partial ordering of events or sets of outcomes is that of "set inclusion."

> DEFINITION. Event E is *contained* in event F, written $E \subset F$, if and only if every element of E is also in F.

The condition defining the larger set or event is automatically satisfied by any outcome in the smaller one, so that set inclusion corresponds to logical implication.

To summarize, the logical equivalents of the operations and relations between events are as follows:

$$
\begin{aligned}
E \text{ and } F &\sim E \cap F,\\
E \text{ or } F &\sim E \cup F,\\
E \text{ implies } F &\sim E \subset F,\\
\text{not } E &\sim E^c.
\end{aligned}
$$

These equivalent expressions, "E or F" and $E \cup F$, for example, will be used interchangeably. Watching for the key words—*and*, *or*, *implies*, and *not*—helps in translating a verbal statement into the algebra of events or sets.

Although the sets defining sample spaces and events can be quite general and abstract, it may be helpful to have in mind a schematic picture associated with each operation. A picture is most feasible when the outcomes are points in the plane. Figure 1-2 shows as event E the points within a triangle, as event F the points within a circle, and then as the intersection EF the points in the part of the triangle that lie within the circle. This is an example of a *Venn diagram.*

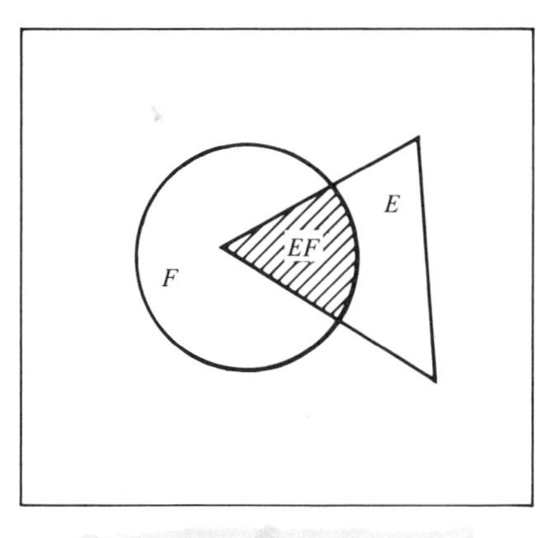

Figure 1-2 Venn diagram.

Example 1-h Let Ω be the sample space of 52 cards, corresponding to the selection of one card from a standard deck. Let events be defined as follows:

E: the card is a spade,
F: the card is a face card,
G: the card is black,
H: the card is an ace.

Then $E \subset G$, and G^c is the set of red cards (26 in number). The event $H + G$ consists of all the spades and clubs plus the aces of hearts and diamonds—28 cards in all. And the event $EFG = EF$ consists of three cards, the king, queen, and jack of spades. ◄

Example 1-i Consider as a sample space the set of points in a plane. These points are defined by rectangular coordinates (x, y). And consider these events:

E: the set of points (x, y) such that $0 < x < 1$ (a vertical strip of unit width),
F: the set of points (x, y) such that $0 < y < 1$ (a horizontal strip of unit width),
G: the set of points (x, y) such that $x^2 + y^2 < 1$ (the interior of a unit circle).

Then EF is the unit square, with vertices at $(0, 0)$, $(1, 1)$, $(0, 1)$, and $(1, 0)$; G^c is the set of points outside or on the unit circle; and EFG is the set of points within the circle that lie in the first quadrant. ◄

1.3 Probability

A model for an experiment of chance, called a *probability model*, consists of a sample space together with an assignment of a probability to each event of interest. But this assignment cannot be completely arbitrary if it is to incorporate the notion that probability is a "*proportion* of the time" in a long sequence of trials. As an objective quantity, $P(E)$ is the limiting relative frequency of occurrence of E, and as such it has certain characteristics that relative frequencies or proportions have. In particular, its value is nonnegative and cannot exceed 1:

$$0 \leq P(E) \leq 1.$$

Also, the whole sample space is an event, one that always occurs, so its relative frequency in any set of trials is 1. And the empty set is an event that never occurs—its relative frequency in any set of trials is 0. Hence

$$P(\Omega) = 1, \qquad P(\varnothing) = 0.$$

If events E and F are disjoint, then clearly, at any stage in a long sequence of trials, the frequency (up to that point) of the union $E \cup F$ is the sum of the frequencies of E and F. This additivity carries over to relative frequencies, obtained by dividing frequencies by the number of trials; and so in a probability model it is postulated that probabilities are additive over disjoint events:

Additivity:

$$\text{If } EF = \varnothing, \quad \text{then } P(E \text{ or } F) = P(E) + P(F).$$

These properties, defining probability as a nonnegative, additive measure on events, normed so that the sample space has probability 1, are often taken as *axioms* in the mathematical development of probability models.

The property of additivity gives at once an expression for the probability of a complement. For, since E and E^c are disjoint, and $E + E^c = \Omega$, it follows that

$$P(E) + P(E^c) = P(\Omega) = 1,$$

or finally,

$$\boxed{P(E^c) = 1 - P(E).}$$

This is often useful for obtaining $P(E)$ when $P(E^c)$ is easier to compute directly, or conversely.

Another useful relation, which follows from the relation

$$E = EF + EF^c$$

(see Problem 4), is the *law of total probability*:

$$\boxed{P(E) = P(EF) + P(EF^c).}$$

In words, the probability of E is the probability of the part of E that is in F plus the probability of the part of E that is in F^c.

Example 1-j Consider once more the selection of a card from a deck of playing cards. Whatever probability model is appropriate for the method of selection used, it must be, for instance, that

$$P(\text{red}) = 1 - P(\text{black})$$

and that

$$P(\text{king}) = P(\text{red king}) + P(\text{black king}). \quad ◄$$

1.4 Discrete Sample Spaces

In the case of a discrete sample space, a probability model can be constructed by first assigning a probability to each possible outcome, up to a total of 1, and then computing from these probabilities the probabilities for more general events.

Probability model for a discrete space:

(i) Sample space: $\Omega \equiv \{o_1, o_2, \ldots\}$.
(ii) Probabilities: $p_i = P(o_i)$, where

$$p_i \geq 0 \quad \text{and} \quad \sum p_i = 1.$$

As before, an event is a subset of the possible outcomes; and in the spirit of the additivity property that probabilities have, the probability of an event is here defined to be the sum of the probabilities of its outcomes.

Probability of the event E:

$$P(E) \equiv \text{sum of } p_i\text{'s for } o_i\text{'s in } E.$$

From this definition it follows that $P(E) \geq 0$, that $P(\Omega) = 1$, and that probability is additive over disjoint events.

Example 1-k A gambler feels that in a poker hand of five cards, the chances are 4 in 10 for a "pair," 1 in 20 for "two pair," 1 in 50 for "three of a kind," 1 in 250 for a "straight," 1 in 500 for a "flush," 1 in 1000 for a "full house," 1 in 4000 for "four of a kind," and 1 in 100,000 for a "straight flush." Thus *his* model for the experiment of drawing five cards consists of the following outcomes and corresponding probabilities:

Outcome	Probability
Straight flush	.00001
Four of a kind	.00025
Full house	.001
Flush	.002
Straight	.004
Three of a kind	.02
Two pair	.05
Pair	.4
None of above	.52274

[The hands are listed in order of preference. They are defined in Chapter 2 (Problem 16), where their a priori or ideal probabilities are computed.] The probability of the outcome "none of above" was obtained by subtracting from 1 the sum of the other probabilities. Some sample computations for other events are:

P(hand no better than 2 pair) = .4 + .05 + .52 = .97,
P(flush or better) = .00326,
P(better than pair but not as good as a full house)
= .05 + .02 + .004 + .002 = .076. ◀

1.5 A Priori Models

There are certain kinds of experiments of chance for which considerations of symmetry, even prior to experimentation, suggest a particular probability model. People would generally agree on the model that *should* represent the experiment if it is properly carried out. Such a model is termed an *a priori model.*

Tossing a coin, and tossing a regular polyhedron—of which the cube or die is the most common instance—are such experiments. When the die, say, is tossed and allowed to fall on a flat surface, one of the sides is "up." The six sides are taken to be the outcomes of the experiment, and symmetry will suggest to most people that no one side should come up any more often than any other in a long sequence of tosses. The a priori model then assigns equal probabilities to the six sides in the sample space—an amount 1/6 for each, if they are to add up to 1. Whether this model represents an actual toss could only be determined by tossing infinitely often, and this would take a long time, indeed. The relative frequencies would have to converge to 1/6 for each side.

A tossing experiment with this property is referred to as ideal, or as "fair." What people do toward achieving this ideal is to toss the die vigorously and with a good spin, so that no side can be favored deliberately, perhaps by shaking it well in a container before casting it onto a green-felt tabletop.

Another type of experiment for which one can construct an a priori model is that in which an object is selected from a group of objects in a way that the man on the street thinks of as "at random." Picking a card from a shuffled deck, picking a draft number from a rotating drum containing 366 capsules, one for each day of the year, and picking from a list of voters the name of a person to whom to send a questionnaire—these are examples of the type of experiment called *random selection*. To be considered fair, one would want all cards or capsules or names to have the same chance of being chosen. That is, one would strive to emulate a mathematical or ideal selection in which the probabilities postulated for the possible outcomes are all equal. Thorough mixing or shuffling and blind selection are essential.

To select an object *at random* will mean, by definition of the phrase, that in the mathematical model for the selection all available objects have the same probability. It is understood that in carrying out the experiment there will be an attempt to achieve experimental conditions that make this particular model correct. The random sampling model is useful not only in lotteries and card games but also in sampling from actual finite populations for the purpose of learning about the characteristics of the population.

The third type of a priori model to be considered is that for a spinning pointer or wheel. The roulette wheel is one that is constructed with 38 (or 37) compartments of equal size, into one of which a small ball falls when it slows down after being spun around the wheel. It would be fair if these compartments are equally likely outcomes, each with probability 1/38 (or 1/37). The wheel of fortune is fitted with equally spaced pegs, so that when it is spun and gradually slows down, a pawl pivoted near the edge falls down between two pegs and causes the wheel to stop in one of the finitely many positions determined by the pegs. In Russian roulette the chamber of a revolver, with six positions, is spun and comes to rest with one of the chambers in the firing position; these are presumed to be equally likely. The spinner that is furnished with certain children's games is a pivoted pointer that stops, after being spun in a horizontal plane, in one of several sectors; these sectors are often constructed with equal central angles so as to be equally likely as regions in which the pointer might stop.

In all of these a priori, discrete models, the list of outcomes is finite, and all outcomes have the same probability: they are *equally likely*. If there are m equally likely outcomes in Ω, each has probability $1/m$. Moreover, the probability of any event is just that common probability $(1/m)$ times the number of outcomes in the event. In summary:

> Given a sample space Ω with m equally likely outcomes $o_1, o_2, \ldots, o_m$:
>
> $$P(o_i) = \frac{1}{m}, \qquad \text{for each outcome;}$$
>
> $$P(E) = \frac{\text{number of outcomes in } E}{\text{number of outcomes in } \Omega}.$$

Example 1-l Let the six faces of a die be coded, as usual, by the integers from 1 to 6.

Then

$$P(i) = \frac{1}{6}, \qquad i = 1, 2, \ldots, 6,$$

and the event that the outcome is an odd number, say, has probability

$$P(\text{odd}) = \frac{\text{number of odd faces}}{\text{number of faces}} = \frac{3}{6}. \quad \blacktriangleleft$$

1.6 Populations

The term *population* is often used for a sample space, a carryover from those situations in which the experiment consists of drawing a person from a certain population of people. The possible outcomes of the experiment are the individuals in the population. The term *population* is also applied to a group of animals, or of inanimate things—the population of chance tickets in a lottery, for instance.

When the interest lies in some characteristic of the individuals in a population, such as eye-color, weight, or blood type, the outcomes may be thought of as the values or categories of that characteristic, rather than as the individuals themselves. The term *population* is then used to denote the sample space of value-outcomes. Thus, if the sex of an individual is what is of interest, the population can be thought of as consisting of the sample space {male, female}.

The term *population* is frequently used to mean something conceptual, a collection of individuals or of individual values that are thought of as possible when an experiment is performed, but which are not tangible—not seen or observed until realized as a result of an actual performance of the experiment. Thus, when a coin is tossed, one can imagine that one is "selecting" from a container with two outcomes, heads and tails (or equally well, from a container with equal numbers of heads and tails). In this sense, the sample space {heads, tails} is a population.

In the case of making a measurement, or observing the result of some ongoing operation such as production line, only those individual outcomes actually observed are manifest; and again the population—the set of possible outcomes—is conceptual. If it is conceptual, there is no harm (and possibly some simplification) in thinking of it as infinite in some instances. For example, the number of incoming calls at a telephone exchange in an hour's time can be any of the numbers 0, 1, 2, ... Practically speaking, the number may have some finite limit, but ideally we may take the population to be the infinite set of nonnegative integers. Continuous quantities such as time and distance are modeled most simply using the infinite population or sample space of real numbers on an interval.

When an individual is drawn from an actual population of N individuals in such a way that all are equally likely, the probability of each one is $1/N$, and the probability of an event is the proportion of outcomes that are in the event. If it is a value or category of some characteristic that is of interest, the individual values or categories will have probabilities defined by their proportions in the population of individuals.

Example 1-m In a group of 100 freshmen, 73 are eighteen, 22 are seventeen, 4 are nineteen, and 1 is sixteen years old. Selecting one freshman from the group at random is represented by assigning probability .01 to each. The sample space can be taken to be the set of possible ages, and these outcomes have probabilities as follows:

Outcome (age)	16	17	18	19
Probability	.01	.22	.73	.04

◄

Problems

1. Give an appropriate sample space for each of these experiments.

(a) A coin is tossed repeatedly until heads first appears.
(b) The number of smokers in a class of 30 students is determined.
(c) A voter is picked from a population of voters and queried as to his political party affiliation.
(d) A person is weighed on a scale in the student health service.
(e) A marksman fires a rifle at a target.
(f) A light bulb is burned until it fails.

2. A penny and a nickel are tossed together, and the sample space is taken to consist of four outcomes: {hH, hT, tH, tT}. (The notation hT means heads on the penny and tails on the nickel.)

(a) How many different events can be defined on this sample space?
(b) Give a list of outcomes defining each of these events:

E: both coins fall heads.
F: at least one coin falls heads.
G: the penny falls heads.

(c) Give a list of outcomes for each of these: $E + G$, EFG, F^c, where E, F, and G are as in part (b).

3. In the sample space for the toss of a die, with outcomes identified by the number of dots on the upturned face, define these events:

E: the outcome is not 6.
F: the outcome is an even number.
G: the outcome is less than 3.

List the outcomes in each of the following:

(a) EG.
(b) $F + G$.
(c) EG^c.

4. Show the following set identities, and study each one with the aid of a Venn diagram.

(a) $E + E^c = \Omega$.
(b) $EE^c = \varnothing$.
(c) $(EF)^c = E^c + F^c$.
(d) $E = EF + EF^c$.
(e) $E + F = EF^c + E^cF + EF$.

5. Observe that the events on the *right* in each of Problems 4(d) and (e) are disjoint, and so deduce formulas for $P(E)$ and $P(E + F)$ by means of the additivity property. Obtain a similar formula for $P(F)$ by decomposing F as in 4(d), and finally relate the probabilities $P(E)$, $P(F)$, and $P(E + F)$ by a single formula.

6. Show that if $E \subset F$, then $P(F) = P(E) + P(FE^c)$. Use this to deduce that $P(E) \leq P(F)$.

7. A card is drawn at random from a standard deck. Determine the probability that the card drawn is

(a) Not a diamond.
(b) An honor card (A, K, Q, J, 10).
(c) A red honor card.
(d) A spade but not an honor card.

8. The 38 positions on a roulette wheel are numbered 0, 00, 1, 2, . . . , 36. The zero and double zero are green; half of the rest are black and half red.

(a) What is the probability of red?
(b) What is the probability that the result is not black?
(c) What is the probability that the result is an odd number?

9. In a population of 100 students, 25 are juniors, half of the 20 seniors smoke, 24 of the 30 freshmen do not smoke, 8 sophomores smoke, and one third of the smokers are seniors. What is the probability that a student picked at random is

(a) A smoker?

(b) A sophomore?
(c) A junior nonsmoker?
(d) Not a senior and does not smoke?

10. Table XIII of Appendix B gives the results of a sequence of trials of an experiment with 10 equally likely outcomes, identified as the digits 0, 1, . . . , 9.

(a) Devise two (physical) experiments each of whose a priori model would be one with 10 equally likely outcomes.
(b) Check the relative frequencies of the 10 digits in a sequence of 200 consecutive entries from Table XIII. Are they equal? (Should they be?)
(c) If you wanted to toss a coin but had no coin, explain how Table XIII might be used as a pseudo-coin.

11. One of three boxes contains a prize. A contestant chooses one box at random, and then the Master of Ceremonies opens one of the other boxes to reveal that it is empty. Assuming that the Master of Ceremonies would not open the box with the prize, should the contestant trade the box she picked for the unopened box if given the chance?

2 Counting

As discussed in Chapter 1, the way to calculate the probability of an event in a model with equally likely outcomes is to count the number of outcomes in the event and the number of outcomes in the sample space, and then divide one number by the other. Counting is simple enough when there is a brief listing of all possible outcomes, but it is often convenient to have at one's command certain techniques for obtaining a count without the labor of a complete enumeration. These will now be developed.

2.1 A Basic Principle

Often, a complicated experiment consists of performing consecutively certain "subexperiments," or it can be thought of in this way. The outcomes of a composite experiment, although arbitrarily designated, may be conveniently and naturally defined in terms of the outcomes of the component experiments, as follows. For the case of two experiments, $\mathscr{E}$ and $\mathscr{F}$, with outcomes $\{e_1, \ldots, e_m\}$ and $\{f_1, \ldots, f_n\}$, respectively, the outcomes of the experiment "$\mathscr{E}$ and then $\mathscr{F}$" are the pairs (e_i, f_j)

with $i = 1, \ldots, m$ and $j = 1, \ldots, n$. It is clear that there are mn such pairs:

$$\begin{array}{c}(e_1, f_1), \ldots, (e_1, f_n), \\ \vdots \\ (e_m, f_1), \ldots, (e_m, f_n).\end{array}$$

Although the notation just used assumed implicitly that the structure of $\mathscr{F}$ is the same no matter what happens when $\mathscr{E}$ is performed, the same counting scheme applies when the sample space for $\mathscr{F}$ depends on the outcome of $\mathscr{E}$, provided that the *same number* of outcomes are possible for $\mathscr{F}$ in every case.

> Basic counting principle: If experiment $\mathscr{E}$ has m distinct outcomes, and if for each of these outcomes $\mathscr{F}$ has n distinct outcomes, then the experiment "$\mathscr{E}$ and then $\mathscr{F}$" has mn distinct outcomes.

The mn outcomes of the composite experiment are *ordered pairs*. This is so even if the structure of $\mathscr{F}$ does not depend on $\mathscr{E}$, in which case the experiments could just as well be actually carried out simultaneously, or in reversed order. But in any case, keeping track of order is essential for the count to be correct. In particular, when $\mathscr{F}$ has the same outcomes as $\mathscr{E}$, as in the case of tossing two coins, or tossing one coin twice, the outcome of heads on one coin or toss and tails on the other coin or toss is counted as different from the outcome of heads on the other and tails on the one.

Example 2-a A license number is to consist of three digits followed by three letters, in sequence. How many distinct license numbers are possible using this scheme?

The license number can be constructed by choosing the numbers and letters in succession, from those available. The first number is one of 10 digits, as is the second and also the third. The first letter is one of 26, as is the second and also the third. The total number of sequences is then the product

$$10 \cdot 10 \cdot 10 \cdot 26 \cdot 26 \cdot 26 = 17{,}576{,}000.$$

[It might occur to one that a sequence could be constructed in some other order—for example, by picking first the third letter, then the second digit, then the second letter, and so on. However, when constructed in this order, the count is the same, because the same factors (three 10's and three 26's) would appear in the product.] ◀

In constructing a priori probability models it will be useful to have the following fact about composite experiments, to be derived in Chapter 4, but perhaps intuitively clear even at this point:

> Suppose that $\mathscr{E}$ has m equally likely outcomes. If, no matter how $\mathscr{E}$ turns out, a subsequent experiment $\mathscr{F}$ has n equally likely outcomes, then the mn outcomes of the composite experiment "$\mathscr{E}$ followed by $\mathscr{F}$" are equally likely.

It should be emphasized that the equal likelihood of $\mathscr{F}$'s outcomes refers to the situation after the outcome of $\mathscr{E}$ has become known. Moreover, the list of outcomes in $\mathscr{F}$ may depend on the outcome of $\mathscr{E}$, as long as there are just n of them in each case. In the first example to follow, the list of outcomes of $\mathscr{F}$ will actually be the same, no matter how $\mathscr{E}$ turns out; but in the second example, the list (although not its size) will depend on what happened in $\mathscr{E}$.

Example 2-b A die is tossed once and then tossed again. The first toss, $\mathscr{E}$, has six outcomes, and no matter which of them occurs the second toss, $\mathscr{F}$, has the same six possible outcomes. If $\mathscr{E}$ is "ideal," its outcomes are equally likely; and if $\mathscr{F}$ is also an ideal toss (unaffected by the outcome of $\mathscr{E}$), its outcomes are equally likely. Then the 36 pairs $(1, 1), (1, 2), \ldots, (6, 5), (6, 6)$ in the sample space of the two successive tosses are equally likely. Knowing then that each outcome (i, j) has probability $1/36$, one can compute the probabilities of various events. For instance,

$$P(\text{sum} = 5) = P[(1,4), (2,3), (3,2), (4,1)] = \frac{4}{36}.$$

In this example, of course, since the first experiment is assumed to have no bearing on the second, they could be performed simultaneously, as in the toss of two dice. ◄

Example 2-c A ball is selected at random from a bag containing four white and three black balls, and then a second ball is selected at random from the six that remain. There are $7 \cdot 6 = 42$ outcomes in the sample space, and these are equally likely because of the phrase "selected at random" that describes each subexperiment. For the event

$$E\colon\quad \text{both balls are white}$$

to occur, it must be that a white ball (one of four) was picked at the first draw and then a white ball (one of three remaining) was picked on the second draw. Thus

$$P(E) = \frac{4 \cdot 3}{7 \cdot 6} = \frac{2}{7}. \quad ◄$$

2.2 Permutations

The basic multiplication principle can be used to count the number of possible arrangements or permutations of objects in a row or sequence. If there are n distinct objects to be arranged, the arrangement can proceed by selecting one object at a time for each position in the sequence. There are n choices for the first position, $n - 1$ for the second position, and so on. Multiplication yields this formula:

> Number of permutations of n distinct objects in a row:
>
> $$n! \equiv n(n-1)(n-2)\cdots 3 \cdot 2 \cdot 1.$$

If only r of n available objects are to be used in an arrangement, there would be only r factors, starting with n and proceeding through successively smaller integers to $n - r + 1$:

> Number of permutations of n things r at a time:
>
> $$(n)_r \equiv n(n-1)\cdots(n-r+1).$$

Clearly, $(n)_n = n!$.

Example 2-d There are $(4)_2 = 4 \cdot 3$, or 12, distinct arrangements of two letters taken from the list A, B, C, D. Listing these 12 arrangements systematically shows why the multiplication principle works:

$$\begin{array}{cccc} AB & BA & CA & DA \\ AC & BC & CB & DB \\ AD & BD & CD & DC. \end{array}$$ ◄

Suppose, in the process of forming an arrangement of objects by choosing one at a time from those available, that the choice at each step is made "at random," that is, the available objects all have the same chance of being chosen. Then, according to a result from the preceding section, the various permutations are *equally likely*.

Example 2-e A reporter called a mathematics department in a state university to inquire whether it might be considered unusual that if the names of the six candidates for three judgeships were arranged at random, the incumbent judges' names would appear first.

There are $6! = 720$ ways of arranging the six names, and if they are arranged at random, each permutation has probability $1/720$. The question is: In how many of these are the incumbents listed first? An arrangement of six names with incumbents listed first can be accomplished by arranging their names within the first three positions (one of $3! = 6$ ways) and then arranging the names of the three others (one of $3! = 6$ ways); all this can be done in $(3!)(3!) = 36$ ways. And then

$$P(\text{incumbents first}) = \frac{(3!)(3!)}{6!} = \frac{1}{20}. \quad \blacktriangleleft$$

It is often necessary to count arrangements of objects some of which are to be considered indistinguishable. That is, even though $n!$ is the number of arrangements, some of these may appear to be the same; and what is desired is the number N of different-looking arrangements. Suppose that k of the n objects are "alike." Then with the other objects fixed in a given configuration, these k can be permuted in $k!$ ways without changing the appearance of the arrangement. A particular one of the $n!$ arrangements can be formed by first putting the unlike objects in their given configuration (one of N ways) and then arranging the superficially like objects (one of $k!$ ways). Thus

$$N \cdot k! = n! \quad \text{or} \quad N = \frac{n!}{k!}.$$

Reasoning along similar lines yields this result:

Number of permutations of n objects, k_1 of one type, k_2 of another type, and so on: $$\frac{n!}{k_1!k_2!\cdots}.$$

Example 2-f A mother has six cookies, two chocolate chip, three raisin, and one peanut butter. In how many distinct ways can she pass them out to six children so that each gets one?

If one thinks of the children as coming to her in sequence, the question boils down to how many ways the six cookies can be arranged, those of the same kind being indistinguishable. The number is

$$\frac{6!}{2!3!} = 60.$$

If the cookies were distinguishable, one of the $6! = 720$ arrangements would be $C_1C_2R_1R_2R_3P$. These 720 arrangements can be found by starting with the 60 in which the C's and R's are not distinguished, and in *each* of them arranging subscripts 1 and 2 on the C's and subscripts 1, 2, and 3 on the R's:

$$720 = 60 \times 2! \times 3!. \quad \blacktriangleleft$$

2.3 Combinations

A group or bunch or selection of k objects taken from n available objects, without regard to order, is called a *combination*. The number of possible combinations can be counted by considering an equivalent problem of arrangement. Suppose that n objects are of two types, say, k black and $n - k$ white ones. The number of distinct arrangements or color patterns is given by

$$\frac{n!}{k!(n-k)!} = \frac{n(n-1)\cdots(n-k+1)}{k!}.$$

But each such pattern or arrangement could be formed as follows: From the n positions in the sequence to be formed, select k positions and put the black objects in these positions (in any way, since it does not matter which black objects go where). Then put the whites in the remaining $n - k$ positions (in any way). There are as many patterns or arrangements as there are ways of selecting a group of k sequence positions from n. This number is denoted by $\binom{n}{k}$.

> Number of combinations of n things k at a time:
>
> $$\binom{n}{k} \equiv \frac{n!}{k!(n-k)!} = \frac{(n)_k}{(k)_k}, \qquad k = 0, 1, \ldots, n.$$

The better formula for hand computation is the last one: Starting with n on top of, and k below, the fraction bar, write out k factors decreasing in steps of 1.

Example 2-g A hockey team includes three centers and six wings. How many different forward lines (line = one center plus two wings) can the coach make up from these nine players?

The number of ways of picking a center is $\binom{3}{1} = 3$, and for each of these three choices the number of ways of picking two wings is

$$\binom{6}{2} = \frac{6!}{2!4!} = \frac{6 \cdot 5}{2 \cdot 1} = 15.$$

The number of ways of forming the complete line is then $3 \times 15 = 45$. ◄

Example 2-h How many different subcommittees can be formed from the six members of a governing board?

There are either one, two, three, four, five, or six members on a subcommittee, and the total number is the sum of the number of subcommittees with one, the number with two, and so on:

$$\binom{6}{1} + \binom{6}{2} + \binom{6}{3} + \binom{6}{4} + \binom{6}{5} + \binom{6}{6}$$
$$= 6 + 15 + 20 + 15 + 6 + 1$$
$$= 63.$$

Although obtained by counting combinations, this number could be found (as is often the case) by another approach. Form a subcommittee by taking each board member in turn and deciding whether that member is to be on the subcommittee; there are two choices for each of the members, so the total number would be the product of these 2's: $2 \times 2 \times 2 \times 2 \times 2 \times 2 = 64$. However, this figure includes as a "subcommittee" one with no members (for *each* board member, decide *not* to use him on the subcommittee); and so this one case, counted as one of the 64, should be subtracted to yield (as before) 63. ◄

Some properties of $\binom{n}{k}$ are worth noting. First, for $k = 1$, there are just n ways of selecting 1 from n objects. For $k = n$, there is only one way of selecting n from n—that is, use them all. Finally, selecting k to use in a combination automatically leaves behind $n - k$ that are not used; that is, there are as many ways of selecting k as there are of selecting $n - k$. Indeed, $\binom{n}{k}$ can be thought of as the number of ways of partitioning n objects into two groups, one with k and one with $n - k$.

Properties of $\binom{n}{k}$:

$$\binom{n}{1} = n,$$

$$\binom{n}{n} = 1;$$

$$\binom{n}{k} = \binom{n}{n-k}.$$

The last relation, expressing a kind of symmetry, is useful in hand computation, as in this illustration:

$$\binom{100}{98} = \binom{100}{2} = \frac{100 \cdot 99}{2 \cdot 1} = 4950.$$

If this had been written out as 98 factors above and below the line, 96 of them would cancel anyway, leaving the two shown.

The experiment of drawing k from n objects, which has $\binom{n}{k}$ outcomes, can sometimes be performed so that each combination has the same chance of occurring as any other. The selection is then said to be *random*.

Model for *random selection* of k objects from n:

The $\binom{n}{k}$ possible combinations are *equally likely*, each occurring with probability $1\Big/\binom{n}{k}$.

To strive for this in practice, shuffle the objects thoroughly and reach in and grab k of them. (If this is not feasible directly, each object could be represented by a slip of paper in a hat.)

The model for a random selection applies also if objects are drawn one at a time, without replacing any, in such a way that at each draw the available objects are equally likely. This follows from the fact (given in Section 2.2) that the $(n)_k$ permutations of k things from n are equally likely. For, each combination can be thought of as an event consisting of the $k!$ permutations of the objects in that combination, with probability

$$P(\text{particular combination}) = k! \times \frac{1}{(n)_k} = \frac{1}{\binom{n}{k}}.$$

Example 2-i Five cards are drawn at random from a standard deck of playing cards. Whether drawn in a bunch at random, or drawn one at a time at random, the $\binom{52}{5}$ combinations are equally likely. The probability that a hand of five cards has a certain property is then the number of hands with that property divided by $\binom{52}{5}$.

For instance, the probability that a hand contains two cards of one denomination and three of another (e.g., two 6's and three jacks), called a "full house," is computed by counting the number of such hands and dividing by $\binom{52}{5}$:

$$P(\text{full house}) = \frac{13 \cdot 12 \cdot \binom{4}{2} \cdot \binom{4}{3}}{\binom{52}{5}} \doteq .00144.$$

The $13 \cdot 12$ in the numerator is the number of ways of selecting a denomination for the pair and a denomination for the three. This selection results in two piles of four cards each; to complete the hand one must select two cards from the first pile and three from the other. ◄

★ 2.4 Binomial and Multinomial Coefficients

The quantities $\binom{n}{k}$ may have been recognized as those occurring in binomial expansions. For example, for $n = 4$, they are 1, 4, 6, 4, and 1. This coincidence is easily explained.

In multiplying out the n factors defining the power $(x + y)^n$, one finds that terms are formed by taking either an x or a y from each factor. There are 2^n terms, each a product of n x's and y's. *Like* terms can be grouped; and there are as many terms of the form $x^k y^{n-k}$ as there are ways of selecting k factors of $(x + y)^n$ from which to use the x, the y's then coming from the remaining $n - k$ factors. These like terms combine to yield

$$\binom{n}{k} x^k y^{n-k},$$

with this result:

Binomial theorem:

$$(x + y)^n = \sum_0^n \binom{n}{k} x^k y^{n-k}.$$

The quantities $\binom{n}{k}$ are often called *binomial coefficients*. Notice that with $x = y = 1$ in the binomial theorem, it is seen again, as in Example 2-g, that

$$\binom{n}{0} + \binom{n}{1} + \cdots + \binom{n}{n} = 2^n.$$

The same kind of reasoning leads to the *multinomial theorem*. For the case of the nth power of a trinomial $(x + y + z)^n$, the expansions consist of 3^n terms, each a product of n factors, the factors in a given term being an x, y, or z from each factor of $(x + y + z)^n$. Again, like terms can be collected, and there are as many terms of the form $x^q y^r z^s$, where $q + r + s = n$, as there are ways of arranging n things, q of one kind, r of a second kind, and s of a third kind: $n!/(q!r!s!)$. Thus

Trinomial theorem:

$$(x + y + z)^n = \sum_{q+r+s=n} \frac{n!}{q!r!s!} x^q y^r z^s.$$

The trinomial coefficient here can be thought of also as the number of ways of picking q factors from which to use x, then r factors from those remaining from which to use y. That is, for $q + r + s = n$,

$$\frac{n!}{q!r!s!} = \frac{n!}{q!(n-q)!} \frac{(n-q)!}{r!(n-q-r)!} = \binom{n}{q}\binom{n-q}{r}.$$

Problems

1. In how many ways can:

(a) Seven banquet speakers be seated along one side of the head table?

(b) A hostess place five name placecards around a round table?
(c) Four keys be put in a flat leather key case?
(d) Four keys be put on a key ring?
(e) Four keys be put in a flat leather key case if the car key must be on the left?

2. In how many ways can four distinct balls be put in six distinct containers:

(a) Without restriction?
(b) So that no container gets more than one?
(c) So that the first container gets exactly one ball?

3. An opera company performed a random opera, described as follows in the program: "Briefly, there is a plot in seven parts which combines the underlying structure of the heroic myth, i.e. the awakening of the hero, his mission and its achievement, with the core structure of all popular opera, i.e. love and its requital (comedy) or denial (tragedy). For each step of this sequence there are eight possible scenes which are described in the booklet accompanying your program. The scene to be played for each step of the progression will be selected prior to that step by one of the three Wheels of Fortune. The other two wheels will select one of the sixteen environments in which the scene is to be played, and one of the eight musical periods which will provide the spirit for the musical improvisation. For the mathematically-minded, the chance that any single step of the plot will have the same scene, the same environment and the same musical style on successive nights is approximately 1000 to 1, and the odds against the same combination of seven scenes with the same environments and musical styles is approximately one sextillion to one."

(a) How many different possible combinations of scene, environment, and style are there at each step?
(b) Check the "sextillion to one" odds quoted.

4. Assuming 365 equally likely days, determine the probability that no two, among six persons at a party, have the same birthday. (HINT: Treat like Problem 2, with 6 balls and 365 containers.)

5. A hatcheck girl loses track and passes out the four hats she has at random. What is the probability that none of the four owners gets the right hat?

6. Eight glasses are filled, four with Coke and four with Pepsi. A taster selects four he thinks to be those with Coke.

(a) How many different selections can he make?
(b) How many of these selections include at least three of the Cokes?
(c) What is the probability, if the taster really cannot distinguish and selects at random, that he picks at least three Cokes?

7. How many connecting cables are needed in order that any two of nine offices in a building can communicate directly?

8. If n countries in a bloc exchange ambassadors, how many ambassadors are involved?

9. A committee of 4 is selected at random from a group consisting of 10 labor and 5 management representatives. What is the probability that the committee selected includes:

(a) Two representatives from labor and two from management?
(b) At least one representative from each group?
(c) The chairman of the labor delegation and the chairman of the management delegation?

10. I have two pairs of green, two pairs of blue, and one pair of red socks mixed up in a drawer.

(a) Dressing in the dark I pick two socks at random. What is the probability that they are of the same color?
(b) If I pick three at random, what is the probability that at least two of the three are of the same color?

11. You are one of 10 finalists in a contest tied for first. Three are selected by lot from the 10 to receive the top prizes. What is your chance of being one of the winners?

12. In comparing a sample of three observations called X's and a sample of five called Y's, one technique involves putting them in order and looking at the pattern of X's and Y's. Suppose the possible patterns are equally likely.

(a) How many patterns of three X's and five Y's are possible?
(b) What is the probability that the X's will be together on one end or the other?
(c) What is the probability that there will be just one Y to the left of the middle?
(d) What is the probability that there will be exactly two Y's to the left of the middle?

13. W. A. Mozart prepared material for a "musikalisches Würfelspiel," or musical dice game. He wrote out 176 bars of music, 11 for each of 16 steps in the construction of a random minuet. According to the outcome of the first toss of a pair of dice, a prepared bar is selected to be the first measure of the minuet; the dice are tossed again for the selection of a bar (from 11 for that step) to serve as measure 2; and so on. (The complete minuet called for a repeat of the first 8 bars, the bar used in the eighth step being provided with a "first ending" and a "second ending" version.)

(a) The present publisher's brochure claims 11^{16} distinct minuets to be possible, but inspection shows that the 11 bars from which measure 8 is to be selected are identical, and that 10 of the 11 for measure 16 are identical. So, then, how many distinct minuets are possible?
(b) Are the minuets counted in part (a) equally likely?
(c) If the odd bar for measure 16 is assigned to the dice total of 10, what is the probability that the composition does not end with that bar?

14. By expressing in terms of factorials, show that $\binom{n}{k} + \binom{n}{k-1} = \binom{n+1}{k}$.

(a) Use this to obtain $\binom{8}{1}, \binom{8}{2}, \ldots, \binom{8}{7}$ by combining $\binom{7}{0} = 1$, $\binom{7}{1} = 7$, $\binom{7}{2} = 21$, and $\binom{7}{3} = 35$.

(b) Starting with $\binom{1}{0} = 1$ and $\binom{1}{1} = 1$, use the identity to calculate each $\binom{2}{k}$, then each $\binom{3}{k}, \ldots$ [starting each sequence with $\binom{n}{0} = 1$ and closing with $\binom{n}{n} = 1$].

★ **15.** Determine the following coefficients:

(a) Of x^3y^5 in $(x + y)^8$.
(b) Of x^2y^4z in $(x + y + z)^7$.
(c) Of $x^2y^2z^2w^2$ in $(x + y + z + w)^8$.

16. Example 1-k gave a gambler's probabilities of various poker hands (5 cards from a deck of 52). Example 2-i shows that the probability of a full house is about .00144 (rather than the gambler's value of .001). Check the other probabilities.

[NOTE: "Pair" means two of one denomination, and the other three of three different denominations. "Three of a kind" means three of one denomination—but not a full house or four of a kind, "Flush" means five cards of one suit—but not a straight flush. "Straight" means five in sequence (e.g., 8, 9, 10, J, Q)—but not a straight flush. "Straight flush" means five in sequence in the same suit.]

3

Discrete Random Variables

The outcome of an experiment of chance with a discrete sample space whose possible outcomes are taken to be numerical is called a *discrete random variable.* The number of raisins in a raisin cookie, the number of leaves on a tree, and the number of defective items in a carton of manufactured items are examples of random variables with a discrete value space.

Often the experiment of chance can be considered as more basic than the determination of a number. For instance, the randomness in the number of raisins in a cookie lies in the idea that the cookie was drawn at random from a box of cookies or from a cookie factory's production line. The outcome could be thought of as the cookie itself, and the number of raisins is a *function* of what cookie is drawn. Similarly, a person's height, to the nearest inch, is random if the person is drawn by some chance mechanism from a given population of people. The height is a *function* of what person is drawn, and one may want to think of the population as the sample space whose elements are the individual people. Thus one is led to the following:

DEFINITION. A real-valued function $X(\omega)$ defined for each outcome ω in a sample space is said to be a *random variable*. It is a *discrete* random variable if the set of its possible values is finite or countably infinite.

If the outcomes in the sample space are taken to be numbers, then the simple function $X(\omega) = \omega$ is a random variable.

In summary, and paraphrasing the above definition, a random variable is a variable whose value is determined by an experiment of chance.

Example 3-a The result of measuring a student's achievement in a course can be thought of as a random variable. It may be that the student is taken from a population of students by a random device, so that the possible outcomes ω are basically the students in the population. The achievement is then a function of ω as assigned according to the outcome of an examination. Such scores are ordinarily rounded off, so the variable is discrete. However, it may be desirable to think of the measurement process—even for a particular student—as an experiment of chance, one whose outcomes ω are integers from 0 to 100, say. The achievement is then $X(\omega) = \omega$. The elements of "chance" here (i.e., factors that would be difficult to model deterministically) might be the room temperature, the state of hunger, the mental attitude, the way in which questions are interpreted or misinterpreted—factors that may enter in both the taking and the grading of the examination. ◄

3.1 Distributions in the Value Space

If a sample space itself has numerical outcomes, the assignment of probabilities to outcomes automatically assigns probabilities to the values of $X(\omega) = \omega$, since the values *are* the outcomes. More generally, if a random variable $X(\omega)$ is defined on an arbitrary sample space, any assignment of probabilities to outcomes ω automatically defines probabilities for the various possible values of $X(\omega)$. This is because the statement $X(\omega) = k$, saying that the function $X(\omega)$ has been determined to have the value k, defines an *event* or set of outcomes ω:

$$\{X(\omega) = k\} = \text{set of } \omega\text{'s assigned the value } k \text{ by } X(\omega).$$

Thus, if the sample space is discrete:

$$P(X(\omega) = k) = P(\text{set of } \omega\text{'s such that } X(\omega) = k),$$

which is just the *sum* of the probabilities of the individual ω's that are assigned the value k by $X(\omega)$.

Suppose that $X(\omega)$ has possible value $x_1, x_2, \ldots$. *Each* of these has a probability attached to it, because they have been given probabilities either directly by a particular model, or indirectly as a consequence of assigned probabilities of the ω's in a sample space. The attachment of a probability to each x_i is another instance of the notion of "function," and the name *probability function* is sometimes used here.

Probability function of the random variable $X(\omega)$:

$$f(x) \equiv P(X(\omega) = x),$$

defined for each possible value x.

It is clear that if *all* possible values of X are taken into account, the total probability is now distributed among the values x so that *their* probabilities add up to 1. This fact and the fact that probabilities are nonnegative together characterize a probability function:

Basic properties of a probability function:

$$\text{(i) } f(x) \geq 0, \qquad \text{all } x.$$
$$\text{(ii) } \textstyle\sum_{\text{all } x} f(x) = 1.$$

Any such function is a probability model on the sample space of values. It defines what is referred to as the *distribution* of $X(\omega)$, meaning by this the way in which the total probability 1 is allotted to its various possible values.

Because the interest usually lies in the value space of a random variable, and in the way probability is distributed there, the reference to the more basic outcomes ω is often suppressed. Thus one writes X in place of the more cumbersome $X(\omega)$, as for instance in defining the probability function: $f(x) = P(X = x)$.

The probability function of X will sometimes be given by means of a *formula* for computing $f(x)$ from x. In simple cases it can be given by means of a *table* of its values; this is possible because of the assumed discreteness, and feasible if the list of values is not too long.

Example 3-b In a draft lottery, it has been customary to draw a capsule from a drum containing 366 capsules, each with a slip of paper in it marked with one of the integers from 1 to 366. Let $X(\omega)$ denote the first number drawn; this variable has one of the discrete set of values $1, 2, \ldots, 366$. If the drawing is a "random selection," then all numbers in this list have the same chance of being drawn:

$$f(x) = \frac{1}{366}, \qquad x = 1, 2, \ldots, 366.$$

The distribution described by this probability function is said to be *uniform* over the set of values $\{1, 2, \ldots, 366\}$. ◄

Example 3-c When two dice are tossed as in Example 2-b, the 36 outcomes $\omega = (i, j)$ are equally likely. One may be (and in some games often is) interested in the *sum* of the two numbers in the pair:

$$X(i, j) \equiv i + j.$$

This function or random variable can have as its value any integer from $2 = 1 + 1$ up to $12 = 6 + 6$. The probabilities 1/36 for each (i, j) automatically define probabilities for the values of X. For example,

$$f(4) \equiv P(X = 4) = \frac{\text{number of pairs with } i + j = 4}{36} = \frac{3}{36},$$

since the three pairs (1, 3), (2, 2), and (3, 1) have sum 4. Similar computations yield the probability table:

x	2	3	4	5	6	7	8	9	10	11	12
$f(x)$	1/36	2/36	3/36	4/36	5/36	6/36	5/36	4/36	3/36	2/36	1/36

This table defines $f(x)$; but it is possible, albeit a little artificial, to give a formula that yields these same probabilities:

$$f(x) = \frac{6 - |x - 7|}{36}, \qquad x = 2, \ldots, 12. \quad ◄$$

Pairs or, more generally, vectors of random variables are often encountered. They arise naturally when variables are considered "together" and defined on the same sample space.

Each member of a pair (X, Y) could be studied as a random variable in its own right, but the two variables X and Y may be related in interesting and relevant ways. Such relationships can only be analyzed in the context of a model for the two together, in what is called their *joint* distribution. This is defined by a joint probability function that gives the probability of each possible pair of values:

> Joint probability function for (X, Y):
> $$f(x, y) = P(X = x \quad \text{and} \quad Y = y).$$

When probabilities are defined initially in the sample space, this is the probability of the event $\{X(\omega) = x \text{ and } Y(\omega) = y\}$.

Again, it is sometimes feasible to give a probability function for (X, Y) by means of a formula, but it can often be conveniently given in a table. It is natural to use a two-way table corresponding to ways of pairing each possible X-value with each possible Y-value. Thus if X can take values $x_1, x_2, \ldots$ and Y can take values $y_1, y_2, \ldots,$ then values for (X, Y) are pairs (x_i, y_j). Some of these pairs may occur with probability zero, but it does not hurt to include them. A table with values of X across the top and values of Y along the left side, say, and probabilities $f(x_i, y_j)$ opposite y_j and below x_i serves to define the joint probability function.

Example 3-d Manufactured items of a certain type can have two types of flaw; let X denote the number of one type and Y the number of the other type. One item is selected at random from a group of six items, listed, with numbers of flaws, as follows:

Item	Probability	X	Y
ω_1	1/6	0	2
ω_2	1/6	2	0
ω_3	1/6	1	1
ω_4	1/6	1	2
ω_5	1/6	2	1
ω_6	1/6	0	0

Since, as it happens, no two items are alike with regard to (X, Y), this table gives the values of $f(x, y)$. An equivalent two-way table is the following:

		X 0	1	2	
Y	0	1/6	0	1/6	1/3
	1	0	1/6	1/6	1/3
	2	1/6	1/6	0	1/3
		1/3	1/3	1/3	

Notice that some pairs [e.g., (1, 0)] have zero probability, since they cannot occur. The table has been extended to give sums of the rows and columns; these sums are probabilities for X and Y separately. For instance, since

$$\{Y = 2\} = \{X = 0 \text{ and } Y = 2\} + \{X = 1 \text{ and } Y = 2\} + \{X = 2 \text{ and } Y = 2\},$$

it follows that

$$P(Y = 2) = f(0, 2) + f(1, 2) + f(2, 2) = \frac{1}{6} + \frac{1}{6} + 0. \quad \blacktriangleleft$$

As in this example, the distribution of Y is found in the row totals in the right margin, and of X, in the column totals in the lower margin of a two-way table for the probability function $f(x, y)$. The name *marginal distribution* is applied to each of these distributions of X by itself and Y by itself. Notice that the marginal probabilities, although determined *by* the joint probabilities, do not determine them. For instance, in Example 3-d, the same *marginal* totals would have resulted had the joint probabilities all been equal to 1/9.

3.2 Expected Value

It is natural to want to think of a varying quantity in terms of some kind of typical value. And a "typical" value ought to be in the middle of the distribution, in some sense. One interpretation of "middle" is given by the *expected value*.

Example 3-e Suppose that a game pays $1 if an ordinary, fair die lands 6, otherwise nothing. The number of dollars won is a random variable:

$$X(6) = 1,$$

$$X(1) = X(2) = X(3) = X(4) = X(5) = 0.$$

The probability function for this variable is given in the following table:

x	$f(x)$
0	5/6
1	1/6

Consider a long sequence of trials, resulting in f_1 6's ($X = 1$) and f_0 non-6's ($X = 0$). The total reward in dollars is

$$f_1 \times 1 + f_0 \times 0,$$

and, divided equally among the n trials, this would amount to

$$\frac{f_1}{n} \times 1 + \frac{f_0}{n} \times 0 \doteq f(1) \times 1 + f(0) \times 0 = \frac{1}{6},$$

as the average value per trial. This latter idealized average has been called the "mathematical expectation" of the game. It is what a fair gambling house using an ideal or mathematical die would charge for one play of the game. [It is a curiosity, not at all obvious at this point, and somewhat off the track, that although one might expect a gambling house that makes this fair charge per game to come out even in the long run, it will in fact probably end up (after a long time) either way ahead or way behind. To be sure of not ending up behind (after a long time), the gambling house would have to charge slightly more than what is "fair."] ◄

The idea in the example above is easily generalized to any discrete random variable in the following definition.

DEFINITION. The *expected value* or *mean value* of a discrete random variable X is defined as the weighted sum

$$E(X) \equiv \sum xf(x).$$

Although Σ means to add, and we can only add numbers in a sequence (e.g., $\Sigma_i a_i$), it will make life simpler if we are permitted to adopt the abbreviated notation Σa. It is to be understood that the a is a generic name for values in a countable list, and that the Σ means to sum over that list. Thus $E(X) = \Sigma xf(x)$, where x refers to a list $x_1, x_2, x_3, \ldots$.

The words "average," "expected," and "mean," are commonly used in various, equivalent forms:

Average, or average value
Mean, or mean value
Expectation
Expected value, or expected number
Population mean, or population average

We may, in the same problem, speak of a random variable, of its distribution, or of the population of values. The term *expected value* is most often used when referring to the random variable, as suggested in the notation $E(X)$. But we also speak of this as the *mean value* of X, or as the mean value of the distribution of X, and then commonly denote it by μ or by μ_X. When referring to the sample space and its distribution as a population, we usually employ the term *population mean*, to distinguish it from (and by analogy with) the mean of a sample from the population, a notion that will be defined in Chapter 9. When the random variable is integer-valued, we may say "average number" or "expected number," as in "the expected number of children." Here we mean, more precisely, the expected value of the random variable X, the number of children, even though this expected number need not itself be an integer.

Example 3-f The usually accepted model for the outcome of the toss of two coins assigns equal probabilities to the four outcomes. Let X denote the number of heads:

ω	$P(\omega)$	$X(\omega)$
HH	1/4	2
HT	1/4	1
TH	1/4	1
TT	1/4	0

The probability function for X is then as given in the following table, along with products whose sum is $E(X)$:

x	$f(x)$	$xf(x)$
0	1/4	0
1	1/2	1/2
2	1/4	1/2
		$1 = E(X)$

The expected number of heads showing is 1. It is significant and useful to observe that $E(X)$ can also be computed from the ω-table:

$$E(X) = 2 \times \frac{1}{4} + 1 \times \frac{1}{4} + 1 \times \frac{1}{4} + 0 \times \frac{1}{4} = \sum X(\omega)P(\omega). \quad \blacktriangleleft$$

The last computation in Example 3-f illustrates an important fact, namely, that the expected value of $X(\omega)$ can be computed *either* from the distribution of probability among the values x_i, or from the distribution of probability among the outcomes ω:

$$E(X) = \sum xf(x) = \sum X(\omega)P(\omega).$$

A simple but useful application of this formula is to the case of a *constant* random variable, that is, a function $X(\omega)$ that has the same value k for each ω:

$$E(k) = \sum kP(\omega) = k\sum P(\omega) = k.$$

The average of a constant is that constant.

Another application of the computation of $E[X(\omega)]$ from the ω-probabilities is in showing the *additivity* of the averaging process. If two random variables $X(\omega)$ and $Y(\omega)$ are defined for each outcome ω, their *sum* is also a random variable:

$$Z(\omega) \equiv X(\omega) + Y(\omega).$$

The mean value of the sum is

$$\begin{aligned} E[Z(\omega)] &= \sum Z(\omega)P(\omega) \\ &= \sum [X(\omega) + Y(\omega)]P(\omega) = \sum X(\omega)P(\omega) + \sum Y(\omega)P(\omega) \\ &= E(X) + E(Y). \end{aligned}$$

Thus no matter how X and Y are related, the average of their sum is the sum of the averages.

Additivity of expected values:

$$E(X + Y) = E(X) + E(Y).$$

Example 3-g Random variables X and Y defined in Example 3-d were seen there to have these joint probabilities:

	0	1	2
0	1/6	0	1/6
1	0	1/6	1/6
2	1/6	1/6	0

The sum $Z = X + Y$ has values and probabilities, computed from the joint probabilities, as follows:

z	$P(Z = z)$	$zP(Z = z)$
0	1/6	0
2	1/2	1
3	1/3	1

The mean value is the sum of the products in the last column,

$$E(Z) = \sum zP(Z = z) = 2.$$

Each of the marginal distributions assigns probability equally among the values 0, 1, 2, with expected value

$$E(X) = E(Y) = 0 \cdot \frac{1}{3} + 1 \cdot \frac{1}{3} + 2 \cdot \frac{1}{3} = 1.$$

Clearly, $E(Z) = E(X) + E(Y)$. ◄

3.3 Expected Value of a Function

A real-valued function $g(x)$ defined for each value x of a random variable X is a random variable, being a function defined on the probability space of X-values. It is denoted by $g(X)$ or, more simply, by a new name, Y. The expected value of $Y = g(X)$ can be computed by translating the earlier formula for $E[X(\omega)]$, replacing ω by X and $X(\omega)$ by $g(x)$:

Expected value of a function of X:

$$E[g(X)] = \sum g(x)f(x).$$

To be sure, one could calculate $E(Y)$ as a sum of its values each weighted by probabilities in the Y-space, but the summation given is usually more direct, avoiding the need for a calculation of the Y-probabilities themselves.

The simplest function of X is a constant—a "function" in only a trivial sense: If $g(X) \equiv k$, then

$$E(k) = \sum kf(x) = k \sum f(x) = k.$$

As we saw in the previous section, the average value of a constant is that constant.

The function $g(X) = kX$, a constant multiple of X, is often encountered:

$$E(kX) = \sum kxf(x) = k \sum xf(x) = kE(X).$$

Thus a constant multiplier factors out of the averaging process. This and the property of additivity may be summarized as follows:

Linearity of the averaging process:

$$E(aX + bY) = aE(X) + bE(Y),$$

$$E(aX + b) = aE(X) + b.$$

An important linear function of X is the deviation of X from its mean, $Y \equiv X - \mu$, or $X - E(X)$. According to the property of linearity,

$$E(X - \mu) = E(X) - \mu = 0.$$

That is, *the average deviation of a variable about its mean is zero,* the terms corresponding to x's to the right of $E(X)$ exactly canceling the contribution from x's to the left of $E(X)$.

A real-valued function $g(x, y)$ defined for each pair (x, y) of values of a pair of random variables (X, Y) is itself a random variable—a function defined on the probability space of (x, y)-values. The expected value of $Z \equiv g(X, Y)$ is again computable from the earlier formula for $E[X(\omega)]$, replacing ω by (x, y) and $X(\omega)$ by $g(x, y)$, to obtain a formula that does not require computation of probabilities for Z.

> Expected value of a function of (X, Y):
>
> $$E[g(X, Y)] = \sum g(x, y)f(x, y),$$
>
> the sum extending over all possible pairs (x, y).

Expectations of this type will be encountered in describing aspects of a joint distribution in Section 3.6.

The summation used in defining $E[g(X, Y)]$ can be done in any order, as long as all possible points (all possible points of positive probability) are taken into account. If the possible pairs are defined by pairing possible X-values $x_1, \ldots, x_m$ with possible Y-values $y_1, \ldots, y_k$, as in a two-way table of the probability function, then the summation can be accomplished systematically, first adding over individual rows and then adding the column of row totals, or conversely:

$$\begin{aligned} E[g(X, Y)] &= \sum_{j=1}^{k} \left[\sum_{i=1}^{m} g(x_i, y_j) f(x_i, y_j) \right] \\ &= \sum_{i=1}^{m} \left[\sum_{j=1}^{k} g(x_i, y_j) f(x_i, y_j) \right]. \end{aligned}$$

Example 3-h Consider again the joint distribution of Example 3-g, repeated in Table 3.1. A second table of values of the particular function $g(x, y) \equiv xy$ is also shown (Table 3.2) along with Table 3.3 of products $g(x, y)f(x, y)$:

Table 3.1 Probabilities

$y \backslash x$	0	1	2
0	1/6	0	1/6
1	0	1/6	1/6
2	1/6	1/6	0

Table 3.2 Values of xy

$y \backslash x$	0	1	2
0	0	0	0
1	0	1	2
2	0	2	4

Table 3.3 Products

$y \backslash x$	0	1	2	Sum
0	0	0	0	0
1	0	1/6	2/6	3/6
2	0	2/6	0	2/6
Sum	0	3/6	2/6	5/6

The final entry, 5/6, is the value of $E(XY)$. (The tables are given to show the computation, but of course it can be done mentally from just the first table of probabilities.) Having seen that $E(X + Y) = E(X) + E(Y)$, one might have been tempted to expect that $E(XY) = E(X)E(Y)$. This example shows that such is not generally the case. ◄

3.4 Expectation as a Center of Gravity

Most people have had, formally or informally (as on a teeter-totter), some experience with distributions of mass. A discrete mass distribution on a line puts, say, m_1 units at x_1, m_2 at $x_2, \ldots,$ and m_k at x_k. A probability distribution, similarly, puts p_1 units of probability at x_1, p_2 at x_2, and so on. The only difference is that $\sum p_i = 1$, whereas $m \equiv \sum m_i$ may not be 1. The analogy is between probabilities p_i and *relative* masses m_i/m, or proportions of the whole.

The center of gravity of a finite system of point masses on a line is

$$\text{c.g.} \equiv \frac{\sum m_i x_i}{\sum m_i} = \sum x_i \left(\frac{m_i}{m}\right).$$

This arithmetic is precisely that which is involved in computing the expected value of a discrete probability distribution with probabilities $p_i = m_i/m$. Interpreting probability as mass endows the expected value with properties of a center of gravity. In particular, an expected value is a *balance* point, expressed in the property $E(X - \mu) = 0$. This means that the expected value, like a balance point, must lie in the midst of the distribution—not off to one side—a property that can be used as a rough check of an expected value computation.

Use of the term *moments* is carried over from the field of mechanics. The mean or center of gravity is called the *first moment* of the distribution. More generally, the average of the *k*th power of a variable is its *k*th *moment*:

$$\mu'_k \equiv E(X^k).$$

The averages of powers of deviations about an arbitrary value $x = a$ are called moments "about" that value. Thus the *k*th moment about the mean is

$$\mu_k \equiv E[(X - \mu)^k].$$

(The *first* moment about the mean is always zero, as was shown in Section 3.3.)

3.5 Variance

The second moment about the mean will be large if there is a heavy concentration of mass or probability far from the mean. It is a measure describing dispersion or variation about the mean:

Variance of a random variable:

$$\begin{aligned} \operatorname{var} X &= E[(X - \mu)^2] \\ &= E(X^2) - \mu^2. \end{aligned}$$

The second expression given here for variance is readily obtained from the first by expanding the square:

$$(X - \mu)^2 = X^2 - 2\mu X + \mu^2$$

and the averaging term by term:

$$E[(X - \mu)^2] = E(X^2) - E(2\mu X) + E(\mu^2).$$

The expected value of a constant is that constant: $E(\mu^2) = \mu^2$, and the constant 2μ factors out of the middle term:

$$E(2\mu X) = 2\mu E(X) = 2\mu^2.$$

Collecting terms yields the given formula.

The unit of variance is the square of the unit of X itself. The square root of the variance, on the other hand, has the same unit as X and can be interpreted as a kind of average deviation about the mean (called a root-mean-square or *r.m.s.* average):

Standard deviation of a random variable:

$$\sigma \equiv \sqrt{\operatorname{var} X} = \sqrt{E[(X-\mu)^2]}$$
$$= \sqrt{\sum (x-\mu)^2 f(x)}.$$

Example 3-i The probabilities for the six faces of a fair die are equal; so if X denotes the number of points showing, then

$$P(X = i) = \frac{1}{6}, \qquad \text{for } i = 1, 2, \ldots, 6.$$

The mean value is

$$E(X) = \sum_1^6 if(i) = \frac{1}{6}(1 + 2 + \cdots + 6) = \frac{7}{2},$$

and the expected square:

$$E(X^2) = \sum_1^6 i^2 f(i) = \frac{1}{6}(1^2 + 2^2 + \cdots + 6^2) = \frac{91}{6}.$$

The variance is then computed as follows:

$$\operatorname{var} X = E(X^2) - (EX)^2 = \frac{91}{6} - \left(\frac{7}{2}\right)^2 = \frac{35}{12}.$$

The standard deviation is the square root:

$$\sigma = \sqrt{\operatorname{var} X} = \sqrt{\frac{35}{12}} = 1.708.$$

The largest absolute deviation about the mean is $6 - 3.5 = 2.5$, and the smallest is 0.5. The standard deviation is a kind of average of the absolute deviations. ◀

3.6 Covariance

A "product moment" or "mixed moment" that describes the joint variation of a pair of random variables (X, Y) is their *covariance*:

> *Covariance* of two random variables:
> $$\begin{aligned}\text{cov}(X, Y) &\equiv E[(X - \mu_X)(Y - \mu_Y)] \\ &= E(XY) - \mu_X\mu_Y.\end{aligned}$$

The second of these expressions for covariance, the expected product minus the product of the expectations, is obtained from the first formula by expanding the product of the deviations and averaging term by term.

The products $(X - \mu_X)(Y - \mu_Y)$ will be positive when X and Y are on the same side of their means. The covariance will therefore be positive if X and Y tend to be large together and small together. And conversely, the covariance is negative when the relationship is contrary—X tends to be large when Y is small, and vice versa. Thus cov (X, Y) is a measure of concordance between X and Y. However, as such, it suffers from being sensitive to changes of scale, a flaw that will be remedied in Chapter 16 in defining a correlation coefficient.

Example 3-j A bag contains five beads, of which two are red, two black, and one white. Let X denote the number of red and Y the number of black beads in a random selection of three. Then, for instance,

$$f(1, 2) = P(X = 1 \text{ and } Y = 2) = P(1 \text{ red and } 2 \text{ black})$$
$$= \frac{\binom{2}{1}\binom{2}{2}\binom{1}{0}}{\binom{5}{3}} = \frac{2}{10}.$$

Similar computations will complete the probability table for (X, Y):

		X = 0	X = 1	X = 2	
Y	0	0	0	1/10	1/10
	1	0	4/10	2/10	6/10
	2	1/10	2/10	0	3/10
		1/10	6/10	3/10	

The means are equal (by symmetry):

$$E(X) = E(Y) = 1 \times .6 + 2 \times .3 = 1.2,$$

and the average product is

$$E(XY) = 1 \times .4 + 2 \times .2 + 2 \times .2 = 1.2,$$

so that

$$\text{cov}\,(X, Y) = 1.2 - (1.2)^2 = -.24.$$

The *negative* covariance is expected, for if the sample includes a large number of red beads, the number of black beads tends to be small. ◄

Example 3-k Consider the bivariate distribution that has equal probabilities at these points: (0, 0), (0, 1), (1, 1), (1, 2), (2, 0), (2, 1). The product sum is

$$\sum x_i y_i = 0 + 0 + 1 + 2 + 0 + 2 = 5,$$

and so (because each probability weight is 1/6), $E(XY) = 5/6$. The means are easily found to be $EX = 1$ and $EY = 5/6$, so that

$$\text{cov}\,(X, Y) = E(XY) - \mu_X\mu_Y = \frac{5}{6} - 1 \times \frac{5}{6} = 0.$$

The variables are uncorrelated—not related, in the special way suggested by covariance. (They *are* related in this sense: If the value of one variable is known, the value of the other variable may be restricted. For instance, if $Y = 2$, then X cannot be 0.) ◄

The covariance will be encountered for the present only in considering the variance of a sum. The variance of $X + Y$ is computed from the joint distribution of X and Y as follows:

$$\begin{aligned}
\text{var}\,(X + Y) &= E\{[X + Y - (\mu_X + \mu_Y)]^2\} \\
&= E\{[(X - \mu_X) + (Y - \mu_Y)]^2\} \\
&= E[(X - \mu_X)^2 + 2(X - \mu_X)(Y - \mu_Y) + (Y - \mu_Y)^2] \\
&= \text{var}\,X + \text{var}\,Y + 2\,\text{cov}\,(X, Y).
\end{aligned}$$

Thus the variance is *not* additive, in general. It *is* additive if the variables have a zero covariance; such variables are said to be *uncorrelated*.

The formula for the variance of a sum is easily extended to any finite number of variables.

Variance of a sum:

$$\operatorname{var}\left(\sum_1^n X_i\right) = \sum_1^n \operatorname{var} X_i + \sum_{i \neq j} \operatorname{cov}(X_i, X_j).$$

That is, the variance of a sum is the sum of the variances *plus* the sum of covariances of all pairs of different terms, and there are $n(n - 1)$ such pairs in all. [Observe that $\operatorname{cov}(X_2, X_3) = \operatorname{cov}(X_3, X_2)$ but that *each* of these occurs in the sum.] If all of the covariances involved are zero, the variables are said to be *pairwise uncorrelated*. In this case the variance is additive:

If $X_1, \ldots, X_n$ are pairwise uncorrelated,

$$\operatorname{var}\left(\sum_1^n X_i\right) = \sum_1^n \operatorname{var} X_i.$$

Problems

1. Three coins are tossed so that the eight elementary outcomes are equally likely. Let X denote the number of heads.

(a) Give the probability table for X.
(b) Compute $E(X)$.

2. If ω is one of the eight outcomes in tossing three coins, as in Problem 1, let

$$Y(\omega) = \begin{cases} 1, & \text{if the coins all turn up the same,} \\ 0, & \text{otherwise.} \end{cases}$$

Compute $E(Y)$.

3. A carton of 10 includes two defective and eight good items. Let X denote the number of defective items in a selection of three items picked at random without replacement from the 10.

(a) Give the probability function for X as a table of possible values and corresponding probabilities.
(b) Calculate $E(X)$.

4. In the Goren system of contract bridge bidding, "points" are assigned to each card, as follows: ace = 4, king = 3, queen = 2, jack = 1. All others are assigned 0.

(a) Let X denote the number of points assigned to a card selected at random from the deck and compute $E(X)$ and var X.
(b) Let Y denote the number of points in a hand (random selection) of 13 cards, and compute $E(Y)$.

(HINT: The number of points in each of the four hands dealt has the same distribution, and so the same mean.)

5. Let X be the number of correct matches when (as in Problem 5, Chapter 2), the hatcheck girl passes out four hats at random to their four owners: A, B, C, and D.

(a) Determine the distribution of X.
(b) Calculate EX.

Let

$$Y_A = \begin{cases} 1, & \text{if } A \text{ gets his own hat,} \\ 0, & \text{if } A \text{ does not get his own hat.} \end{cases}$$

(c) Compute $P(Y_A = 1)$.
(HINT: There are 3! permutations in which A's hat is in the correct position.)
(d) Compute $E(Y_A)$.
(e) Note that the distributions of Y_A, Y_B, Y_C, and Y_D are identical. Use the fact that $X = Y_A + Y_B + Y_C + Y_D$ to verify your answer in part (b).

6. Compute the variances for the following variables:

(a) The random variable X defined in Problem 1.
(b) The random variable X defined in Problem 3.

7. Just one of four similar keys opens my office door, and I try them one at a time. Let X denote the number of keys I have to put in the lock before the door opens.

(a) Construct the probability table for X assuming that the various possible sequences in which keys are tried to be equally likely.
(b) Compute the mean and variance of X.

8. Let X denote the result of picking a digit at random, as in Table XIII [i.e., $X(\omega) = \omega$, where $\omega = 0, 1, 2, \ldots, 9$ and $P(\omega) \equiv 1/10$].

(a) Compute $E(X)$.
(b) Compute $E(|X - E(X)|)$.
(c) Compute $E([X - E(X)]^2)$.

9. A contestant on "Let's Make a Deal" had won \$9000 in prizes and was then offered the option of trading for whatever was behind one of three doors (the choice of door being his): Behind one door was \$2000; behind another, \$5000; and behind the third, \$20,000.

(a) What is the expected value of this option?
(b) Should the contestant take it?

10. In the theory of "utility," an amount of money is assigned a value according to what it is worth to a person under given circumstances. Consider these situations:

(a) A man owes a loan shark $20,000, due the next day with no apparent way to get the money. Assign utility as follows: $u(20{,}000) = 1$, $u(2000) = u(5000) = u(9000) = 0$ and compute the expected utility of the option offered in Problem 9 if he is the contestant. Should he trade?
(b) A man owes a loan shark $9000, due the next day, with no way to get the money. Now $2000 and $5000 are useless, but $9000 is not. Assign

$$u(2000) = u(5000) = 0, \qquad u(9000) = u(20{,}000) = 1$$

and compute the expected utility of the deal. Should he trade?

11. The following frequencies of scores on a certain hole on a golf course have been observed over a period of several years, for professional golfers and for club members.

	Relative frequency	
Score	Professional golfer	Club member
2 (eagle)	.02	.00
3 (birdie)	.16	.03
4 (par)	.68	.22
5 (bogie)	.13	.27
6 (double bogie)	.01	.27
7	.00	.13
8	.00	.05
9	.00	.02
10	.00	.01

Interpreting the given relative frequencies as probabilities, compute the means and standard deviations of pro scores and of member scores. [Par, as in golf, is another kind of "average." It means *standard* or *norm*, and its interpretation is that of a "mode," or most common value, for accomplished players (a standard that is thought of as a goal by run-of-the-mill golfers).]

12. Three coins are tossed so that the eight elementary outcomes are equally likely. Let X denote the number of heads and Y the number of tails showing.

(a) Compute $E(X + Y)$ and $\text{var}(X + Y)$.
(b) Compute $E(X - Y)$ and $\text{var}(X - Y)$.
(c) Compute $\text{cov}(X, Y)$.

(d) Compute var X and var Y, and then verify the formula for var $(X + Y)$ in terms of moments of X and Y.

13. In picking three socks at random from four blue, four green, and two red socks (Problem 10, Chapter 2), let X, Y, and Z denote the number of blue, the number of green, and the number of red socks, respectively, among the three picked.

(a) Construct the probability table for X and Y.
(b) Notice that X and Y have the same distribution. Was this to be expected?
(c) Compute $E(X + Y)$.
(d) Compute $E(Z)$.
(NOTE: $X + Y + Z = 3$.)
(e) Obtain the probability table for Z.
(f) Explain why var Z must equal var $(X + Y)$.
(g) Compute var $(X + Y)$.
(h) Compute cov (X, Y).

14. Using the fact that $X - Y = X + (-Y)$, obtain a formula for var $(X - Y)$ like that obtained for var $(X + Y)$ in the text.

Independence

If a particular probability model for an experiment of chance remains valid when the outcome of another experiment becomes known and whatever that outcome may be, the two experiments are said to be independent. When they are not independent, the betting odds for one experiment *are* influenced by what happened in the other. The study of such relationships, or the absence of them, will use the notion of "conditional" probability.

4.1 Conditional Probability

Consider a single experiment of chance, and two events A and B in the sample space of the experiment. In a given probability model for the experiment, each of these has a probability, called $P(A)$ and $P(B)$, respectively. Suppose now that when the experiment is performed, we learn by some means that B has occurred. Shall we then change our betting odds for A?

The information that B has occurred has the effect of eliminating, as possible outcomes, all outcomes of the original sample space that are not in B. The appropriate model now has B as the sample space. The various outcomes in B still have the same chances relative to each other, because in any sequence of trials their frequencies are unaltered by dropping outcomes not in B. But the probabilities in the new sample space will have to be revised upward from what they were originally, because the outcomes in the reduced list of outcomes that are possible are now more likely. The new probability for A is called $P(A|B)$, read: "the probability of A *given* B."

Example 4-a A die is tossed so that the six sides are equally likely—they occur in about equal proportions in a long sequence of tosses. Suppose it becomes known that the outcome is an *even* number, an event B of probability $P(B) = 1/2$. In light of this, the outcomes 2, 4, and 6 are the only ones possible. However, in a long sequence of tosses, if odd outcomes are ignored, the even ones *still occur in the same proportions*; but these proportions are now about $1/3$ instead of $1/6$:

$$P(2|B) = P(4|B) = P(6|B) = \frac{1}{3}.$$

In this reduced model one can calculate the probability of any event, and this is called a conditional probability to distinguish it from the probability assigned to it in the original model. For instance, let C denote the event $\{3, 6\}$. What is the proper way to bet on C given that B has occurred, that the outcome is even? The outcome 3 is not in B, so only the 6 is possible in the reduced sample space, and its probability is $1/3$:

$$P(C|B) = P(6|B) = \frac{1}{3}. \quad ◄$$

The new probability for an outcome ω in the new sample space B, appropriate when B is known to have occurred, is obtained from the unconditional probability as follows:

$$P(\omega|B) = \frac{P(\omega)}{P(B)}, \qquad \text{for } \omega \text{ in } B.$$

With probabilities revised in this way, the new probabilities of the ω's in B are still in the same ratios to each other [having all been divided by $P(B)$], but they add up to 1:

$$\sum_{\omega \text{ in } B} P(\omega|B) = \sum_{\omega \text{ in } B} \frac{P(\omega)}{P(B)} = \frac{P(B)}{P(B)} = 1.$$

And because $P(\omega|B) \geq 0$, the conditional probabilities given B define a proper probability model for the reduced sample space B.

As in Example 4-a, questions may be asked, given that B occurred, about events A that are not entirely within the reduced sample space B. That is, one may still want to work in the original sample space. But an $\omega \in \Omega$ that is *not* in B is not possible, given that B occurs, so that

$$P(\omega|B) \equiv 0, \qquad \text{if } \omega \notin B.$$

And then for any event A in the original sample space,

$$P(A|B) = \sum_{\omega \in A} P(\omega|B) = \sum_{\omega \in AB} \frac{P(\omega)}{P(B)} = \frac{P(AB)}{P(B)}.$$

This line of reasoning leads to the definition of conditional probability, even in nondiscrete cases, provided that the division is possible:

Conditional probability:

$$P(A|B) \equiv \frac{P(AB)}{P(B)}, \qquad \text{if } P(B) \neq 0.$$

Example 4-b Two cards are selected at random from a standard deck of playing cards. The $\binom{52}{2}$ possible combinations are thus equally likely. If only the colors are of interest, there are three outcomes, with probabilities as follows:

Outcome	Probability
Both black	25/102
Both red	25/102
One of each	52/102

[For instance, $P(BB) = \binom{26}{2} \Big/ \binom{52}{2}$.] If it is known (perhaps from a flash of red in a reflecting surface) that not both cards are black, the probability that both are red is as follows:

$$P(\text{both red}|\text{not both black}) = \frac{P(\text{both red})}{P(\text{not both black})}$$

$$= \frac{25/102}{77/102} = \frac{25}{77}.$$

Observe here that because "both red" is entirely within the event "not both black," the intersection of these two events is the smaller, "both red." ◀

The formula for conditional probability is often given in a form obtained from the defining formula multiplied through by the denominator, to yield the following:

Multiplication rule:

$$P(AB) = P(A|B)P(B).$$

Upon interchanging the roles of A and B, one obtains

$$P(AB) = P(B|A)P(A),$$

so that

$$P(A|B)P(B) = P(B|A)P(A).$$

And then, since

$$\begin{aligned} P(B) &= P(AB) + P(A^cB) \\ &= P(B|A)P(A) + P(B|A^c)P(A^c), \end{aligned}$$

this important result follows:

Bayes' theorem:

$$P(A|B) = \frac{P(B|A)P(A)}{P(B|A)P(A) + P(B|A^c)P(A^c)}.$$

This fundamental formula shows how to reverse the roles of A and B, to obtain $P(A|B)$ from $P(B|A)$. It is at the heart of Bayesian inference, to be discussed in Chapter 13.

Events A and B define two schemes of classification for outcomes, namely, according to whether A or A^c occurs, and according to whether B or B^c occurs. The various probabilities may be presented in a "two-way" table:

	B	B^c	
A	$P(AB)$	$P(AB^c)$	$P(A)$
A^c	$P(A^cB)$	$P(A^cB^c)$	$P(A^c)$
	$P(B)$	$P(B^c)$	1

The ratios of entries in the body of the table to marginal entries define conditional probabilities. If probabilities for A, $B|A$, and $B|A^c$ are given, the table can be completed and used to read out the probabilities for B, $A|B$, and $A|B^c$. Bayes' theorem simply formalizes this process.

Example 4-c A bead is picked at random from a container with beads that are green or yellow, some with a stripe and some not. Suppose that half the beads are striped and that three fifths of the striped beads and two fifths of the nonstriped beads are green,

$$P(S) = P(N) = \frac{1}{2},$$

$$P(G|S) = P(Y|N) = \frac{3}{5}.$$

Then the probability that the bead drawn is striped, given that it is green, is computed as follows:

$$\begin{aligned} P(S|G) &= \frac{P(G|S)P(S)}{P(G|S)P(S) + P(G|N)P(N)} \\ &= \frac{(3/5) \times (1/2)}{(3/5) \times (1/2) + (2/5) \times (1/2)} = \frac{3}{5}. \end{aligned}$$

The corresponding two-way table is constructed by means of the multiplication rule, which gives

$$P(SG) = P(G|S)P(S) = \frac{3}{5} \times \frac{1}{2},$$

$$P(NG) = P(G|N)P(N) = \frac{2}{5} \times \frac{1}{2},$$

or

	G	Y	
S	.3	.2	.5
N	.2	.3	.5

The column totals yield $P(G) = .5$ and $P(Y) = .5$, and the ratios of the G-column entries to $P(G) = .5$ are the conditional probabilities for S and N given G. ◄

Example 4-d Workers A, B, and C make the same type of article, A being responsible for 20 per cent, B for 30 per cent, and C for 50 per cent of the total output. Suppose that 5 per cent of A's output, 10 per cent of B's output, and 4 per cent of C's output have flaws. What is the probability that an article returned as flawed was made by A?

The probability that an article selected randomly from the total output is flawed is

$$P(F) = P(F|A)P(A) + P(F|B)P(B) + P(F|C)P(C)$$
$$= .05 \times .20 + .10 \times .30 + .04 \times .50 = .06.$$

Then, according to Bayes' formula,

$$P(A|F) = \frac{P(F|A)P(A)}{P(F)} = \frac{.05 \times .20}{.06} = \frac{1}{6},$$

in contrast with the unconditional probability 1/5. The knowledge that the article is flawed has changed the odds on A.

The computation of $P(A|F)$ becomes somewhat more transparent, perhaps, if one imagines that of 100 articles, one of the 20 made by A is flawed, three of the 30 made by B are flawed, and 2 of the 50 made by C are flawed. Then if one article selected at random from the 100 is flawed, it is one of $1 + 3 + 2 = 6$ equally likely articles, and

$$P(A|F) = \frac{1}{6}, \qquad P(B|F) = \frac{3}{6}, \qquad P(C|F) = \frac{2}{6}. \quad ◀$$

4.2 Independent Events

It may happen that knowledge of the occurrence of event B does *not* change the probability assigned to A. It then follows that information as to the occurrence of A does not affect the odds on B. For, if

$$P(A|B) = P(A),$$

then

$$P(B|A) = P(A|B)\frac{P(B)}{P(A)} = P(B).$$

In such case the events A and B are said to be *independent.* If they are so, then the multiplication rule becomes $P(AB) = P(A)P(B)$; and this, in turn, implies that the conditional and unconditional probabilities are the same. This symmetric product relation is usually taken as defining independence:

DEFINITION. Events A and B are *independent* if and only if

$$P(AB) = P(A)P(B).$$

Notice that if $P(B)$ or $P(A)$ is zero, then $P(AB)$ is also zero (Chapter 1, Problem 6); so an event of zero probability is independent of every other event, although conditional probabilities, given that event, are not defined.

Example 4-e A card drawn from a standard deck may be classed as red (R) or black, and also as a face card (F) or not a face card. Since there are 6 red face cards, 12 face cards, and 26 red cards,

$$P(RF) = \frac{6}{52} = \frac{12}{52} \times \frac{26}{52} = P(R)P(F).$$

Thus R and F are independent events, and

$$P(R|F) = \frac{6}{12} = \frac{26}{52} = P(R),$$

$$P(F|R) = \frac{6}{26} = \frac{12}{52} = P(F).$$

The proportion of red cards in the deck is the same as the proportion of red cards among the face cards (1/2), and the proportion of face cards in the deck is the same as the proportion of face cards among the red ones (3/13). Observe the proportionality of rows and the proportionality of columns in the two-way probability table:

	R	B	
F	6/52	6/52	12/52
N	20/52	20/52	40/52
	26/52	26/52	52/52

◀

4.3 Composite Experiments

An experiment $\mathscr{E}$ may be composed of two subexperiments, $\mathscr{E}_1$ followed by $\mathscr{E}_2$. The sample space and probabilities for $\mathscr{E}_2$ may depend on what happens when $\mathscr{E}_1$ is performed.

Suppose, as is often the case, a model for $\mathscr{E}$ is to be constructed from a model for $\mathscr{E}_1$ and a model for $\mathscr{E}_2$ that depends on the outcome of $\mathscr{E}_1$. The multiplication rule for arbitrary events is used to define probabilities for each outcome of $\mathscr{E}$ that is a pair consisting of an outcome of $\mathscr{E}_1$ and an outcome of $\mathscr{E}_2$.

Example 4-f A ball is drawn at random from a container with two red and three black balls, and then a second ball is drawn at random from the four that remain. The probability of the outcome RB, red ball on the first draw and black on the second, is given by the multiplication rule:

$$P(RB) = P(\text{red on 1st draw})P(\text{black on 2nd}|\text{red on 1st})$$

$$= \frac{2}{5} \times \frac{3}{4} = \frac{3}{10}.$$

The 2/5 and 3/4 are computed from the assumption that each draw is *random*. ◀

If the n_1 outcomes of $\mathscr{E}_1$ are equally likely, and if, given the outcome of $\mathscr{E}_1$, the n_2 outcomes of $\mathscr{E}_2$ are equally likely, then each outcome of the composite experiment $\mathscr{E}$ has probability

$$P(\omega) = \frac{1}{n_1} \times \frac{1}{n_2} = \frac{1}{n_1 n_2},$$

where the $1/n_2$ is a conditional probability. That is, the $n_1 n_2$ outcomes of $\mathscr{E}$ are equally likely, as asserted in Section 2.2. This result clearly extends to the case of several subexperiments, provided at each stage the outcomes are equally likely *given* what has occurred at previous stages.

Example 4-g A permutation of six objects three at a time is to be formed by selecting an object at random for the first position, then selecting one at random from the remaining five for the second position, and finally selecting one at random for the third position from those remaining. There are then $6 \times 5 \times 4 = 120$ permutations, and these are equally likely. For instance, if the objects are named A, B, C, D, E, and F, the probability of a particular permutation, say AFE, is

$$P(AFE) = \frac{1}{6} \cdot \frac{1}{5} \cdot \frac{1}{4} = \frac{1}{(6)_3} = \frac{1}{120}. \quad ◀$$

Example 4-h One of three closed boxes contains keys to a new car. A contestant in "Let's Make a Deal" chooses one of the boxes. She turns down \$500 for the box she has picked. The quizmaster then says that he will open one of the other boxes; he does so and shows her that it is empty. He then says that her chances are now one in two and offers \$1000 for the box. (She refuses.) Is he right in asserting that the two unopened boxes are equally likely to be the lucky one?

To answer this, it must be understood how he chooses a box to open. First, he would not show her the box with the key, and he knows where it is. Second, if the contestant *does* have the box with the key, the quizmaster has to choose between two empty boxes; suppose he does so at random. Call the box that the contestant picks box No. 1, and let E_i denote the event that the keys are in box i. Then

$$P(\text{opens box } 2|E_1) = \frac{1}{2},$$

$$P(\text{opens box } 2|E_2) = 0,$$

$$P(\text{opens box } 2|E_3) = 1.$$

As far as the contestant knows, the key is equally likely to be in box 1, box 2, or box 3:

$$P(E_1) = P(E_2) = P(E_3) = \frac{1}{3}.$$

Using the multiplication rule to combine the location of the key and the quizmaster's selection, one obtains

$$P(\text{opens } 2) = \frac{1}{2} \times \frac{1}{3} + 0 \times \frac{1}{3} + 1 \times \frac{1}{3} = \frac{1}{2}.$$

The probability desired is then (as a manifestation of Bayes' theorem):

$$P(E_1|\text{opens } 2) = \frac{P(\text{opens } 2|E_1)P(E_1)}{P(\text{opens } 2)}$$

$$= \frac{(1/2) \times (1/3)}{1/2} = \frac{1}{3}.$$

That is, after box 2 is seen to be empty, box 1 still has probability 1/3. However, box 3 now has probability 2/3. ◄

Suppose that composite experiments $\mathscr{E}$ and $\mathscr{F}$ are performed *independently*, in such a way that the outcome of one has no bearing on the other's outcomes and their relative chances. The model for the composite experiment $\mathscr{E}$ followed by $\mathscr{F}$ is again constructed by the multiplication rule, but now the probabilities in $\mathscr{F}$ are unconditional. In this composite model, the conditional probability of an event in $\mathscr{E}$ given an event in $\mathscr{F}$ is the same as in the model for $\mathscr{E}$ alone. Moreover, the model for $\mathscr{F}$ followed by $\mathscr{E}$ is the same as for $\mathscr{E}$ followed by $\mathscr{F}$; so they could even be performed simultaneously.

Example 4-i When three coins are tossed, the outcomes of the combined toss are formed by associating each outcome for each coin with each outcome of the others. They are $2 \cdot 2 \cdot 2 = 8$ in number:

HHH, HHT, HTH, THH, THT, HTT, TTH, TTT.

Their probabilities are computed by the multiplication of probabilities from the models for the individual coins. If each coin is a fair coin, then, for instance,

$$P(\text{HTH}) = P_1(\text{H})P_2(\text{T})P_3(\text{H}) = \left(\frac{1}{2}\right)^3,$$

where, for example, $P_1(\text{H})$ means P (heads on coin 1). And then

$$P_1(H) = P(\text{HHH, HHT, HTH, or HTT})$$
$$= \frac{1}{8} + \frac{1}{8} + \frac{1}{8} + \frac{1}{8} \quad ◀$$

An experiment that is formed by combining the independent, discrete experiments $\mathscr{E}$ with outcomes $\{e_1, e_2, \ldots, e_k\}$, and $\mathscr{F}$ with outcomes $\{f_1, f_2, \ldots, f_m\}$, has km outcomes that are conveniently listed as cells in a two-way table:

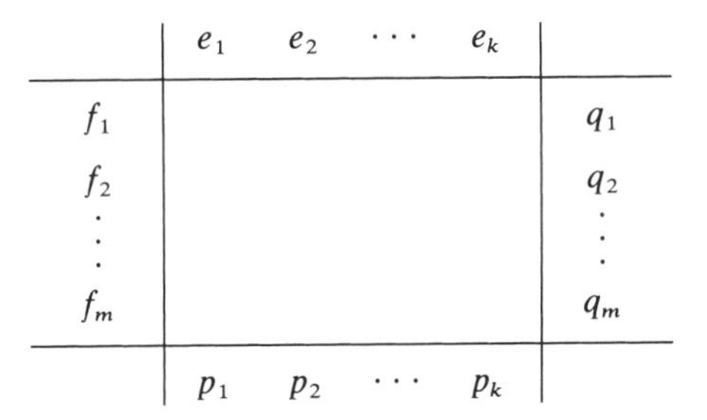

	e_1	e_2	$\cdots$	e_k	
f_1					q_1
f_2					q_2
$\vdots$					$\vdots$
f_m					q_m
	p_1	p_2	$\cdots$	p_k	

If the probabilities $P(e_i) = p_i$ from $\mathscr{E}$ and the probabilities $P(f_j) = q_j$ from $\mathscr{F}$ are put in the margins as shown, then the probability for the pairs (e_i, f_j) may be entered in the appropriate cell as a *product* of marginal probabilities. Then the rows of the table are proportional to the bottom marginal row, and the columns are proportional to the right-hand marginal column.

4.4 Independent Random Variables

Discrete random variables X and Y can be thought of as experiments with discrete sample spaces that happen to consist of *numerical* outcomes. The model for the pair (X, Y) will *incorporate* independence if joint probabilities are defined by multiplication of the marginal probability functions:

> Condition for independence of X and Y:
>
> $$f(x, y) = f_X(x)f_Y(y), \qquad \text{for all } x, y.$$

The model *implies* independence of X and Y if it is defined by a joint probability function that has this property.

Example 4-j Two dice are tossed, the outcome being one of the 36 pairs (i, j) where $i = 1, \ldots, 6$ and $j = 1, \ldots, 6$. If these 36 possible outcomes are *assigned* equal probabilities,

$$f(i, j) = \frac{1}{36},$$

then the marginal probabilities are

$$f_X(i) = \sum_{j=1}^{6} f(i, j) = \frac{1}{6},$$

$$f_Y(j) = \sum_{i=1}^{6} f(i, j) = \frac{1}{6},$$

and clearly these can be multiplied to recover the joint probability $f(i, j)$. Thus X and Y turn out to be independent.

On the other hand, if it is felt, from the way the dice are tossed, that X and Y must be independent, then the model should incorporate this independence, and one *defines* joint probabilities by multiplying probabilities for the X-values and Y-values:

$$f(i, j) = P(X = i, Y = j) \equiv P(X = i)P(Y = j)$$
$$= \frac{1}{6} \times \frac{1}{6} = \frac{1}{36}. \quad ◀$$

When X and Y are independent, they are also uncorrelated. This follows because of the factorization of the joint probability function:

$$\begin{aligned} E(XY) &= \sum_y \sum_x xyf(x, y) \\ &= \sum_y \sum_x xyf_X(x)f_Y(y) \\ &= \left[\sum_y yf_Y(y)\right]\left[\sum_x xf_X(x)\right] \\ &= (EX)(EY). \end{aligned}$$

Since the expected product equals the product of the means, independent variables are uncorrelated—have zero covariance.

The n variables $(X_1, \ldots, X_n)$ are said to be independent if their joint probability function factors

$$P(X_1 = x_1, \ldots, X_n = x_n) = P(X_1 = x_1) \cdots P(X_n = x_n).$$

This implies, in particular, that any *pair* of them would also be independent and so have zero covariance; they are pairwise-uncorrelated. Hence the following:

> If $X_1, \ldots, X_n$ are independent, then
>
> $$\text{var}\left(\sum_1^n X_i\right) = \sum_1^n \text{var } X_i.$$

Example 4-k Two fair dice are tossed so that the results (X, Y) are independent random variables. The distribution of the pair (X, Y) is uniform over the 36 outcomes, and

$$E(X) = \sum iP(X = i) = \frac{7}{2} = E(Y).$$

The distribution of the *sum* $X + Y$ was obtained in Example 3-c. However, the mean and variance of the sum can be computed directly from the means and variances of X and Y without reference to the joint distribution or to the distribution of the sum, as follows:

$$E(X + Y) = E(X) + E(Y) = \frac{7}{2} + \frac{7}{2} = 7,$$

and (because of independence)

$$\text{var } (X + Y) = \text{var } X + \text{var } Y = \frac{35}{12} + \frac{35}{12} = \frac{35}{6}.$$

The standard deviation of the sum is $\sqrt{35/6}$. ◄

Problems

1. A card is drawn at random from a deck of playing cards. Given that it is a heart, what is the appropriate space and what are the probabilities?

2. I have to see a certain businessman. I figure that my chances are only 1 in 5 of getting by the outer secretary; but if I get by her, the odds are even for getting by his private secretary. What are my chances of getting in to see him?

3. Suppose you play the game of Russian Roulette *twice* in a row. Compute your chances for survival if

(a) You spin the chamber between trials.
(b) You do not spin the chamber between trials.

4. I am to play A or B in a game. My chances of beating A are 4/10, and of beating B, 8/10. My chance of playing A is 1/4. What is the probability that I win?

5. A relatively rare disease is to be detected by a device that is not perfect but can err in two ways. It may give a positive reading for a healthy person, and it may fail to give a positive reading when the disease is actually present. Suppose that

$$P(+|D) = .98 \text{ and } P(+|H) = .02.$$

Calculate the probability that a person whose reading is positive is actually diseased: $P(D|+)$, assuming that the incidence of disease in the population is $P(D) = .0005$.

6. If $P(A) = .3$ and $P(B) = .5$, compute $P(AB)$,

(a) If A and B are disjoint.
(b) If A and B are independent.

7. When a card is selected at random from a standard deck, the events R: the card is red, and S: the card is a spade, are disjoint. Are they independent? (Are disjoint events *ever* independent?)

8. A card is drawn at random from a deck of playing cards, and then another card is drawn at random from those that are left. Compute the probability of each of the following events:

(a) Both are hearts.
(b) The first is not a heart but the second is.
(c) The second is a heart.

9. If the probability that a white male lives to age 40 is .93 and that he lives to age 65 is .66, what is the probability that if he reaches age 40 he will live to 65?

10. As in Problem 13 of Chapter 3, let X, Y, and Z denote the number of blue, of green, and of red socks among three picked at random from a drawer with four blue, four green, and two red socks.

(a) From the probability table for X and Y (obtained earlier) deduce that X and Y are not independent.
(b) Compute $P(X = 1|Z = 2)$. Are X and Z independent?

11. A random pair (X, Y) has the distribution defined by the following table:

Y \ X	0	1	2
2	1/12	1/6	1/12
3	1/6	0	1/6
4	0	1/3	0

(a) Are X and Y independent?

(b) Compute $P(X + Y = 3|X = 1)$.
(c) Compute $E(Y|X = 0)$, the mean of the conditional distribution of Y given $X = 0$.

12. Suppose that X' and Y' are independent and have the same marginals as X and Y, respectively, in Problem 11. Compute the following:

(a) $P(X' + Y' = 3|X' = 1)$.
(b) $E(Y'|X' = 0)$.
(c) $P(X' + Y' = 3)$.
(d) $\text{cov}\,(X', Y')$.
(e) $E(X'^2 Y'^2)$.

13. Consider the problem of studying the relationship between degree of education and sentiment on the death penalty for skyjackers. Suppose that the population proportions (or probabilities) are as given in the accompanying table. For a person drawn at random from the population, determine

(a) P(grade school).
(b) P(favor).
(c) P(favor|grade school).
(d) P(college|oppose).
(e) Whether education and sentiment are *independent.* (Explain your answer.)

	Favor	Oppose
Grade school	.15	.05
High school	.20	.10
College	.25	.25

14. A coin is tossed. If it falls heads, a die is tossed once; but if it falls tails, the die is tossed twice. If the total number of points thrown with the die is 4, what is the probability that the die was tossed just once?

15. A "series" string of Christmas tree lights stays lit as long as all eight of its bulbs stay lit. If the probability that a bulb lasts 100 hours is .9, what is the probability that the string stays lit 100 hours?

16. A box contains two balls, an unknown number M of which are black and the rest white. Suppose the chances are *equal* that the box has $M = 0, 1$, or 2 black balls. A ball is drawn at random from the box and found to be black. Now what are the chances that $M = 2$?

5

Bernoulli Populations

An important type of experiment is that in which the sample space has just two outcomes. Some obvious examples are the toss of a coin (heads or tails), determining the sex of a new baby (male, female), testing a product (good or defective), playing a game (win or lose), conducting a mission (succeed or fail), comparing one variable with another (larger or smaller), giving an inoculation (takes or does not take), and so on. More generally, in any given experiment, observing whether an event E occurs or does not occur is a derived experiment with two outcomes.

It is convenient to code the two outcomes 1 and 0, an assignment that defines a random variable, sometimes referred to as a *Bernoulli* variable. Its probability function is defined by a single number p:

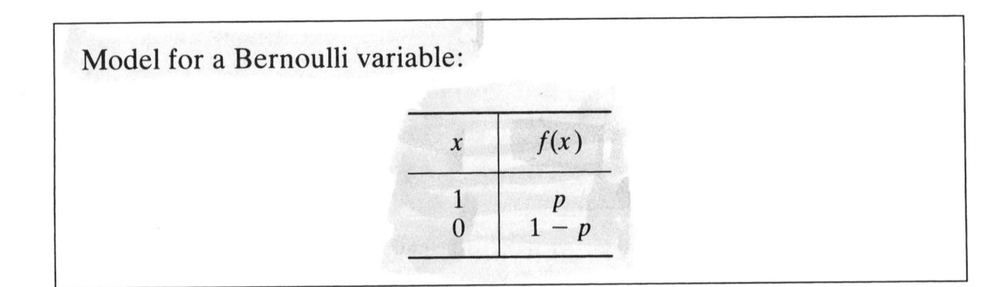

Model for a Bernoulli variable:

x	$f(x)$
1	p
0	$1 - p$

The complementary probability $1 - p$ is often denoted q, so that $p + q = 1$.

The moments of a Bernoulli distribution are easily computed. Observe that if $X = 0$ or 1, then $X^k = X$, and

$$E(X^k) = E(X) = 1 \cdot p + 0 \cdot q = p.$$

The variance is

$$\text{var}\, X = E(X^2) - [E(X)]^2 = p - p^2 = pq.$$

Moments of a Bernoulli variable:

$$E(X) = p, \qquad \text{var}\, X = pq.$$

A commonly encountered situation involving a Bernoulli variable is that of sampling an actual, finite population consisting of individuals or individual items of just two types. Again, the two types can be coded 0 and 1, and the population proportion of type 1 individuals is the probability p that an individual drawn at random is a 1.

5.1 Sampling a Bernoulli Population

Suppose that a Bernoulli experiment is repeated, say n times. Let X_1 denote the result (0 or 1) of the first trial, X_2, the result of the second trial, and so on. The result of the n trials is then a sequence $(X_1, X_2, \ldots, X_n)$, where each element X_i is a Bernoulli variable. In a particular case the outcome is a sequence of particular values, of 0's and 1's. Thus, in a sequence of six trials, the result might be (0, 1, 1, 0, 0, 0), indicating that the event coded 1 occurred in the second and third trials but not in the other four.

A more compact way of recording the results, often sufficient for inference about the population proportion p, is to make a tally mark opposite each possible value as it occurs and then count the marks for each. The sequence (0, 1, 1, 0, 0, 0), for instance, would be recorded and summarized as follows:

x	Tally
1	//
0	////

x	Frequency
1	2
0	4

The second table, giving the frequency of each value in the sample, is called a *frequency distribution.* The frequency of 1's, the number of "successes" in n observations, is a random variable that is often of interest. It has a distribution that depends on the nature of the sampling.

The method of sampling that is mathematically simplest is that in which the observations are *independent* random variables. The probability of a particular sequence of outcomes is then obtained by multiplication. For instance, in six independent trials,

$$P(0, 1, 1, 0, 0, 0) = q \cdot p \cdot p \cdot q \cdot q \cdot q = p^2 q^4.$$

To achieve independence of the trials when sampling from a finite population, it is necessary to restore the population after each draw by replacing the item or individual drawn and then thoroughly mixing or shuffling. Probabilities for succeeding selections are then unaffected by what occurred up to that point.

If objects are drawn from a finite population one at a time at random, but *without* replacement, then successive X_i's in the sequence of observations are no longer independent. However, each X_i, considered by itself, with no knowledge of the other observations, is still a *Bernoulli* variable, having the distribution of the population. This is a consequence of the fact (seen in Chapter 4) that in forming a sequence of n individuals from N by one-by-one random selections, the $(N)_n$ permutations are equally likely. For, suppose there are M individuals of type 1. Then the number of permutations with a type 1 individual in the kth place is just M times the number of ways of arranging $n - 1$ from $N - 1$ things: $M(N - 1)_{n-1}$. Therefore,

$$P(X_k = 1) = \frac{M(N - 1)_{n-1}}{(N)_n} = \frac{M}{N},$$

which is the population proportion of type 1 individuals, or the p of the Bernoulli population.

5.2 The Binomial Distribution

The number of successes (1's) in a sequence of a given number of independent Bernoulli trials is a random variable of considerable interest. (In various applications, it is the number of heads in n coin tosses, the number of defective items in a sample of n, the number of persons among n whose inoculation was effective, and so on.) The number of successes will be denoted Y_n if there are n trials and is just the *sum* of the 0's and 1's in the sequence of outcomes:

$$Y_n = X_1 + X_2 + \cdots + X_n.$$

The possible values of Y_n are $0, 1, \ldots, n$, and the probability of the value k is the sum of the probabilities of the $\binom{n}{k}$ sequences in which there are k 1's and $n - k$ 0's:

Binomial probability function:

$$P(Y_n = k) = \binom{n}{k} p^k q^{n-k}, \qquad k = 0, 1, \ldots, n.$$

The distribution of Y_n is called *binomial* because the individual probabilities are just terms in a binomial expansion, terms that add to 1:

$$\sum_0^n \binom{n}{k} p^k q^{n-k} = (p + q)^n = 1.$$

Example 5-a To test for ESP, one person cuts a deck of cards at random and observes the suit of the card turned up; the subject of the experiment, without seeing the card, tries to identify the suit mentally and names one of the four suits. If the subject does *not* have extrasensory powers, he has one chance in four of being right, at each trial. Suppose there are 10 independent trials, and let Y denote the number of successful identifications in the 10 trials. Then, in the absence of ESP (or of cheating),

$$p(k) = P(Y = k) = \binom{10}{k}\left(\frac{1}{4}\right)^k\left(\frac{3}{4}\right)^{10-k},$$

or in table form:

k	0	1	2	3	4	5	6	7	8	9	10
$p(k)$	.056	.188	.282	.250	.146	.058	.016	.003	.000	.000	.000

(If Y turns out to be 8 in a particular sequence of 10 trials, one might, as will be explained in Chapter 11, have doubts about the assumption of pure chance and conclude that there is something to ESP.) Computation of $E(Y)$ from the table gives 2.493, but these probabilities were rounded off; the actual mean is 2.5, as will be demonstrated next. ◄

The mean and variance of a binomial variable are easily computed by exploiting its structure as a sum of independent Bernoulli variables: $Y_n = \Sigma_1^n X_i$. Recalling that for a Bernoulli X with parameter $p = P(X = 1)$, $EX = p$ and var $X = pq$, one obtains

$$E(Y_n) = E\left(\sum_1^n X_i\right) = \sum_1^n E(X_i) = np.$$

and (because of the independence)

$$\text{var}\,(Y_n) = \text{var}\left(\sum_1^n X_i\right) = \sum_1^n \text{var}\,X_i = npq.$$

The formula $\mu = np$ for the mean is especially appealing and therefore easy to remember—the number of trials times the probability of success at each trial.

Moments of a binomial variable: $$E(Y_n) = np,$$ $$\text{var}\,Y_n = npq.$$

Notice that if $p = 0$, the variance is zero; this is not surprising because then every sample sequence is all zeros, and $Y_n = 0$ with probability 1. Similar reasoning applies to $p = 1$, where var Y_n is also 0.

A binomial distribution with $p = 1/2$ is symmetric about its mean, $n/2$, because the probabilities are proportional to the binomial coefficients:

$$p(k) = \binom{n}{k}\left(\frac{1}{2}\right)^k\left(\frac{1}{2}\right)^{n-k} = \binom{n}{k}\left(\frac{1}{2}\right)^n = \binom{n}{n-k}\left(\frac{1}{2}\right)^n.$$

If $p > 1/2$, the larger values of Y_n have greater probabilities, and the distribution is said to be *skewed* to the right. If $p < 1/2$, it is skewed to the left.

In principle, binomial probabilities can be calculated directly from the binomial formula. Some hand calculators have factorial and x^y keys that facilitate the computation. But hand computation can be tedious, and so a table is provided (Table X, Appendix B) which gives probabilities for $n = 4, \ldots, 10$ and a selection of values of p. For larger values of n, methods of approximate calculation of binomial probabilities will be taken up in Chapters 7 and 8.

5.3 Hypergeometric Probabilities

When sampling without replacement from a finite Bernoulli population of N objects, including M 1's and $N - M$ 0's, the result is a sequence $(X_1, \ldots, X_n)$ in which each X_i is 0 or 1. The probability of a *particular* sequence of n 0's and 1's can again be calculated as a product, but the factors will have to be conditional probabilities. Alternatively, one can exploit the equal likelihood of the $(N)_n = \binom{N}{n} n!$ permutations of n objects taken from N, and count the number among them with the given configuration of 0's and 1's. Thus the number of permutations giving rise to the sequence 0, 1, 1, 0, 0, 0 is the number of ways of arranging 2 from M 1's and 4 from $N - M$ 0's:

$$(M)_2(N - M)_4 = \binom{M}{2}\binom{N - M}{4} 2!4!,$$

so that

$$P(0, 1, 1, 0, 0, 0) = \frac{\binom{M}{2}\binom{N - M}{4} 2!4!}{\binom{N}{6} 6!} = \frac{\binom{M}{2}\binom{N - M}{4}}{\binom{6}{2}\binom{N}{6}}.$$

As in the case of sampling with replacement, the probability of a given number of 1's in a sequence of n trials is the sum of the probabilities of the various sequences with that many 1's. Thus, in the case $n = 6$ above, there are $\binom{6}{2}$ sequences having two 1's (and four 0's), each with the same probability, as given above; the probability of two 1's is then $\binom{6}{2}$ times that common value:

$$\binom{6}{2} \times \frac{\binom{M}{2}\binom{N - M}{4}}{\binom{6}{2}\binom{N}{6}}.$$

If Y_n is again used to denote the number of 1's,

$$Y_n = X_1 + \cdots + X_n,$$

then the probability function of Y_n is given, in general, as follows:

Hypergeometric probability function:

$$P(Y_n = k) = \frac{\binom{M}{k}\binom{N-M}{n-k}}{\binom{N}{n}}, \qquad k = 0, 1, \ldots, n.$$

The list of possible values given here is sometimes too inclusive. That is, if n is larger than M or $N - M$, one may run out of individuals of one type or the other. This will show up in the formula for probability as a combination symbol in which the upper element is less than the lower one $\left[\text{e.g., } \binom{4}{6}\right]$. If it is understood that such combination symbols are *defined* to be 0, then the formula works for $k = 0, 1, \ldots, n$.

The same formula is obtained, of course, if the sampling is thought of as a random selection of n out of N, the $\binom{N}{n}$ possible combinations being equally likely. Of these, there are $\binom{M}{k}\binom{N-M}{n-k}$ combinations in which k are taken from the M 1's and $n - k$ are taken from the $N - M$ 0's.

Example 5-b Three articles are selected from a lot of 10 articles, 2 defective and 8 good, by picking one at a time at random and without replacement. Let Y denote the number of defectives among the three drawn. The probability function of Y is as given in the following table:

y	$P(Y = y)$
0	$\binom{8}{3}\Big/\binom{10}{3} = 7/15$
1	$\binom{2}{1}\binom{8}{2}\Big/\binom{10}{3} = 7/15$
2	$\binom{2}{2}\binom{8}{1}\Big/\binom{10}{3} = 1/15$

The mean number of defectives is

$$E(Y) = 0 \times \frac{7}{15} + 1 \times \frac{7}{15} + 2 \times \frac{1}{15} = \frac{3}{5},$$

and the mean square:

$$E(Y^2) = 0^2 \times \frac{7}{15} + 1^2 \times \frac{7}{15} + 2^2 \times \frac{1}{15} = \frac{11}{15}. \quad \blacktriangleleft$$

The mean and variance of a hypergeometric distribution are calculable in general by using the fact that the variable Y_n is a sum of Bernoulli variables, with $p = M/N$. Thus

$$E(Y_n) = \sum_1^n E(X_i) = np = n \cdot \frac{M}{N},$$

as in the binomial model, and

$$\text{var } Y_n = \text{var}\left(\sum_1^n X_i\right) = \sum_1^n \text{var } X_i + \sum_{i \neq j} \text{cov }(X_i, X_j)$$
$$= n\sigma^2 + n(n-1)\,\text{cov }(X_1, X_2).$$

The last expression follows because the individual X_i's have a *common* Bernoulli distribution with mean p and variance pq, and because (by similar reasoning) the *joint* distribution of every pair is the same. Now if $n = N$, this formula for variance must reduce to 0, because there is no variation in Y_n computed from samples that consist of the whole population. That is,

$$0 = N\sigma^2 + N(N-1)\,\text{cov }(X_1, X_2).$$

This determines the covariance as $-\sigma^2/(N-1)$, and therefore (with $\sigma^2 = pq = (M/N)[1 - (M/N)]$)

$$\text{var } Y_n = n\sigma^2 + n(n-1)\frac{-\sigma^2}{N-1} = n\sigma^2\left(1 - \frac{n-1}{N-1}\right)$$
$$= n\frac{M}{n}\left(1 - \frac{M}{N}\right)\frac{N-n}{N-1}.$$

That is, the variance is a factor $(N - n)/(N - 1)$ times the binomial variance npq that would apply if there were replacement:

Moments of a hypergeometric variable:

$$E(Y_n) = n\frac{M}{N},$$

$$\text{var } Y_n = n\frac{M}{N}\left(1 - \frac{M}{N}\right) \cdot \frac{N-n}{N-1}.$$

Observe that the finite population factor $(N - n)/(N - 1)$ disappears (approaches 1) as N becomes infinite. Indeed, as N and M become infinite in fixed ratio, hypergeometric probabilities approach binomial probabilities:

$$\frac{\binom{M}{k}\binom{N-M}{n-k}}{\binom{N}{n}} = \frac{\dfrac{M(M-1)\cdots(M-k+1)}{k!}\,\dfrac{(N-M)\cdots(N-M-n+k+1)}{(n-k)!}}{\dfrac{N(N-1)\cdots(N-n+1)}{n!}}$$

$$= \binom{n}{k}\underbrace{\frac{M}{N}\cdot\frac{M-1}{N-1}\cdots\frac{M-k+1}{N-k+1}}_{k \text{ factors}}\underbrace{\frac{N-M}{N-k}\cdots\frac{N-M-n+k+1}{N-n+1}}_{n-k \text{ factors}}$$

$$\to \binom{n}{k}p\cdot p\cdots p\cdot q\cdots q = \binom{n}{k}p^k q^{n-k}$$

Because a quantity approaching a limit must eventually get close to that limit, it follows that a hypergeometric probability can be approximated by using the binomial formula if N is sufficiently large:

> Binomial approximation to the hypergeometric if $N \gg n$:
>
> $$\frac{\binom{M}{k}\binom{N-M}{n-k}}{\binom{N}{n}} \doteq \binom{n}{k}\left(\frac{M}{N}\right)^k\left(1-\frac{M}{N}\right)^{n-k}, \qquad k = 0, 1, \ldots, n.$$

Example 5-c A jury of 12 members is drawn at random from a voter list of 1000 persons, 700 white and 300 black. What is the probability that the jury will be all white?

The actual probability is hypergeometric:

$$P(0 \text{ black, } 12 \text{ white}) = \frac{\binom{300}{0}\binom{700}{12}}{\binom{1000}{12}} = \frac{700}{1000}\cdot\cdots\cdot\frac{689}{989},$$

but since 1000 is much larger than 12, this may be approximated by a binomial probability with $n = 12$ and $p = 300/1000$:

$$P(0) = \binom{12}{0}(.3)^0(.7)^{12} = \left(\frac{700}{1000}\right)^{12} \doteq .0138.$$

Observe that this would be the same for any larger population, with the same proportion of blacks and whites, as well. For instance, if the 12 were drawn from a population of 140 million whites and 60 million blacks, the probability of no blacks among the 12 drawn would be approximated by the same binomial expression, with the value .0138. ◀

★ 5.4 Inverse Sampling

The binomial and hypergeometric distributions give the probabilities for the number of "successes" in a given number of Bernoulli trials, with and without replacement, respectively. A Bernoulli population can also be sampled, one at a time (with or without replacement), *until a specified number of successes is achieved.* Sampling with this rule for stopping is referred to as *inverse* sampling.

Suppose first that the Bernoulli trials are independent (sampling with replacement). If the sampling stops when precisely k successes have been observed, the number of trials required is a random variable; call it N. The possible values of N are $k, k + 1, k + 2, \ldots$. The event $N = r$ occurs if and only if there are $k - 1$ successes in $r - 1$ trials followed by a success in the rth trial:

$$P(N = r) = \left[\binom{r-1}{k-1}p^{k-1}q^{r-k}\right] \times p.$$

For given k (a parameter of the distribution), these define a probability function:

> Negative binomial probability function:
>
> $$f(r) = \binom{r-1}{k-1}p^k q^{r-k}, \qquad r = k, k + 1, \ldots.$$

If $k = 1$, the sampling goes on only until the *first* success, and the probabilities for N, the number of observations required, are

$$p, pq, pq^2, pq^3, \ldots,$$

corresponding to values $N = 1, 2, 3, \ldots$. Observe that these probabilities are terms in a *geometric series*:

$$p + pq + pq^2 + \cdots = p(1 + q + q^2 + \cdots)$$
$$= p \cdot \frac{1}{1 - q} = 1.$$

For this reason the distribution of N in this special case ($k = 1$) is called a *geometric* distribution.

In the more general case, the probabilities are terms in the negative binomial expansion

$$1 = p^k \cdot \frac{1}{p^k} = p^k(1 - q)^{-k} = p^k \sum_{n=0}^{\infty} \frac{(-k)(-k - 1)\cdots(-k - n + 1)}{n!}(-q)^{-n}$$
$$= p^k \sum_{r=k}^{\infty} \binom{r - 1}{k - 1} q^{r-k}.$$

This expansion, showing that the probabilities for $N = k, k + 1, \ldots$ add up to 1, gives rise to the name of the distribution.

The mean of N is obtainable by observing that it is a sum of geometric variables:

$$N = N_1 + N_2 + \cdots + N_k,$$

where N_1 is the number of trials needed to get the first success, N_2 is the number of additional trials to get the second success, and so on. Each N_i is geometric, with mean

$$E(N_i) = \sum_1^{\infty} npq^{n-1} = p \sum_1^{\infty} \frac{d}{dq} q^n$$
$$= p \frac{d}{dq}(q + q^2 + \cdots)$$
$$= p \frac{d}{dq}\left(\frac{q}{1 - q}\right) = \frac{1}{p}.$$

Hence

$$E(N) = \frac{1}{p} + \cdots + \frac{1}{p} = \frac{k}{p}.$$

For example, if $p = 1/4$ it takes, on the average, $4k$ trials to get k successes.

Example 5-d A caller decides to make as many phone calls as necessary to obtain 20 responses favorable to a certain proposition on the ballot. How many calls will he make if one out of three in the population is favorable?

The number of calls required is *random*, with a distribution that is approximately negative binomial:

$$P(r \text{ calls required}) = \binom{r-1}{19}\left(\frac{1}{3}\right)^{20}\left(\frac{2}{3}\right)^{r-20}, \qquad r = 20, 21, \ldots.$$

(The approximation is good if the population is large compared to the sample size.) The mean number of calls needed to obtain the first favorable response is $1/p = 3$, and to obtain 20 is $20/p = 60$. ◀

When sampling from a *finite* population to reach a given number of successes, the number of necessary trials cannot exceed the population size. A general formula (available in more advanced texts) will not be presented, but the following example illustrates how probabilities can be computed.

Example 5-e Suppose I need 4 batteries and have a pile of old batteries that is a mixture of 6 good ones and 14 that are no good. How many tests do I have to perform to locate 4 good batteries?

The number of tests required, of course, is random. The probability that it takes 8 tests, for instance, is

$$p(8) = \frac{\binom{6}{3}\binom{14}{4}}{\binom{20}{7}} \times \frac{3}{13} = .0596.$$

The second factor (3/13) is the conditional probability that a good battery is located at the eighth test, *given* that 3 have been found in the first 7 tests. The first factor is the probability that 3 good ones are located in the first 7 tests. ◀

Problems

1. A basketball player makes 75 per cent of his free-throw attempts.

(a) What is the expected number of successes in 12 tries?
(b) What is the probability that he makes 10 of 12 attempts?
(c) What is the probability that he makes at least 10 of 12 attempts?
(d) What is the probability that his first basket comes on his third toss?

2. In my class of 40 students I sometimes think that each student decides whether or not to attend class according to the toss of a coin. If this were so, what would be the mean and standard deviation of the class size?

3. A TV commercial reports that 54 of 100 drivers prefer the ride and handling of car M over car O. If the cars are of equal with regard to these qualities, it might be assumed that the test subject's indication of a preference is like the toss of a coin, with $p = P\,(\text{prefer } M) = 1/2$. If p is really $1/2$, compute the mean number of drivers who would say they prefer car M, and also the standard deviation. (How does the standard deviation compare with the observed deviation from the mean? In view of this do you think the commercial has a strong argument?)

4. In Table XIV in Appendix B are listed a sequence of observations on a particular experiment for which the probability of a positive outcome and of a negative outcome are both $1/2$. The observations are given in groups of five.

(a) Give a table of values and probabilities for X, the number of $-$'s in five observations.
(b) What is the probability of at most two minus signs ($X \leq 2$)?
(c) Count the number of $-$'s in each of 100 groups of five on a page, and make a tabulation of the results in the form of a frequency of $X = k$ for each of $k = 0, 1, 2, 3, 4, 5$. Divide by 100 and compare these relative frequencies with the probabilities in part (a).

5. If you look through a shuffled deck of playing cards, the probability of finding at least one king next to a queen is about .49. (The same would be true for any pair of prechosen denominations.) What are the chances of succeeding in this trick at least five out of eight shuffles?

6. Give the probability distribution for Y, the number of black jelly beans among 3 drawn at random from a bag of 50, just 8 of which are black.

7. A town council consists of seven members, five liberal and two conservative. A committee of *three* is selected by picking from a hat in which each member is represented by a slip of paper with the member's name on it. Let X denote the number of conservatives on the committee.

(a) Compute $E(X)$.
(b) Determine the probability table for X.

8. Let Z denote the number of students who favor coed dorms in a random selection of 100 from 2000 dorm residents. If the sentiment were evenly divided in the population of 2000, what would be the mean and standard deviation of Z? (Would these be different if the selection were done one at a time *with* replacement?)

9. Evaluate approximately (without the aid of a calculator):

$$\frac{\binom{500}{4}\binom{500}{4}}{\binom{1000}{8}}.$$

10. We catch four fish from a pond, tag them, and release them back into pond. Later, we catch four more. Let X = number of tagged fish in second catch. Problem:

(a) What is the probability distribution of X and $E(X)$ if there are 10 fish in the pond?
(b) Suppose there are N fish (an unknown number) in the pond. Give a formula for $E(X)$ in terms of N.

(APPLICATION. In general one would not know the number of fish in the pond and would want to *estimate* this on the basis of the observed value of X.)

11. Evaluate:

(a) $\displaystyle\sum_0^8 k \frac{\binom{500}{k}\binom{500}{8-k}}{\binom{1000}{8}}$

(b) $\displaystyle\sum_0^{20} k^2\binom{20}{k}(.05)^k(.95)^{20-k}$

(c) $\displaystyle\sum_{12}^{\infty} n\binom{n-1}{11}2^{-n}$

★ **12.** A man plays Russian Roulette repeatedly. How long does he have to live? Assume one chance in six of expiration at a given trial, and give the distribution of the number of trials he will carry out, along with the expected value of this number.

★ **13.** If X is the number of (independent) trials of a Bernoulli experiment needed to obtain a success, it is geometric with parameter p, where p is the probability of success at a given trial.

(a) Compute $E(X^2 - X) = pq \sum_2^\infty n(n-1)q^{n-2}$.
(b) From the answer to part (a), compute var X.
(c) If $Y = X_1 + \cdots + X_r$, where the X's are independent in parts (a) and (b), then Y is the number of trials needed to get r successes. Compute var Y, exploiting the independence of the X's.

6

Continuous Random Variables

Physical variables such as distance, time, weight, force, and concentration are measured on the scale of real numbers, a continuous scale with uncountably many values. It is true that in making and recording measurements of such variables one is forced to make discrete approximations—to round off to the nearest centimeter, or tenth of a second, or whatever is dictated by the precision of the recording device. Nevertheless, it is often the case that a continuous model is simpler than a discrete model that would take into account the realities of measurement processes. The feature of continuous models that makes them different from discrete ones is that individual values—individual outcomes—cannot have positive probabilities. The definition and development of a continuous model will have to proceed somewhat differently.

6.1 Recording and Presenting Data

In recording the results of a number of trials of a continuous experiment, one might simply write down the value of the measured quantity after each trial, obtaining thereby a list of numbers—the outcomes, in the order of their occurrence. As pointed out previously, these numbers are rounded-off numbers; for not only can we not measure with infinite precision, we cannot even write down a real number with infinite precision—unless it should happen to be rational.

Listed in Table 6.1 are cholesterol levels of 100 males, given to the nearest integer unit of measurement. It is not easy to comprehend the meaning of the data in such a list, and putting the observations in numerical order is a first step of simplification. One way to do this is to make a mark on the scale or axis of values for each observation as it is obtained. With large data sets, such a picture is quickly confused by crowding. To avoid this problem, one can prepare a round-off scheme that is perhaps coarser than that implied in the accuracy of the measuring device. This is done by determining an interval that will include all measurements, dividing it into equal parts called class intervals. Each observation will fall into one of the class intervals. Each class interval may be represented by, say, its middle value, called the *class mark*. Such representation amounts to a round-off scheme, values occurring in a given class interval being rounded off to the class mark.

Table 6.1 Cholesterol level of 100 males

283	248	280	230	258	204	270	307	368	281
165	234	264	274	229	189	245	220	299	270
171	243	310	289	230	220	389	253	189	295
202	294	226	219	293	229	269	208	218	164
245	312	321	253	208	194	218	280	183	294
261	206	285	306	191	258	245	196	255	272
299	325	269	219	258	295	243	198	270	194
208	260	165	236	196	216	271	227	208	272
220	276	276	249	204	198	246	270	306	232
208	271	199	206	323	148	194	194	198	310

The number of observations falling into a given class interval is called its *frequency*, and a list of class intervals and their corresponding frequencies is called a *frequency distribution*. This provides a rather informative, though approximate, summary of the observations, especially useful when there are large amounts of data.

Example 6-a The data of Table 6.1 are summarized in the frequency distribution of Table 6.2, using a class interval width of 31 starting at the value 118.5. Table 6.3 gives frequency distributions for the same data, using different class interval schemes. Observe that although similar in broad outline, these distributions can differ considerably in detail. The fact that different class interval schemes result in different frequency distributions suggests that some care is called for when interpreting such a summary of data. ◀

Table 6.2 Frequency distribution of data in Table 6.1

Class interval	Class mark	Frequency
118.5–149.5	134	1
149.5–180.5	165	4
180.5–211.5	196	24
211.5–242.5	227	17
242.5–273.5	258	27
273.5–304.5	289	16
304.5–335.5	320	9
335.5–366.5	351	0
366.5–397.5	382	2

Table 6.3 Frequency distributions of data in Table 6.1

Class mark	Frequency	Class mark	Frequency
137	1		
162	4	156	4
187	14	177	2
212	18	198	23
237	17	219	12
262	20	240	13
287	15	261	17
312	8	282	11
337	1	303	13
362	1	324	3
387	1	345	0
		366	1
		387	1

The data in a frequency distribution can be represented graphically in what is called a histogram, a picture in which the axis of values is drawn horizontally, and a bar with height proportional to frequency is constructed over each class interval. The use of the bar suggests that the observations represented by a particular bar may have come from anywhere on the class interval.

Example 6-b Two of the frequency distributions in Example 6-a are represented by histograms in Figure 6-1. Although derived from—and purportedly representing—the same set of measurements, they are different in various details because of the different round-off schemes that were employed. ◄

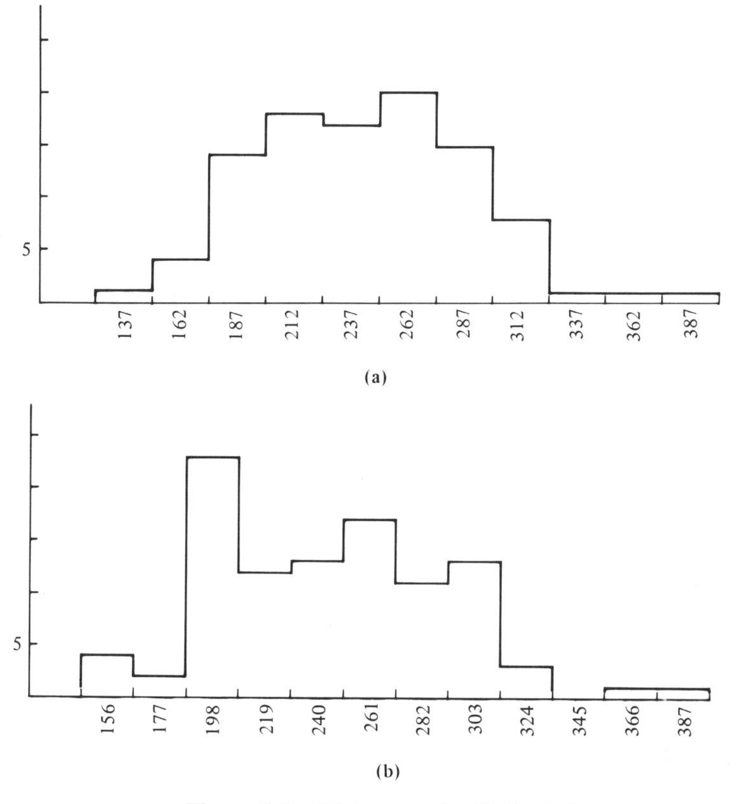

Figure 6-1 Histograms for Table 6-3.

6.2 Density

Just as the probability of an event is an idealization of the notion of relative frequency in a finite sample, and just as a discrete probability distribution is an idealization of a discrete sample frequency distribution, so a suitable model for a continuous random variable is obtained upon idealizing the frequency distribution of a finite sample (or its equivalent histogram) as the sample size is increased without limit.

It would be true, for a given scheme of class intervals, that the relative frequency of a class interval approaches the probability of that interval, as the number of sample observations is increased. However, specification of these class interval probabilities would be inadequate for general probability computations. To see what is appropriate within the intervals, one must take into account what happens, not just the sample size increases, but as the class interval width simultaneously shrinks to zero. A very fine class interval structure will tend to produce a jaggedness in the histogram, but a larger sample size will tend to smooth it out. With n (sample size) and k (number of class intervals) in proper balance, the limiting shape will generally be a smooth curve.

In analyzing limiting tendencies, it is not practical to obtain data from "real" situations, such as the cholesterol measurements in Example 6-a. Rather, it is convenient to simulate a real situation, using a computer to generate data that are like data from a continuous experiment. Of course, owing to the limitations of computers, observations are necessarily discrete; but since they can be generated and recorded to 10 to 20 significant digits, they will appear at least as continuous as real data. The next two examples give results from sampling such artificial populations.

Example 6-c A computer was used to generate data from a continuous distribution equivalent to that describing the average of two successive spins of a pointer with a continuous scale from 0 to 1 on its circumference. Six sets of data were obtained and summarized in frequency distributions.

Sample number	Sample size	Number of class intervals
1	250	20
2	1,000	20
3	4,000	20
4	1,000	100
5	10,000	100
6	40,000	100

In each case a histogram was drawn, shown in Figures 6-2 and 6-3. With 20 class intervals, going from 250 to 4000 observations clearly tends to iron out irregularities. Using 100 class intervals for 1000 observations gives a more detailed account of what happened, but again results in irregularities due to sampling. In the histogram for a sample of 40,000 observations, those fluctuations are smoothed out, leaving mainly the jogs that are unavoidable in the round-off to two decimal places, implicit in the use of 100 class intervals. The reader should have little difficulty imagining that a histogram with, say, 10,000 class intervals and 1,000,000 observations would look to the naked eye like a triangle. This triangle is the ideal that describes this particular continuous experiment. ◄

Example 6-d Data were again generated artificially; this time the individual observations happened to be the average of four successive spins of a pointer on the

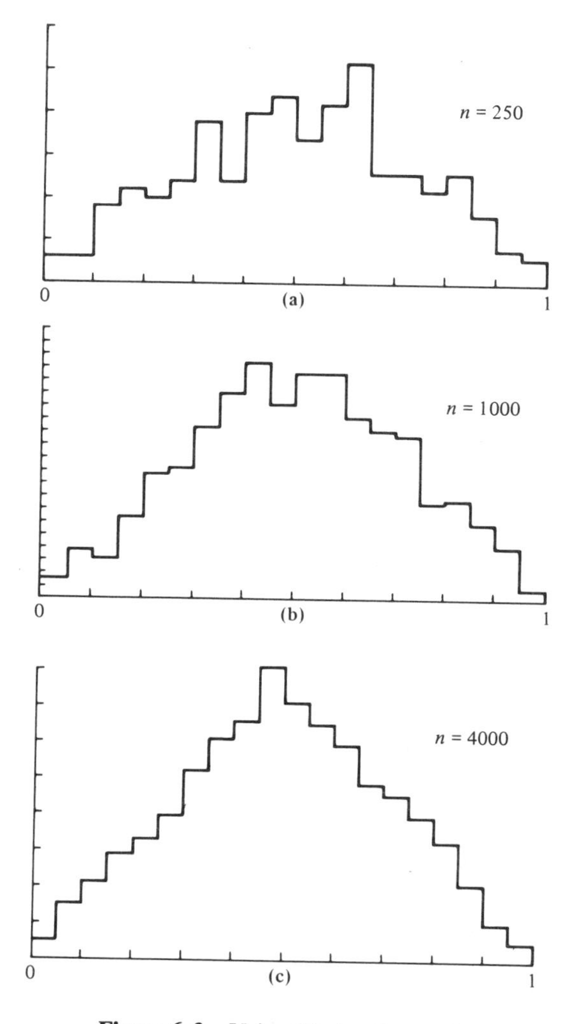

Figure 6-2 Using 20 class intervals.

interval [0, 1]. The sample size was 20,000, and the observations were put into one of 100 class intervals for the purpose of drawing a histogram. The result is given in Figure 6-4. Here it is apparent that the ideal that will emerge when more and more observations are taken, with a corresponding increase in the number of class intervals, is a smooth curve shaped something like a bell. ◄

Although the process of rounding off into class intervals results in histograms with jagged tops, there is no reason to think that the continuous experiment being sampled is that irregular. Experience shows that, even as

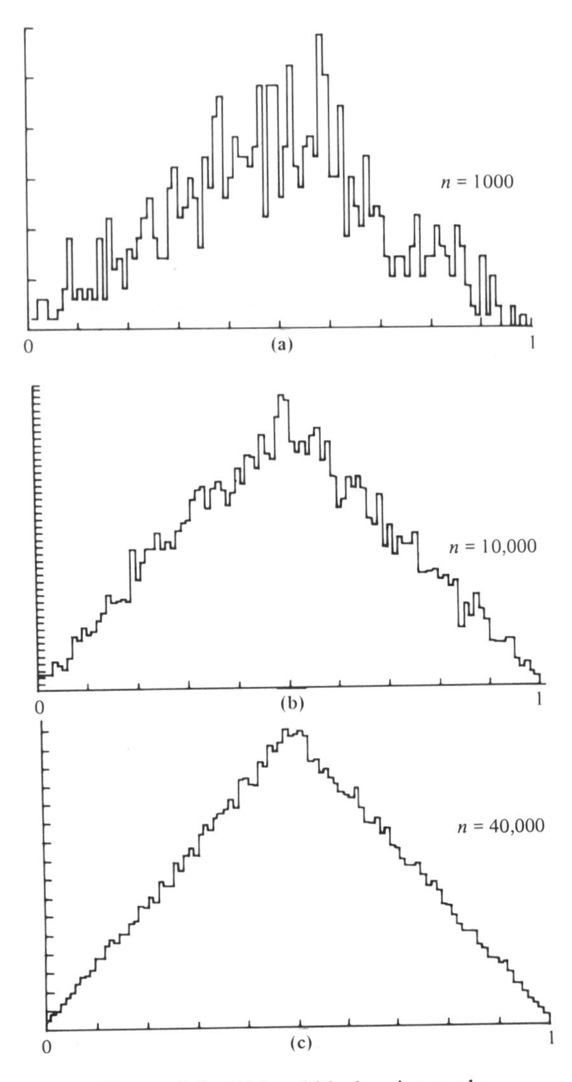

Figure 6-3 Using 100 class intervals.

are the artificial experiments described in Examples 6-c and 6-d, real phenomenon are best described by curves that are mostly smooth.

In a histogram, and so at each stage of such a limiting process, the relative frequency of any interval of values is proportional to the *area* over that interval between the top edge of the histogram and the horizontal axis. In a continuous probability model defined by a curve, or by a function $y = f(x)$ whose graph lies above the horizontal axis, the probability of an event (a set

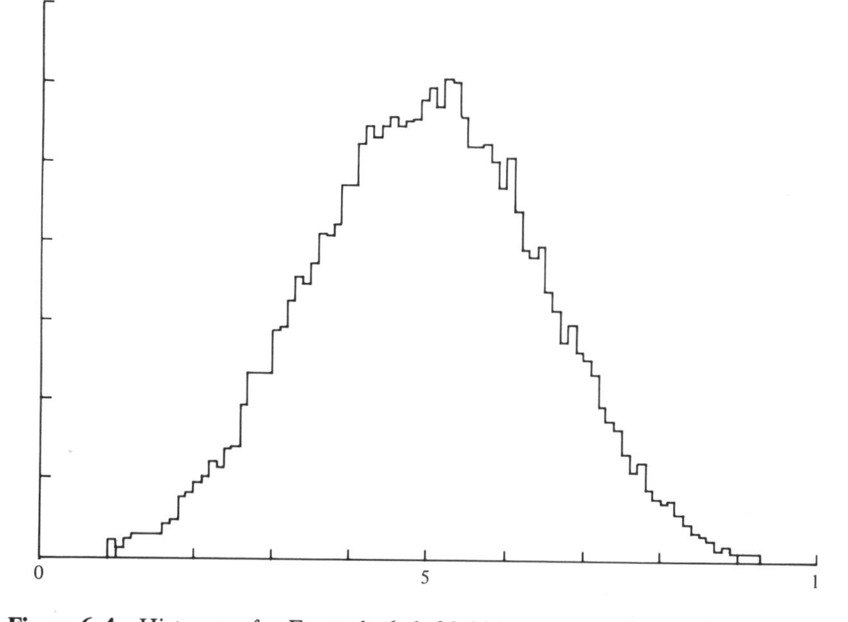

Figure 6-4 Histogram for Example 6-d: 20,000 averages of four pointer spins.

of values on the x-axis) is taken to be proportional to the area under the curve and above that event. If the total area, which would have to be finite, is made equal to 1 by a suitable adjustment of the vertical scale, then *probabilities are equal to areas under the curve.* When scaled in this way, the function $f(x)$ is called a probability *density function*, and the total area under its graph is the total of all the probability in the distribution.

> A probability density function is a nonnegative function with unit area under its graph:
>
> $$f(x) \geq 0, \qquad \int_{-\infty}^{\infty} f(x)\,dx = 1.$$

The density function defines a probability distribution for the continuous quantity being measured. If this continuous random variable is denoted by X, reference to X may be made using a subscript when needed for clarity: $f_X(x)$. Probabilities of events (sets of values) are calculated as areas—as

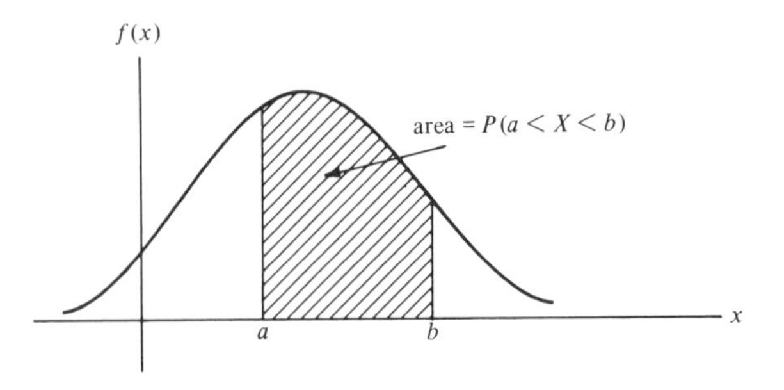

Figure 6-5 Probability as area.

definite integrals. In particular, for an *interval* $a < x < b$, one has the following (see Figure 6-5):

> Probability from density:
>
> $$P(a < X < b) = \int_a^b f(x)\,dx$$
> $$= \text{area under } f(x) \text{ between } a \text{ and } b.$$

Most events to be encountered will be either intervals or unions of intervals.

Example 6-e Considerations of symmetry suggest an a priori model for the "spinning pointer"—a pivoted pointer with a uniform number scale around the path of its tip with which to identify the stopping point when it is spun. If one thinks of this as an idealization of a wheel of fortune, as the number of stopping points becomes infinite, it is clear that the proper model is defined by a density function that is *constant* over the interval of values used. If the scale extends from 0 to 1, the density function is constant on $0 < x < 1$; and since the area under the density curve is to be 1, that constant value is 1:

> A priori model for a spinning pointer:
>
> $$f(x) = 1, \qquad \text{for } 0 < x < 1.$$

A random variable represented by this model is said to have a *uniform distribution*. Figure 6-6 shows the uniform density on the interval $0 < x < 1$, and indicates the rectangle whose area is the probability of the event $.2 < X < .6$; this probability or area is just the width of the interval from .2 to .6 (because the rectangle's height is 1):

$$P(.2 < X < .6) = \int_{.2}^{.6} dx = .4.$$

Table XIII in Appendix B gives numerous observations on this particular random variable; the user can round these observations to any desired number of decimal places. ◄

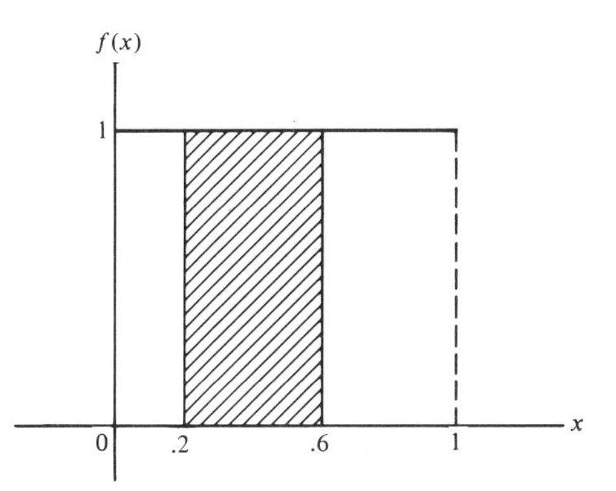

Figure 6-6 Density for a spinning pointer.

Example 6-f Suppose that X has the density function defined by $f(x) = 1 - |x - 1|$ for $0 < x < 2$, shown in Figure 6-7. This turns out to be the model for the sum of two independent spins of the pointer described in the preceding example. (It would be instructive for the student to plot, say, 100 successive sums of pairs of numbers from Table XIII to see this triangular distribution developing.) Again, probability is calculated as an area or an integral. For example,

$$P\left(|X - 1| > \frac{1}{2}\right) = 1 - P\left(\frac{1}{2} < X < \frac{3}{2}\right) = 1 - \int_{1/2}^{3/2} (1 - |1 - x|)\, dx$$

$$= 1 - \left[\text{area between } \frac{1}{2} \text{ and } \frac{3}{2}\right] = .25.$$

This area is indicated in Figure 6-7. ◄

Although the probability of a particular value x is zero, the density function $f(x)$ can be thought of as proportional to the probability "near" x, in the following sense. In a small interval $(x, x + dx)$, the probability is equal

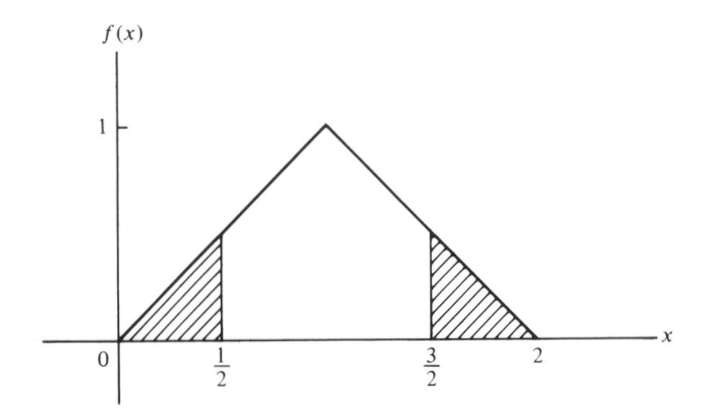

Figure 6-7 Triangular density (Example 6-f).

to the area under $f(x)$ between x and $x + dx$; but this is approximately the same as the area of a rectangle of height $f(x)$ and base dx (see Figure 6-8):

$$P(x < X < x + dx) = \text{area under } f(x) \text{ between } x \text{ and } x + dx$$
$$\doteq f(x)\,dx.$$

This differential approximation defines the *probability element.* (The inequalities given and the representation in Figure 2-8 assumes $dx > 0$. Similar reasoning applies if $dx < 0$.)

$$\text{Probability element} \equiv f(x)\,dx$$
$$\doteq \text{probability in } (x, x + dx).$$

The term *density* is apropos in describing $f(x)$, since this function is quite analogous to the density function used to describe the distribution of mass

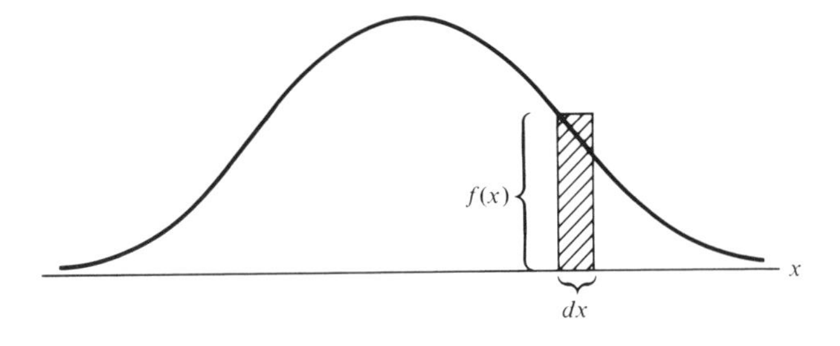

Figure 6-8 $P(x < X < x + dx) = f(x)\,dx.$

along a line. At any point x_0 the average density in an interval dx, including x_0, is defined as the amount of probability (mass) per unit interval:

$$\text{average density near } x_0 \doteq \frac{f(x_0)\,dx}{dx}.$$

In the limit as $dx \to 0$, the approximation becomes exact, so that the "point density" at x_0, the limiting value of the average density, is just $f(x_0)$:

$$\text{density at } x_0 \equiv \lim_{dx\to 0} (\text{average density near } x_0) = f(x_0).$$

6.3 The Distribution Function

Although a continuous probability model is defined in terms of a given density function, it is the *area* under the density curve that is of primary interest, since probabilities are defined as areas. Thus, for a given model, it would be more helpful to have a table of areas for events of possible interest, rather than a table of values of the density function. To be sure, areas might well be computable (perhaps numerically) from the density function values; nevertheless, the more directly useful tables are those that give areas.

Rather than give areas or probabilities for a large number of intervals $a < X < b$, it is more efficient to give areas of one-sided "intervals" of the form $X \le \lambda$, for a suitable sprinkling of values of λ. From these one can compute the probability of an interval $a < X \le b$. For, since

$$\{X \le a\} + \{a < X \le b\} = \{X \le b\},$$

it follows, by the addition law, that

$$P(X \le a) + P(a < X \le b) = P(X \le b),$$

and hence

$$P(a < X < b) = P(a < X \le b) = P(X \le b) - P(X \le a).$$

(Bear in mind again that when dealing with continuous distributions, there is no distinction between $<$ and $\le$, since individual values have zero probability.)

The probability $P(X \le \lambda)$ depends, of course, on the value λ. It is a *function* of λ, called the *distribution function* of X. It is also termed the *cumulative* distribution function, abbreviated c.d.f.

Distribution function (c.d.f):

$$F(\lambda) \equiv P(X \leq \lambda).$$

The symbol $F(\cdot)$ will always mean the c.d.f., although sometimes, for clarity, a subscript will be attached to make explicit the random variable that is involved: $F_X(\lambda)$.

In terms of this function, the probability of an interval is a difference, as illustrated in Figure 6-9:

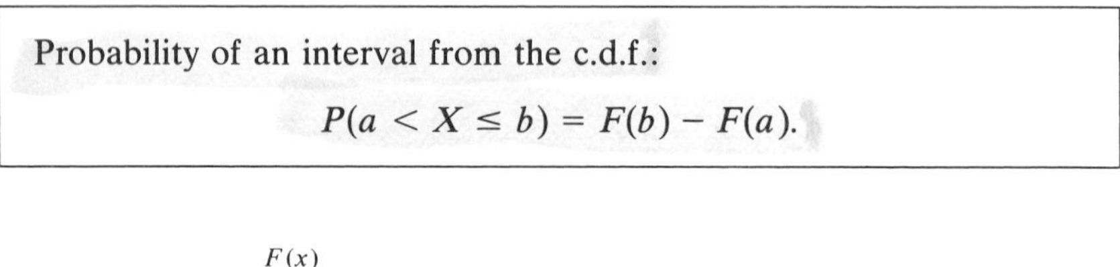

Probability of an interval from the c.d.f.:

$$P(a < X \leq b) = F(b) - F(a).$$

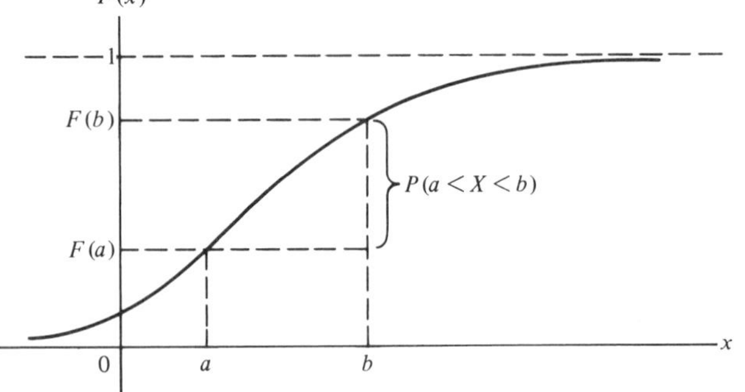

Figure 6-9 Probability of an interval from the cumulative distribution function.

As is the case with any function, including the density function defined earlier, the letter inside the parentheses is a "dummy" variable. Thus $f(x)$, $f(t)$, and $f(\lambda)$ define the same *function*—the same operations to be performed on x, t, or λ, respectively. Similarly, $F(\lambda)$, $F(x)$, and $F(\alpha)$ would all define the same function—or operation to be performed on λ, x, or α to obtain the corresponding value of the function. Having used capital X as the name of a random variable, one often writes $f(x)$ as its density, using the corresponding lower-case x as the dummy, or $F(x)$ as its c.d.f. Thus,

$$F(x) = P(X \leq x);$$

but the distinction, on the right, between random variable (X) and dummy (x) needs to be kept clear.

Because the distribution function $F(x)$ is a probability, for any given x, it is expressible as an area, or definite integral, of the density function:

$$F(x) = P(X \leq x) = \text{area under } f(x) \text{ to the left of } x$$
$$= \int_{-\infty}^{x} f(u)\, du.$$

(Notice that because x is used as the dummy in defining the c.d.f., another dummy u is needed in the integral to distinguish between upper limit and the variable of integration.) This relation giving $F(x)$ in terms of $f(x)$ is shown in Figure 6-10.

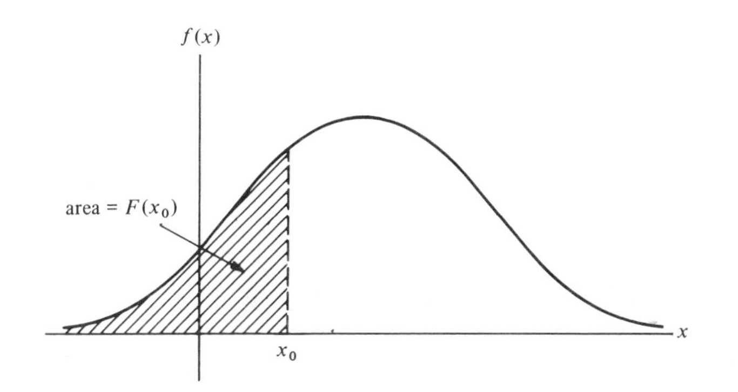

Figure 6-10 The cumulative distribution function from the density.

The integral giving the c.d.f. $F(x)$ in terms of the density $f(x)$ is a particular indefinite integral of $f(x)$, sharing with any indefinite integral the property that its derivative is the function integrated:

$$f(x) = F'(x) = \frac{d}{dx}\int_{-\infty}^{x} f(u)\, du.$$

(This fact is the essential content of what is called in calculus the Fundamental Theorem of Integral Calculus.) Thus the *slope* of $F(x)$ at any point is the density of the distribution at that point. The distribution concentrates probability heavily at points where $F(x)$ is steep [large density $f(x)$] and spreads it thinly where $F(x)$ rises slowly (small density).

The distribution function, giving the accumulated probability from the left up to the point x, is a nondecreasing function of x; for probability (which is nonnegative) can only be *added* on in going from left to right. The function starts at $-\infty$ with the value 0 and rises to the value 1 at ∞. These essential properties are summarized as follows.

Properties of a c.d.f.:

(i) $0 \leq F(x) \leq 1$.

(ii) $F'(x) \geq 0$.

(iii) $F(-\infty) = 0, \quad F(\infty) = 1$.

[Writing "at ∞" and $F(\infty)$ abbreviates the more precise expressions involving limits. Thus $F(\infty) \equiv \lim_{x\to\infty} F(x)$, for example.]

Example 6-g Consider the triangular density of Example 6-f, given by the density

$$f(x) = \begin{cases} x, & \text{if } 0 < x \leq 1, \\ 2 - x, & \text{if } 1 \leq x \leq 2, \\ 0, & \text{otherwise} \end{cases}$$

(as in Figure 6-7). The c.d.f. can be calculated in pieces. Surely, $F(x) = 0$ if $x < 0$, and $F(x) = 1$ if $x > 2$, since all the probability is on the interval $0 < x < 2$. Now, for $0 < x < 1$, the probability $P(X \leq x)$ is the area of a 45° right-angle triangle with base x, namely, $x^2/2$; and for $1 < x < 2$, it is 1 minus the area of the triangle to the right of x, namely, $1 - (2 - x)^2/2$:

$$F(x) = \begin{cases} 0, & x < 0, \\ \frac{1}{2}x^2, & 0 < x < 1, \\ 1 - \frac{1}{2}(2 - x)^2, & 1 < x < 2, \\ 1, & x > 2. \end{cases}$$

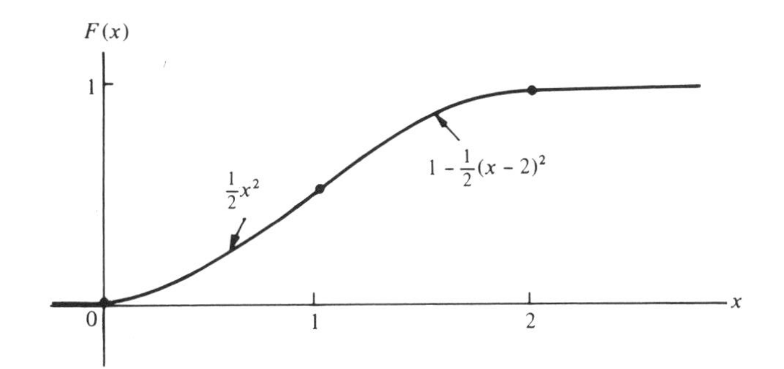

Figure 6-11 The cumulative distribution function for Example 6-g.

This is shown in Figure 6-11. Observe that the steepest slope is at $X = 1$, the point of maximum density. ◄

6.4 Percentiles

As has been seen, a distribution function (c.d.f.) can be used in computing probabilities. A useful table for the functional relationship given by $F(x)$ will give either (a) probabilities $F(x)$ for equally spaced values of x (over the interesting range of values), or (b) values of X corresponding to equally spaced probabilities on the interval 0 to 1. Table I is constructed by method (a); one enters the table at the side (and top) with a value of x and reads in the body of the table the probability $P(X \le x)$, for the particular distribution for which the table is constructed. Tables II and III (among others) give what are called *percentiles.* One enters such a table at a probability that is a multiple of .01 and reads in the body of the table the value of the random variables with that probability to its left. Thus entering at .90 the table entry gives the value $x_{.90}$ such that $P(X \le x_{.90}) \equiv .90$. This value is called the 90th percentile. Ninety per cent of the probability is distributed to the left of the 90th percentile. More generally (see Figure 6-12):

> (100p)th percentile x_p:
>
> $$P(X \le x_p) = p.$$

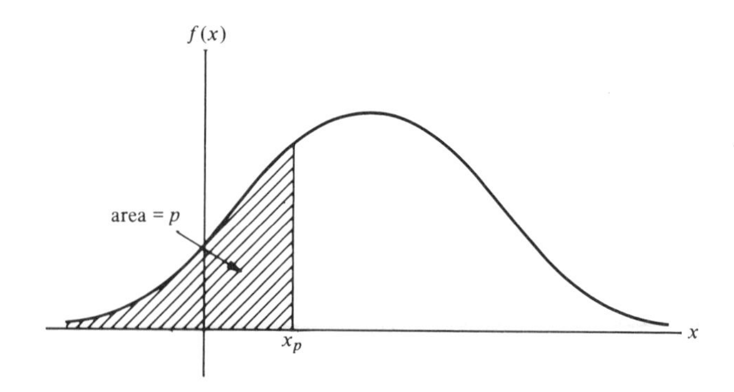

Figure 6-12 Percentile of a distribution.

It is clear that a graph of $F(x)$ could be constructed from either type of table—from type (a) by plotting points $(x, F(x))$, and from type (b) by plotting points (x_p, p).

The 25th, 50th, and 75th percentiles are called, respectively, the 1st, 2nd, and 3rd *quartiles* of the distribution; and the 2nd quartile or 50th percentile is called the *median*:

$$P(X < \text{median}) = P(X > \text{median}) = \frac{1}{2}.$$

In the sense of the definition, the median is a measure of the "middle" or "center" of a distribution: An observation is as likely to be below as above the median. Another measure of "middle" or "location" will be considered in the next section.

6.5 Expected Value

The *expected value* or *mean value* or *average value* or *expectation* of a discrete random variable was defined as the weighted sum of its possible values, the weights being the corresponding probabilities. A continuous random variable has too many possible values for this definition to work, but the same idea will give the proper definition if the "summation" is properly carried out. The probability element $f(x)\,dx$ can be used as a weight for the value x, the products of value and probability element being *integrated* to give the expected value or mean value:

Mean of a continuous variable:

$$\mu \equiv E(X) \equiv \int_{-\infty}^{\infty} x f(x)\,dx.$$

As in the case of discrete distribution, this formula is precisely the formula for *center of gravity* of a continuous mass distribution with mass density function $f(x)$, normalized or scaled so that the total mass is 1.

It is clear from the analogy with mass distributions that the expected value or center of gravity of a *symmetric* distribution is the center of symmetry. That is, if c is such a point:

$$f(c - x) \equiv f(x + c), \qquad \text{for all } x,$$

then the mean value is $E(X) = c$.

Example 6-h Suppose that X has the density $f(x) = 2x$ for $0 < x < 1$. The expected value is computed as follows:

$$EX = \int_0^1 x(2x)\,dx = \left.\frac{2x^3}{3}\right|_0^1 = \frac{2}{3}.$$

This is not, incidentally, the same as the median (.707), which divides the area of the triangle into two equal parts:

$$\int_0^{1/\sqrt{2}} 2x\,dx = \left. x^2 \right|_0^{1/\sqrt{2}} = \frac{1}{2}. \quad ◀$$

The average of a function $g(x)$ of a continuous random variable is defined by analogy with the formula for the discrete case:

$$E[g(X)] \equiv \int_{-\infty}^{\infty} g(x) f(x)\,dx.$$

Having defined expectations for X and for a function $g(X)$, one can go on to various properties and other definitions, just as in the discrete case:

Linearity of $E(\cdot)$:

$$E(aX + b) = aE(X) + b.$$

Balance property of the mean:

$$E(X - \mu) = 0.$$

Definition of *variance*:

$$\text{var}\, X \equiv E[(X - \mu)^2] = E(X^2) - \mu^2.$$

Definition of *standard deviation*:

$$\sigma_X \equiv \sqrt{\text{var}\, X} = \sqrt{E(X^2) - \mu^2}.$$

Example 6-i The mean of the distribution with triangular density $f(x) = 1 - |x - 1|$, for $0 < x < 2$, is the center of symmetry (see Figure 6-7), namely, $\mu = 1$. The expected square is

$$EX^2 \equiv \int_{-\infty}^{\infty} x^2 f(x)\, dx = \int_0^1 x^2 \cdot x\, dx + \int_1^2 x^2(2 - x)\, dx = \frac{7}{6},$$

and so the variance is $7/6 - \mu^2 = 1/6$. ◄

6.6 Several Continuous Variables

The *joint* variation of continuous random variables X and Y is described by a *joint density function*, $f(x, y)$. As in the case of one variable, this density is nonnegative, and its integral over the plane is 1:

Joint density:

$$f(x, y) \geq 0, \qquad \int_{-\infty}^{\infty} \int_{-\infty}^{\infty} f(x, y)\, dx\, dy = 1.$$

This integral is a double integral over the *plane* of values of the pair (X, Y). Probabilities for regions of these values are calculated in terms of the probability element:

$$f(x, y)\,dx\,dy \equiv \text{probability element},$$

by "summing," or integrating this over the region of interest. Such calculations will not be needed here.

If Z is a function of X and Y, the mean value of Z is computed, much as was the function of a single variable, as the weighted sum of its values over the whole plane where the weight is the probability element:

$$E[g(X, Y)] = \int_{-\infty}^{\infty}\int_{-\infty}^{\infty} g(x, y)f(x, y)\,dx\,dy.$$

With these definitions for the distribution of two random variables (X, Y), one can obtain the following extensions of ideas and formulas given earlier for the discrete case:

Additivity of expectation:

$$E(X + Y) = E(X) + E(Y).$$

Condition for independence of X and Y:

$$f(x, y)\,dx\,dy = f_X(x)\,dx\,f_Y(y)\,dy.$$

Covariance of X and Y:

$$\begin{aligned}\text{cov}\,(X, Y) = \sigma_{X,Y} &\equiv E[(X - \mu_X)(Y - \mu_Y)]\\ &= E(XY) - E(X)E(Y).\end{aligned}$$

Variance of a sum:

$$\text{var}\,(X + Y) = \text{var}\,X + \text{var}\,Y + 2\,\text{cov}\,(X, Y).$$

Additivity of variances of uncorrelated variables:

$$\text{var}\,(X + Y) = \text{var}\,X + \text{var}\,Y, \qquad \text{if } \sigma_{X,Y} = 0.$$

Independence of X and Y implies that $\text{cov}\,(X, Y) = 0$.

To illustrate these various relations in specific probability distributions would involve manipulations with double and iterated integrals. All that is really needed in what follows is some intuitive notion of the definitions, and, in the case of uncorrelated variables, the facts of the additivity of expectations and the additivity of variances.

Problems

1. Make a frequency distribution and histogram for the cholesterol data in Table 6.1, using five class intervals. (Is it less erratic than the second histogram in Figure 6-1, which uses 12 class intervals? How about using *two* class intervals—is this wise?)

2. (a) Table XIII gives independent observations on a spinning pointer. Think of each group of five integers as defining a five-decimal digit number on the interval $0 < x < 1$ and make a frequency tabulation and histogram for the first 100 observations. Use a class interval size of .1.
(b) Table XIV gives independent observations from a continuous random variable. Use a class interval size of .5, starting at -3, to make a frequency tabulation and histogram for the first 100 observations in the table.
(c) Would you get the same frequency distribution in part (a), or in part (b), if you started at some other point in the table? (To use the table in simulating actual sampling, one should start at a random point.)

3. Suppose that X is uniformly distributed on $-1 < x < 1$.

(a) Determine the probability that $X > 3/4$, given $|X| > 1/2$.
(b) Obtain the c.d.f. of X.
(c) Given that U is uniform on $0 < x < 1$, as are the numbers in Table XIII of Appendix B, how could you use observations from U to generate a sample of observations from X?

4. Suppose that X is uniformly distributed on $-1 < x < 1$.

(a) Compute $P(X^2 \leq 1/4)$.
(b) Determine $P(X^2 \leq y)$ as a function of y for $0 < y < 1$.
(c) Determine the density function of $Y \equiv X^2$ using the result in part (b).

5. A random variable X has density $f(x) = x/2$, for $0 < x < 2$, and 0 elsewhere.

(a) Determine $P(X > 1)$.
(b) Calculate its mean and variance.
(c) Determine the c.d.f. of X.
(d) Determine the c.d.f. of $Y \equiv X^2$.

6. Let X be uniform on $-1 < x < 1$. Determine $E(X^2)$

(a) Using an integral involving the density of X^2, from Problem 4(c).
(b) Using an integral involving the density of X.

7. Compute σ_X, given that X has the density $6x(1 - x)$ on $0 < x < 1$.

8. Given the c.d.f. $F(x) = \sin x$, for $0 < x < \pi/2$,

(a) Determine the density function.
(b) Determine the median of the distribution.

9. A random variable V is distributed with density function

$$f_V(x) = \frac{1/\pi}{1 + x^2}.$$

(a) Estimate $P(0 < V < .1)$, without integration or tables, by thinking geometrically.
(b) Determine the median of V.
(c) Try calculating $E(V)$ as an integral. What is the stumbling block? (The mean value here is not defined. The distribution is too heavy in the tails.)
(d) Determine the c.d.f. of the distribution.

10. Table I of Appendix B gives values of the c.d.f. of an important continuous distribution. If Z has this distribution, use the table to obtain $P(|Z| > 2)$ and $P(-1.34 < Z < 2.61)$.

11. Table II lists percentiles for several continuous distributions in an important family of distributions called "chi-square," indexed by what is called its "number of degrees of freedom." (This number happens to be the mean of the distribution.) The density function is

$$f(x; n) = (\text{constant})x^{n/2-1}e^{-x/2}, \qquad x > 0,$$

and because the integral is not an elementary function (unless n = 2k, where k is an integer), probabilities must be found by numerical integration or (if one is available) in a table. Use Table II to determine the probability that a chi-square variable with 4 degrees of freedom takes a value between 6 and 15.

12. If X and Y are independent, and if each is uniform on $0 < x < 1$, then $X + Y$ has the triangular distribution of Example 6-i. Verify that in this case var $(X + Y) =$ var X + var Y.

13. Given that X and Y are independent, each uniform on the interval (0, 1), match the following random variables with density function graphs in figures (A)–**(F)**.

(a) $1 - X$.
(b) $X + Y$.
(c) $X - Y$.
(d) $2X$.
(e) $2X - 1$.
(f) $2|X + Y - 1|$.

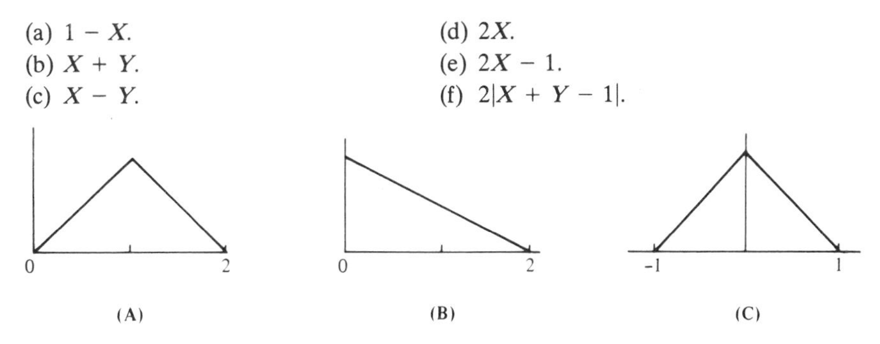

(A) (B) (C)

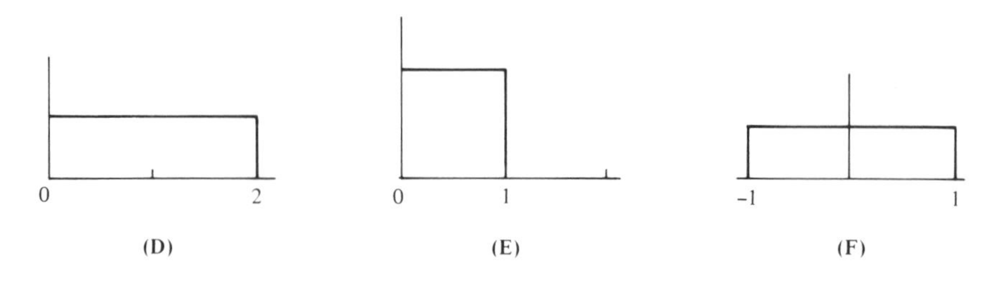

(D) (E) (F)

The Poisson Process

An experiment of chance that continues in time and is observed as it unfolds is sometimes called a *stochastic process*, or a *random process* (or, simply, a *process*). A snapshot or observation at each instant of time is an experiment of chance. If one is observing the time variation of some numerical variable, the instantaneous values make up a family of random variables indexed by time as a parameter. An observation on the complete process is a *function* of time.

One important process is the *Poisson process*, used to describe a wide variety of phenomena that share certain characteristics, phenomena in which some kind of happening takes place sporadically over a period of time in a manner that is commonly thought of as "at random." Examples of happenings for which the model is found useful include arrivals at a service counter, flaws in a long manufactured tape or wire, clicks or counts recorded by a Geiger counter near a radioactive substance, and breakdowns or failures of a piece of equipment or component (when failures are corrected as they occur and the equipment is put back into operation). Just to have a convenient reference term in the general discussion, the happening will be referred to as an "event."†

† This use of the word *event* is not the same as that introduced in Chapter 1 (as a set of outcomes).

The particular model for such processes that goes by the name *Poisson* is that in which the following postulates can be assumed to hold.

Poisson postulates:

1. The numbers of events in nonoverlapping intervals of time are independent random variables.
2. The probability structure is independent of location in time.
3. The probability of exactly one event in a small interval of time is approximately proportional to the width of the interval.
4. The probability of *more* than one event in a small interval of time is negligible.

The terms *approximately* and *negligible* need precise definition in order that a mathematical development can be rigorously based on these postulates. This will be left to Section 7.3.

The description of a process, so far, has been in terms of a *time* variable, although the possibility of wider applicability was hinted in citing the example of flaws in a tape or wire. The variable, to be referred to as t, along which the process evolves, may indeed be a lineal dimension. More generally, it may be a vector of two or three space dimensions, with appropriate modification of the language. Thus the "interval of time" would be an element of area or of volume, with flaws or other occurrences scattered randomly throughout a plane or solid region—as in a sheet of metal or photographic film, or in a container of some fluid. In Postulate 2 one would read just "independent of location," without the restriction "in time." Nevertheless, it will be convenient to use the language of time.

7.1 The Poisson Random Variable

A random variable of particular interest, in connection with a Poisson process, is this:

$$X \equiv \text{number of events in an interval of width } t,$$

for some given t. Its distribution will be derived from the Poisson postulates.

Let the interval from 0 to t be divided into n equal subintervals, of width $h = t/n$. The number n will be allowed to become infinite, so that h can be considered small. According to the postulates, one of two things can be said of each subinterval: Either an event happens, or it does not—discounting the highly improbable contingency that *more* than one event happens. Coding an occurrence with 1 and nonoccurrence with 0 yields a variable that is approximately of the Bernoulli type. The probability of 1 for a particular subinterval is the probability of one event on that subinterval, proportional (approximately) to its width h; let this probability be denoted by λh. The probability of 0 is approximately one minus the probability of 1:

Value	Probability
1	λh
0	$1 - \lambda h$

Notice that Postulate 2 is used in asserting that these same probabilities apply for any of the subintervals of width h.

According to Postulate 1, the n Bernoulli variables defined in this way (one for each subinterval) are independent. So the total number of events in the interval from 0 to $t = nh$ can be thought of as a binomial variable, as the sum of the 1's and 0's for the independent trials or subintervals. This is only approximately so, because the result for each subinterval is only approximately Bernoulli; but it becomes nearer to being correct as n increases without limit. The probability function for X, the number of successes (occurrences of an event) in the n trials (subintervals), is given approximately by the binomial formula:

$$\begin{aligned} P(X = k) &\doteq \binom{n}{k}(\lambda h)^k(1 - \lambda h)^{n-k} \\ &= \binom{n}{k}\left(\frac{\lambda t}{n}\right)^k\left(1 - \frac{\lambda t}{n}\right)^{n-k} \\ &= \frac{(\lambda t)^k}{k!}\cdot\frac{n}{n}\cdot\frac{n-1}{n}\cdots\frac{n-k+1}{n} \\ &\quad \cdot\left(1 - \frac{\lambda t}{n}\right)^{-k}\left(1 - \frac{\lambda t}{n}\right)^n. \end{aligned}$$

As n increases, the first factor is fixed, and the next $k + 1$ factors tend to 1. The last factor (with $x = -\lambda t/n$) is

$$\left(1 - \frac{\lambda t}{n}\right)^n = (1 + x)^{-\lambda t/x} = [(1 + x)^{1/x}]^{-\lambda t} \longrightarrow e^{-\lambda t},$$

since the limit of $(1 + x)^{1/x}$ as $x \to 0$ defines e. The final result is the *Poisson formula*:

$$P(X = k) = \frac{(\lambda t)^k}{k!} e^{-\lambda t}, \qquad k = 0, 1, 2, \ldots.$$

This Poisson formula defines a family of probability models depending on a parameter λ, or equivalently, on the parameter $m \equiv \lambda t$, the only combination in which λ appears. The factors $m^k/k!$, for $k = 0, 1, 2, \ldots$, are terms in the expansion of the exponential function:

$$e^m = 1 + m + \frac{m^2}{2} + \frac{m^3}{3!} + \cdots,$$

so it is clear that $\Sigma P(X = k) = 1$, as must be the case. The mean is easily computed:

$$E(X) = \sum_0^\infty k \frac{m^k}{k!} e^{-m}$$

$$= me^{-m} \sum_1^\infty \frac{m^{k-1}}{(k-1)!} = m.$$

Similarly, one can compute

$$E[X(X-1)] = \sum_0^\infty k(k-1) \frac{m^k}{k!} e^{-m}$$

$$= m^2 e^{-m} \sum_2^\infty \frac{m^{k-2}}{(k-2)!} = m^2,$$

and then $E(X^2) = m^2 + E(X) = m^2 + m$, from which it is evident that the variance equals the mean:

$$\text{var}\, X = E(X^2) - (EX)^2 = (m^2 + m) - m^2 = m.$$

Poisson probability function:

$$f(k; m) = \frac{m^k e^{-m}}{k!}, \qquad k = 0, 1, 2, \ldots.$$

Moments:

$$E(X) = m, \qquad \text{var}\, X = m.$$

In the application to the Poisson process, where X denotes the number of "events" in an interval of length 5, the mean value is $m = \lambda t$, which says that the expected number of events in an interval is proportional to its width. In particular, upon setting $t = 1$, it is evident that the parameter λ of the process is the expected number of events in a unit interval. It is measured in reciprocal time units. For example, if t is a number of *hours*, then λ is the expected number of events *per hour*, so that λt is dimensionless.

Example 7-a Telephone calls are assumed to arrive, at a certain exchange, according to the Poisson "law" at an average of 20 per hour. What is the probability that no calls arrive in a 5-minute interval?

The average number in a 5-minute interval is λt, where $\lambda = 20$ and $t = 5/60$ hour. Alternatively, $\lambda = 1/3$ per minute, and $t = 5$ minutes; but whatever unit is chosen, $\lambda t = 5/3$. And then

$$P(\text{no calls in 5 minutes}) = P(X = 0) = e^{-5/3}\frac{(5/3)^0}{0!} \doteq 0.19. \quad \blacktriangleleft$$

Table XII in Appendix B gives Poisson probabilities for a selected set of values of the mean m (or λt). Probabilities are given in cumulative form, this being the type of probability usually needed in inference problems:

$$P(X \le c) = F(c) = \sum_{k=0}^{c} f(k; m) = \sum_{k=0}^{c} e^{-m}\frac{m^k}{k!}.$$

Probabilities of individual values can be obtained as differences of table entries:

$$P(X = c) = P(X \le c) - P(X \le c - 1) = F(c) - F(c - 1).$$

The Poisson distribution was derived to describe events occurring randomly in continuous time (as in the preceding example), but it is also quite useful in providing approximations to binomial probabilities. It was seen, in deriving the Poisson formula from the postulates, that for large n (and small $h = t/n$),

$$\binom{n}{k}(\lambda h)^k(1 - \lambda h)^{n-k} \doteq \frac{(\lambda t)^k}{k!}e^{-\lambda t}.$$

Setting $p = \lambda h$, so that $\lambda t = pt/h = np$, yields the desired formula:

Poisson approximation to the binomial:

$$\binom{n}{k}p^k(1 - p)^{n-k} \doteq \frac{(np)^k}{k!}e^{-np} \qquad \text{(large } n\text{, small } p\text{).}$$

As in the case of any approximation, it is important to know when it is usable. Experience with actual numbers is a reasonable guide. If $.1 < p < .9$, the normal approximation to be given in Chapter 8 will perhaps be preferred. The following table gives some helpful comparisons of Poisson probabilities and binomial probabilities with the same mean. For the case $p = .1$ the table gives binomial probabilities (n, p) and corresponding Poisson probabilities with mean np, for $k = 0, 1, \ldots, 5$.

Table 7.1 Probability that $X = k$

k	Poisson $m = 5$	Binomial $(50, .1)$	Poisson $m = 2$	Binomial $(20, .1)$	Poisson $m = 1$	Binomial $(10, .1)$
0	.007	.005	.135	.122	.368	.349
1	.033	.029	.271	.270	.368	.387
2	.085	.078	.271	.285	.184	.194
3	.140	.139	.180	.190	.061	.057
4	.175	.181	.090	.090	.015	.011
5	.174	.185	.036	.032	.0031	.0015

Example 7-b One hundred inhabitants of a city of 10,000 are color-blind. What is the probability that a random selection of 200 includes exactly 3 who are color-blind?

The exact probability is hypergeometric:

$$f(3) = \frac{\binom{100}{3}\binom{9900}{197}}{\binom{10{,}000}{200}},$$

which is approximable, since the sample size is small in comparison with the population, by the binomial formula with $p = 100/10{,}000$ and $n = 200$:

$$f(3) \doteq \binom{200}{3}(.01)^3(.99)^{197} = .1814.$$

It is true that this can be evaluated by some hand-held calculators, but it may be sometimes more convenient to approximate it by means of the Poisson formula, with $m = np = 2$:

$$f(3) \doteq \frac{2^3}{3!}e^{-2} = F(3) - F(2)$$

$$= .857 - .677 = .180,$$

where F denotes the Poisson c.d.f. (Table XII). ◄

7.2 The Waiting-Time Distribution

Another random variable associated with a Poisson process is the time from one event to the next one. This is a continuous variable, whose distribution can be obtained with the aid of the Poisson formula.

Let $t = 0$ denote an arbitrary point on the time scale of the process, a point at which a stopwatch is started. Let T denote the time from $t = 0$ to the first event thereafter. The time T will exceed any given $t > 0$ if and only if there are *no events in the interval* $(0, t)$:

$$\begin{aligned} P(T > t) &= P[0 \text{ events in } (0, t)] \\ &= e^{-\lambda t}, \qquad \text{for } t > 0. \end{aligned}$$

The c.d.f. of T is obtained from this by complementation:

$$F_T(t) = P(T \le t) = 1 - e^{-\lambda t}, \qquad t > 0,$$

and its derivative is the density of the distribution of T:

Waiting-time density:

$$f_T(t) = \lambda e^{-\lambda t}, \qquad t > 0.$$

This density for $\lambda = 1$ is pictured in Figure 7-1, along with the corresponding c.d.f.

It is to be observed that the exponential distribution obtained for the waiting time applies no matter how long it happened to have been, before the stopwatch was started, that the most recent event occurred. This is particularly significant if the event is the failure of a unit or system. It means that if the unit or system is operating at a given time, a *new* unit of the same type would have the same future life distribution as the used one currently in operation; that is, the "preventive maintenance" of replacing a used unit is of no use as long as it is still operating. The exponential model for system failure is usually not applicable when failure is caused by wear-out; but it is often useful for describing the operating life when failure is caused by random environmental conditions—shocks, voltage drops, road hazards, and so on. In such contexts it is not surprising that maintenance by systematic replacement of units is in vain.

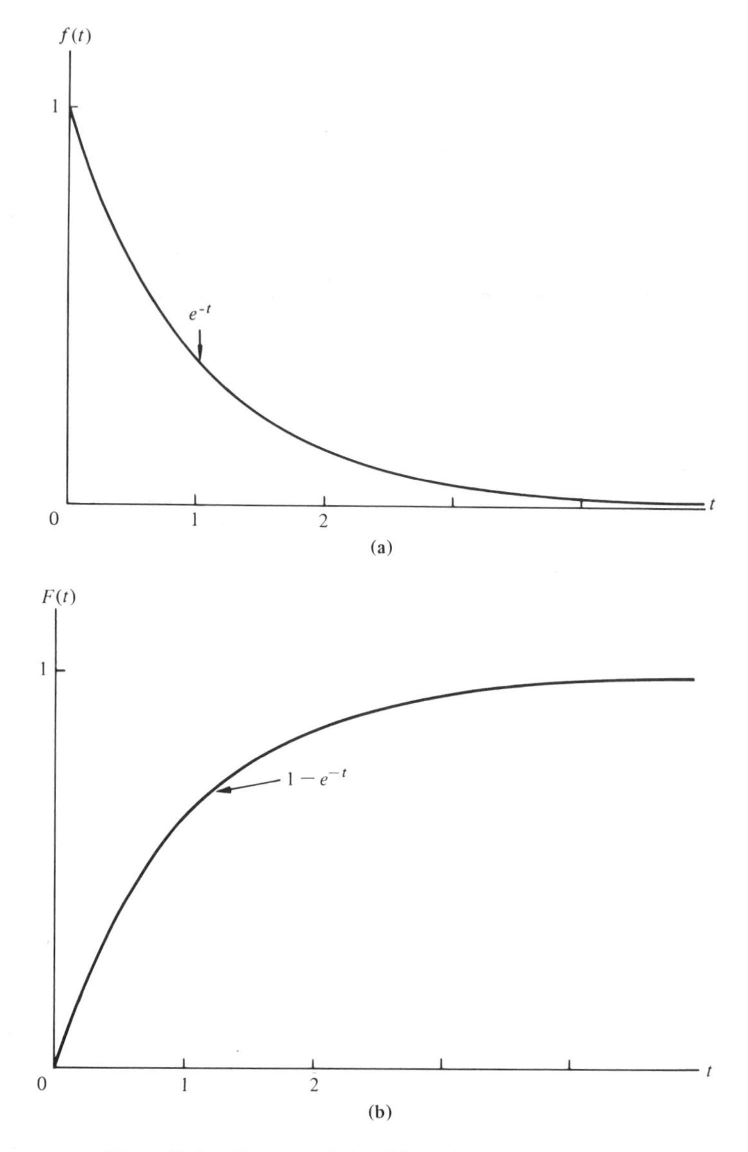

Figure 7-1 Exponential waiting-time density and c.d.f.

The integral that gives the mean waiting time can be evaluated by an "integration by parts":

$$E(T) = \int_0^\infty t\lambda e^{-\lambda t}\,dt = \frac{1}{\lambda}\int_0^\infty ue^{-u}\,du = \frac{1}{\lambda}.$$

This is the reciprocal of the average number of events per unit time. Thus if there are 10 events per hour on the average, the mean time to wait for an event (starting at any point on the time scale) is 1/10 hour, or 6 minutes.

The time from an arbitrary $t = 0$ out to the rth subsequent event is the sum of r independent waiting times, each having the exponential waiting-time distribution. The distribution of this sum will not be derived here, but of course the *mean* time to the rth event would be the sum of the r mean times between events, namely, r/λ.

It has been seen that for a given Poisson process, the waiting times are independent with identical exponential distributions. It is not difficult to show, conversely, that if the waiting times are independent and have the same exponential distributions, the Poisson postulates are satisfied.

Example 7-c Suppose the time between arrivals at a service counter is exponential with mean 12 minutes. If successive times are independent, the number of arrivals in an hour is Poisson with mean $\lambda t = 60/12 = 5$, and then, for example,

$$P(10 \text{ or fewer arrivals in 1 hour}) = e^{-5} \sum_{0}^{10} \frac{5^k}{k!} = .986. \quad ◄$$

The waiting time in a Poisson process is a continuous analog of the "time" or number of trials to the first success in a sequence of independent, Bernoulli trials. The probability of having to wait for n trials is geometric, and upon referring to the derivation of the Poisson formula, it is seen that the geometric probability formula converges to the exponential probability element:

$$pq^n = \frac{\lambda t}{n}\left(1 - \frac{\lambda t}{n}\right)^n = \lambda\left[\left(1 - \frac{\lambda t}{n}\right)^{-n/(\lambda t)}\right]^{-\lambda t}\left(\frac{t}{n}\right) \longrightarrow \lambda e^{-\lambda t}\, dt.$$

Similarly, the time to the rth arrival or failure is the analog of the time to the rth success, and so the distribution of the sum of r independent exponential variables is the continuous analog of the negative binomial distribution.

The Poisson process is thus seen to be the continuous-time version of the Bernoulli process, and both are characterized by the property that the future waiting time, after any point, does not depend on how long one has been waiting.

Example 7-d If $p = .01$ in a Bernoulli experiment, the probability that one would have to wait 100 trials or more to get the first success is

$$\sum_{100}^{\infty} pq^n \doteq \int_{1/\lambda}^{\infty} \lambda e^{-\lambda t}\, dt = \frac{1}{e} \doteq .368.$$

[The series can also be summed directly, to yield $(.99)^{100} = .366$. The lower limit $1/\lambda$ is the t-value that corresponds to $n = 100$.] ◄

★ 7.3 An Alternative Derivation

The Poisson formula can be derived from the Poisson postulates using a differential approach, one that can be applied to a variety of related situations. A more precise formulation of Postulates 3 and 4 is required. Since it is the differential with respect to the time variable that is considered, the following notation will be employed, for a Poisson process with parameter λ:

$$P_n(t) = P(n \text{ events in time } t).$$

Observe once again that the time invariance of Postulate 2 permits definition of this function of the interval width, independent of its location on the time axis.

The postulates are now given as follows:

3. For small h, $P_1(h) = \lambda h + o(h)$.
4. For small h, $\Sigma_2^\infty P_n(h) = o(h)$,

where $o(h)$ denotes a residual or error, a function of h, that is of "smaller order" than h, in this sense:

$$\lim_{h\to 0} \frac{o(h)}{h} = 0.$$

Thus in Postulate 3,

$$\lim_{h\to 0} \frac{P_1(h) - \lambda h}{h} = 0.$$

This is a formal and precise way of saying that the probability of one event in time h is essentially proportional to h. Similarly, in terms of a limit, Postulate 4 reads

$$\lim_{h\to 0} \frac{1}{h} \sum_2^\infty P_n(h) = 0.$$

The differential technique for deriving $P_n(t)$ is to consider an increment in t, from t to $t + h$, and to express what takes place in the slightly wider interval $(0, t + h)$ in terms of what occurs in the two portions $(0, t)$ and $(t, t + h)$. Clearly, if there are *no* events in $(0, t + h)$, then there are none in $(0, t)$ and none in $(t, t + h)$, and conversely:

$$\begin{aligned} P_0(t + h) &= P[0 \text{ events in } (0, t)]P[0 \text{ events in } (t, t + h)] \\ &= P_0(t)P_0(h), \end{aligned}$$

the factorization being permitted by the independence assumed in Postulate 1.

Using Postulate 3 and transposing a term, one obtains

$$P_0(t + h) - P_0(t) = -\lambda h P_0(t) + o(h),$$

from which the derivative is obtained as the limit of the increment quotient:

$$P_0'(t) = \lim_{h \to 0} \frac{-\lambda P_0(t) + o(h)}{h} = -\lambda P_0(t).$$

Because $P_0(0)$ must be 1 (the probability of no events in time 0), it follows upon integration that

$$P_0(t) = e^{-\lambda t}.$$

Similar reasoning yields

$$P_1(t + h) = P_0(t)P_1(h) + P_1(t)P_0(h),$$

which, by means of Postulates 3 and 4, leads to the following first-order linear differential equation for $P_1(t)$:

$$P_1'(t) = -\lambda P_1(t) + \lambda P_0(t).$$

Using the $P_0(t)$ determined previously, one obtains

$$P_1(t) = \lambda t e^{-\lambda t}.$$

And for general n, it is seen similarly that

$$P_n'(t) \equiv -\lambda P_n(t) + \lambda P_{n-1}(t),$$

with solution

$$P_n(t) = \frac{(\lambda t)^n}{n!} e^{-\lambda t}.$$

This is the Poisson formula obtained heuristically in Section 7.1.

To illustrate the point made earlier that the differential technique may be used in other situations, consider the following. A machine has exponential life distribution with parameter λ, and when it breaks down a repairman takes a length of time to repair the machine that is also exponential but with parameter μ. Let 0 denote the state that the machine is operating, and 1 the state that it is being repaired. Let $P_n(t)$ be the probability that it is in state n. Then for an increment from t to $t + h$, there follows

$$P_0(t + h) = P_0(t)e^{-\lambda h} + P_1(t)(1 - e^{-\mu h}).$$

For, if the machine is operating (state 0) at $t + h$, then either it was operating at t and survives an additional h units, or it was down (state 1) at t and was repaired in time less than h units. Subtraction of $P_0(t)$ and division by h yields

$$\frac{1}{h}[P_0(t + h) - P_0(t)] = P_0(t)\frac{e^{-\lambda h} - 1}{h} + P_1(t)\frac{1 - e^{-\mu h}}{h}$$

with limit, as $h \to 0$:

$$P_0'(t) = -\lambda P_0(t) + \mu P_1(t).$$

Similarly, it may be shown that

$$P_1'(t) = -\mu P_1(t) + \lambda P_0(t),$$

and these two linear differential equations for $P_0(t)$ and $P_1(t)$ can be solved by standard methods.

★ 7.4 Randomness of Events

If one looks at a record of a Poisson process, he would expect, in view of the basic postulates, to find events sprinkled along the time axis in what might be thought of as a random manner. One way of justifying this expectation is to consider the conditional distribution of events in a particular interval *given* that there are a certain number, say n, in that interval.

Suppose first that an interval $(0, t)$ is known to contain *one* event. Where is it likely to be? Let its location (relative to $t = 0$) be denoted by T. The conditional c.d.f. of T is

$$\begin{aligned} P[T \le t \mid \text{one event in } (0, b)] &= \frac{P[T \le t \text{ and 1 event in } (0, b)]}{P[\text{1 event in } (0, b)]} \\ &= \frac{P[\text{1 event in } (0, t)]P[\text{0 events in } (t, b)]}{P[\text{1 event in } (0, b)]} \\ &= \frac{(\lambda t e^{-\lambda t})(e^{-\lambda(b-t)})}{\lambda b e^{-\lambda b}} = \frac{t}{b} \end{aligned}$$

if $0 < t < b$. That is, the position of the event is *uniformly* distributed, with density $1/b$, on the interval $(0, b)$.

More generally, the following can be demonstrated: If it is given that n events have occurred on the interval $(0, b)$, these locations are distributed on that interval in the same way as would be the results of n independent spins of a uniform pointer with scale $0 < t < b$ around its circumference.

★ 7.5 Reliability

The probability that a unit operates without failure to time t is denoted by $R(t)$ and is called the unit's *reliability* function. If L denotes the length of life or time to failure, then

> Reliability function:
> $$R(t) = P(L > t) = 1 - F_L(t).$$

If the time to failure has the *exponential* density $\lambda e^{-\lambda t}$, then

$$R(t) = \int_t^\infty \lambda e^{-\lambda u}\,du = e^{-\lambda t}.$$

The logarithmic derivative of the reliability is called the *hazard*:

$$H(t) \equiv -\frac{d}{dt}[\log R(t)] = \frac{-R'(t)}{R(t)}.$$

This can be interpreted as the rate of "dying" among such units relative to the number still operating. If the time to failure is exponential, the hazard is λ, a constant; and conversely, if the hazard is constant, it follows (upon integration) that the unit has an exponential failure pattern.

The mean life of a unit can be computed in terms of its reliability function:

$$E(L) = \int_0^\infty tF_L'(t)\,dt = -\int_0^\infty tR'(t)\,dt.$$

Integrating by parts, with $u = t$, $dv = -R'(t)\,dt$, and $v = -R(t)$, one obtains

$$E(L) = -tR(t)\Big|_0^\infty + \int_0^\infty R(t)\,dt.$$

If $E(L)$ is finite, then $R(t)$ goes to zero faster than $1/t$, and the first term vanishes; then

$$E(L) = \int_0^{\infty} R(t)\,dt.$$

Sometimes a system is so constituted that its reliability can be determined from a knowledge of the reliability of its component units. The simplest combination of units are those termed *series* and *parallel* combinations.

If units are placed in a system so that the system fails if any one of the unit fails, they are said to be in series. The reliability of the system is computed in terms of unit reliabilities $R_1(t), \ldots, R_n(t)$ as follows:

$$\begin{aligned} R(t) = P(\text{system life} > t) &= P(\text{all units live to at least } t) \\ &= \prod P(i\text{th unit lives to at least } t) \\ &= \prod R_i(t). \end{aligned}$$

If the units are placed in *parallel*, so that the system fails only when all units have failed, the system reliability is

$$\begin{aligned} R(t) &= 1 - P(\text{system life} \le t) \\ &= 1 - P(\text{all units fail before } t) \\ &= 1 - \prod P(i\text{th unit fails before } t) = 1 - \prod [1 - R_i(t)]. \end{aligned}$$

In summary, then:

Series system reliability:

$$\prod R_i(t).$$

Parallel system reliability:

$$1 - \prod [1 - R_i(t)].$$

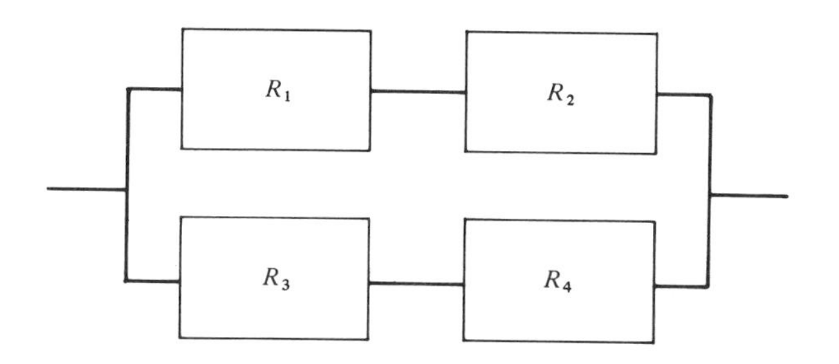

Figure 7-2 Schematic diagram for Example 7-e.

Example 7-e Suppose a system is composed of four units, arranged as shown in Figure 7-2. That is, the system fails if some unit in the upper path *and* some unit in the lower path fails. The reliabilities in the upper and lower paths are

$$R^*(t) = R_1(t)R_2(t),$$

$$R_*(t) = R_3(t)R_4(t).$$

and so the system reliability is

$$1 - (1 - R_1R_2)(1 - R_3R_4).$$

If, for example, each unit has an exponential life with mean 1 hour: $R_i(t) = e^{-t}$, then

$$R(t) = 1 - (1 - e^{-2t})(1 - e^{-2t}),$$

and the mean system life in hours is

$$E(L) = \int_0^\infty R(t)\,dt = \int_0^\infty (2e^{-2t} - e^{-4t})\,dt = \frac{3}{4}. \quad ◀$$

Problems

1. Customers arrive at a service counter according to a Poisson law, at an average of 30 per hour. Determine the probability that

(a) Just one customer arrives during a given 1-minute period.
(b) At least two customers arrive during a 1-minute period.
(c) No customers arrive in a 5-minute period.

2. Flaws appear at random on sheets of photographic film on the average of two per sheet. Assuming a Poisson distribution, what is the probability that a sheet has no flaws?

3. Breakdowns of a certain piece of airborne equipment are found to occur according to a Poisson law, with an average of one breakdown per 10-hour period of operating time.

(a) What is the probability of no breakdown during a 15-hour mission?
(b) What is the mean number of breakdowns in 100 hours of operating time?

4. A company will accept a shipment of 1000 items if no more than one defective item is found in a random selection of 20 items.

(a) Give an exact hypergeometric formula for the probability that the shipment is accepted when it actually contains 50 defectives.
(b) Give a binomial approximation for the probability in part (a).
(c) Use Table XII to obtain an approximate numerical value for the probability in parts (a) and (b).

5. A certain type of inoculation has been found to be about 99.5 per cent effective. What is the probability that it is not effective for more than one out of a group of 400 given the inoculation?

6. Three service counters, for three different types of customers, experience arrivals at an average of 10 per hour, 20 per hour, and 5 per hour, respectively. What is the probability that if the three service people take a 6-minute break together, no customer will arrive for service of any type? (Assume indepent, Poisson arrivals.)

7. Rat hairs are distributed, in a certain brand of peanut butter, according to a Poisson law, at an average of 10 per 2 pounds.

(a) What is the probability that a 1-pound jar contains two or more?
(b) Given that a 1-pound jar contains at least one, what is the probability that there are more in that jar?
(c) What is the probability that of the 1-pound jars in a case of 24, none is free of rat hairs?

8. In a certain application, an electric fuse, which fails when overloaded, has a time to failure which is assumed to be exponentially distributed with mean 6 months. What is the probability that it will last 12 months?

9. Evaluate the following, in terms of e:

(a) $\sum_{0}^{\infty} k^2 2^k / k!$.

(b) $\binom{100}{3}(.94)^{97}(.06)^3$.

10. Let L denote the length of life of a certain unit in a given environment, and suppose that the probability that it lasts at least x hours is e^{-2x}.

(a) Determine the c.d.f. and the density function of L.

(b) Determine the probability that the unit lasts a total of at least $x + 1$ hours, *given* that it is still operating after 1 hour.

11. In the Poisson process of Problem 1 ($\lambda = 30$/hour):

(a) Give the density function of the time to the next customer after an arbitrary time $t = 0$.

(b) Determine the mean time to the next customer, and the mean time until the tenth customer arrives (after $t = 0$).

12. In Problem 6, determine the mean time to the arrival (after $t = 0$) of the first customer of any type.

13. In a system of k identical units in series, each with an exponential life distribution with mean $1/\lambda$, determine the mean life of the system.

14. Let W be the time to the first arrival, after an arbitrary starting time $t = 0$, in a Poisson process with mean λ arrivals per unit time.

(a) Use this integral formula to evaluate var W: $\int_0^\infty x^2 e^{-ax}\,dx = 2/a^3$.

(b) Obtain a formula for the variance of the time to the rth arrival, exploiting its structure as a sum of independent interarrival times.

8

Normal Distributions

A particular type of continuous distribution of considerable importance is that which is called *normal* or (especially in the physical sciences and engineering) *Gaussian*.

The normal distribution is important for several reasons: (1) Random variables occurring in practice can often be represented rather well by means of a normal distribution; (2) large-sample statistics often turn out to be approximately normally distributed; and (3) the form of the normal density lends itself to mathematical manipulations that make possible a precise analysis of inference procedures for normal populations—procedures that happen to work well even for some nonnormal populations.

8.1 The Standard Normal Distribution

The function $e^{-z^2/2}$ is nonnegative; it goes rapidly to 0 as $|z| \to \infty$, so the area under its graph is finite. It can be shown that

$$\int_{-\infty}^{\infty} e^{-z^2/2}\, dz = \sqrt{2\pi}$$

so that a density function is obtained upon division by $\sqrt{2\pi}$. This density defines what is called the *standard normal distribution*.

> Standard normal density:
>
> $$\frac{1}{\sqrt{2\pi}}e^{-z^2/2}, \qquad -\infty < z < \infty.$$

The function $e^{-z^2/2}$ is symmetric about $z = 0$, having the same value at $-z$ as at $+z$, and clearly has a maximum at $z = 0$. The usual method of calculus involving the second derivative shows that there are points of inflection at $z = \pm 1$. The graph is shown in Figure 8-1.

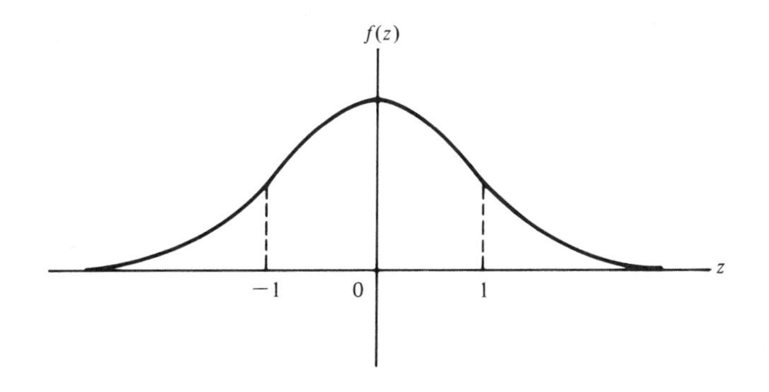

Figure 8-1 Standard normal density function.

Let Z denote a random variable having the standard normal distribution. Because the distribution is symmetric about $z = 0$, its mean value is this center of symmetry: $EZ = 0$. The variance is

$$\operatorname{var} Z = E(Z^2) = \frac{1}{\sqrt{2\pi}}\int_{-\infty}^{\infty} z^2 e^{-z^2/2}\, dz = 1,$$

which can be verified by an integration by parts. That is, the standard deviation is 1, the distance from the mean to the points of inflection.

The term *standard* refers to the parameter values $\mu = 0$ and $\sigma^2 = 1$, to a distribution centered at the origin and scaled so that the unit of measurement is 1 standard deviation.

The integral of the density defines the c.d.f.:

> Standard normal c.d.f.:
>
> $$\Phi(z) \equiv P(Z \leq z) = \int_{-\infty}^{z} \frac{1}{\sqrt{2\pi}} e^{-u^2/2}\, du.$$

This integral is not "elementary" in the sense of calculus, but a table of values can be developed by numerical integration. One such table is given as Table I of Appendix B. Probabilities of intervals or simple combinations of intervals can be computed by referring to that table.

Example 8-a Suppose that Z is standard normal. The probability

$$P(|Z - 1| > 1.42) = 1 - P(-.42 < Z < 2.42)$$

can be expressed in terms of $\Phi(z)$ as

$$1 - [\Phi(2.42) - \Phi(-.42)] = 1 - .9922 + .3372 = .3450. \quad ◄$$

8.2 The Normal Family

A distribution is said to be normal if it can be transformed into a standard normal distribution by a linear transformation—that is, by a shift of origin and a change of scale. The following transformation makes any random variable into a "standardized" variable with mean 0, variance 1.

> Standardizing transformation:
>
> $$Z = \frac{X - \mu}{\sigma} \qquad (EZ = 0, \quad \text{var}\, Z = 1).$$

So the standard normal distribution is the basis for a whole family of distributions, namely, distributions of $X = \mu + \sigma Z$, where Z is standard normal and μ and σ are any real numbers with $\sigma > 0$. Clearly, X has mean μ and standard deviation σ, and its shape is in essence the same (except for uniform stretching and shifting) as that of Z:

General normal density:

$$f_X(x) = \frac{1}{\sigma\sqrt{2\pi}} e^{-(x-\mu)^2/(2\sigma^2)}$$

This density is symmetric about $x = \mu$, and the distance from μ to the points of inflection ($z = \pm 1$ on the standardized scale) is just one sigma; they are at $\mu \pm \sigma$. (See Figure 8-2.)

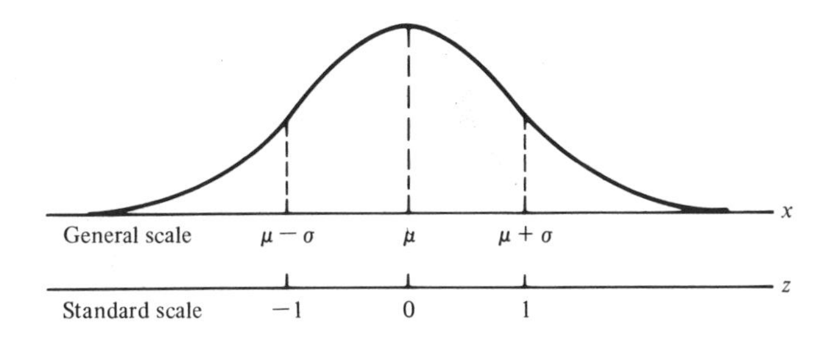

Figure 8-2 General normal density with corresponding standardized scale.

Because of the simple, linear relationship between a general normal distribution and a standard normal one, it is possible to calculate probabilities for the former using the table for the latter. Since the point $x = \mu + z\sigma$ on the general scale corresponds to the point z on the standard scale, the relative area to the left of x is the same as to the left of $z = (x - \mu)/\sigma$.

Determining probabilities for a normally distributed X from the standard normal c.d.f. (Table I):

$$P(X \le x) = \Phi\left(\frac{x - \mu}{\sigma}\right),$$

where $\mu = E(X)$ and $\sigma^2 = \text{var } X$.

That is, the point x is located relative to the mean, as so many sigmas to the right, and that number (the number of σ's) is the z used to enter Table I for $\Phi(z)$. (See Figure 8-3.)

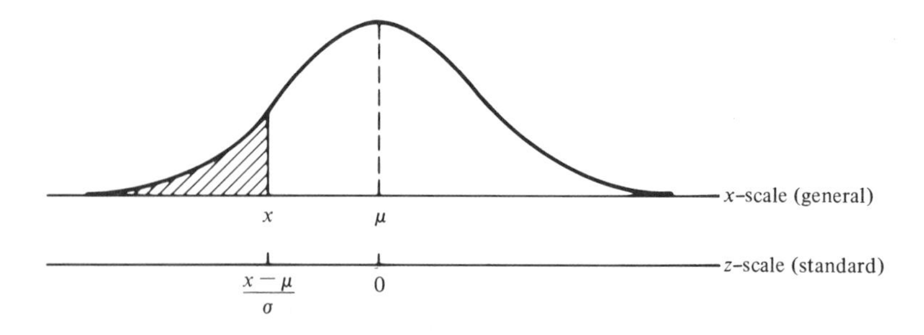

Figure 8-3 General normal probabilities using the standard normal scale.

Example 8-b Scores in a certain examination are found through long experience to be distributed in such a way as to be well represented by a normal distribution, with mean 100 and standard deviation 20. The probability that a score falls between 90 and 110 is then computed as follows with the aid of Table I:

$$\begin{aligned} P(90 < X < 110) &= F_X(110) - F_X(90) \\ &= \Phi\left(\frac{110 - 100}{20}\right) - \Phi\left(\frac{90 - 100}{20}\right) \\ &= .6915 - .3085 = .3830. \quad ◄ \end{aligned}$$

It is convenient to have a shorthand way of referring to a normal distribution. The phrase "X is $\mathcal{N}(\mu, \sigma^2)$" will mean that X has a normal distribution with mean μ and variance σ^2.

8.3 The Central Limit Theorem

Many useful statistics are sums of independent random variables; and it is essential, in interpreting them, to know their patterns of variation from sample to sample—their distributions, in other words. These will generally be quite complicated, but for large samples the situation is simplified by an important fact: The *sum* of a large number of independent observations on a random variable is *approximately normally* distributed.

Before stating this formally and accurately, we recall some facts about sums. Let $X_1, X_2, \ldots, X_n$ be independent, all having the same distribution with mean μ and variance σ^2; that is, the X's are independent observations on the same numerical experiment. Then the expected value and variance are additive:

Let $X_1, \ldots, X_n$ have common mean μ and common variance σ^2. Then

(i) $E(\sum X_i) = \sum E(X_i) = n\mu$.

(ii) $\text{var}\,(\sum X_i) = \sum \text{var}\, X_i = n\sigma^2$, when X's are independent.

If S_n denotes the sum $\sum X_i$, then the *standardized sum* is

$$\frac{S_n - E(S_n)}{\sqrt{\text{var}\, S_n}} = \frac{S_n - n\mu}{\sqrt{n\sigma^2}}.$$

The following important result, given without proof, states that the limiting distribution of the standardized sum is $\mathcal{N}(0, 1)$.

CENTRAL LIMIT THEOREM. If for each integer n the observations $X_1, \ldots, X_n$ are independent with a common distribution having mean μ and finite variance σ^2, then the sum $S_n \equiv X_1 + \cdots + X_n$ is asymptotically normal, in the following sense:

$$\lim_{n\to\infty} P\left(\frac{S_n - n\mu}{\sqrt{n\sigma^2}} \le z\right) = \Phi(z),$$

where $\Phi(z)$ is the standard normal c.d.f.

The practical significance of such a statement involving a limit is that for large n, the limitand is close to the limit:

$$P\left(\frac{S_n - n\mu}{\sqrt{n\sigma^2}} \le z\right) \doteq \Phi(z).$$

The Central Limit Theorem (CLT) can then be given in an equivalent form that is more directly useful:

Practical version of the CLT:
Let $X_1, \ldots, X_n$ be independent, having a common distribution with mean μ and finite variance σ^2. Their sum $S_n \equiv X_1 + \cdots + X_n$ is distributed approximately as $\mathcal{N}(n\mu, n\sigma^2)$, and probabilities can be approximated using Table I:

$$P(S_n \le y) \doteq \Phi\left(\frac{y - n\mu}{\sqrt{n\sigma^2}}\right).$$

The larger the n, of course, the better the approximation, but no general rule can be given as to how large n should be. It would depend on (1) how precise an answer is desired, and (2) how nearly normal the common distribution is. If the common distribution is itself normal, then S_n is exactly normal. For most situations, an n of 10 or 20 would be sufficiently large for a reasonable approximation.

Example 8-c An elevator sign reads "maximum weight 2700 lb, capacity 17 persons." If the population of passengers has mean 150 pounds and standard deviation 25 pounds, what is the probability that a capacity crowd will overload the elevator?

The total weight of 17 persons is a random variable with mean $17 \times 150 = 2550$ and standard deviation $25\sqrt{17} = 103$. Therefore (even without knowing the precise shape of the population distribution),

$$P(\text{total weight} > 2700) \doteq 1 - \Phi\left(\frac{2700 - 2550}{103}\right)$$

$$= 1 - \Phi(1.46) = .07. \quad ◀$$

The *average* of $X_1, \ldots, X_n$ is important:

$$\bar{X} \equiv \bar{X}_n \equiv \frac{1}{n}\sum X_i,$$

and it is noteworthy that *the standardized* $\bar{X}_n$ *is the same as the standardized sum*:

$$\frac{S_n - n\mu}{\sqrt{n\sigma^2}} = \frac{n(\bar{X}_n - \mu)}{\sqrt{n\sigma^2}} = \frac{\bar{X}_n - \mu}{\sigma/\sqrt{n}}$$

where $\mu = E(\bar{X}_n) = E(X_i)$, and $\sigma_{\bar{X}_n} = \sigma/\sqrt{n}$. Thus probabilities for $\bar{X}_n$ are also calculable, approximately, for large n, using Table I:

$$P(\bar{X}_n \leq v) \doteq \Phi\left(\frac{v - \mu}{\sigma/\sqrt{n}}\right).$$

Example 8-d Consider a population of students whose IQ's have an average of 100 and a standard deviation of 20. A student drawn at random from the population will have an IQ that is random, with the distribution of the population. If the population is very large, it may be assumed that the IQ's of 50 students drawn at random are independent and so constitute a random sample from the population. The average $\bar{X}$ of the 50 IQ's will then be approximately normal, with

$$E(\bar{X}) = \mu = 100, \qquad \sigma_{\bar{X}} = \frac{20}{\sqrt{50}} = 2.828.$$

Probabilities such as the following may be computed with the aid of Table I:

$$P(\bar{X} > 105) \doteq 1 - \Phi\left(\frac{105 - 100}{2.828}\right) = .039.$$

This calculation is valid no matter what the shape of the IQ distribution, and the beauty of this is that the population distribution does not need to be specified in detail. ◄

8.4 Approximating Binomial Probabilities

The binomial formula, the formula (from Chapter 5) for the probability of k successes in n independent trials of a Bernoulli experiment with parameter p, is as follows:

$$f(k) = \binom{n}{k} p^k (1 - p)^{n-k}.$$

When n is large, this can be quite tedious to compute. A method of approximating such probabilities with the aid of the Central Limit Theorem will now be explained.

Consider the sum $Y_n = X_1 + \cdots + X_n$ of n observations on X, a Bernoulli population with parameter p:

	Value	Probability
X	0	$1 - p$
	1	p

Mean: $E(X) = p$,
Variance: $\text{var } X = p(1 - p)$.

The sum Y_n (being the sum of 0's and 1's) is just the number of 1's among the n trials—the number of successes. Its mean and variance are

$$EY_n = n\mu_X = np,$$

$$\text{var } Y_n = n\sigma_X^2 = np(1 - p).$$

The Central Limit Theorem asserts (as a particular case) that the sum Y_n is approximately normal with these moments, which fact we write as follows:

$$Y_n \sim \mathcal{N}(np, np(1 - p)).$$

Equivalently, the mean $\bar{X}_n = Y_n/n$, which is the relative frequency or sample proportion of 1's, is approximately normal:

$$\bar{X}_n \sim \mathcal{N}\left(p, \frac{p(1 - p)}{n}\right).$$

Thus (with $q = 1 - p$, as usual)

$$P(\text{number of successes} \le k) \doteq \Phi\left(\frac{k - np}{\sqrt{npq}}\right)$$

or, equivalently,

$$P(\text{proportion of successes} \le r) \doteq \Phi\left(\frac{r - p}{\sqrt{pq/n}}\right).$$

To work in terms of the *number* of successes (an integer), use the first relation. To work in terms of the *proportion* of successes (a number between 0 and 1), use the second.

Example 8-e For a fair coin, $p = 1/2$ and $pq = 1/4$. If, for instance, $n = 100$, then $np = 50$ and $npq = 25$, and

$$\begin{aligned} P(50 \text{ heads in } 100 \text{ trials}) &= P(Y_n \le 50) - P(Y_n \le 49) \\ &\doteq \Phi\left(\frac{50 - 50}{\sqrt{25}}\right) - \Phi\left(\frac{49 - 50}{\sqrt{25}}\right) \\ &= .5000 - .4207 = .0793. \end{aligned}$$ ◄

In approximating the discrete binomial distribution function by the continuous normal c.d.f., it is possible to improve the degree of success by making what is called a *continuity correction.* Figure 8-4 shows the sort of thing that is going on. The approximating normal curve tends to cut through the "risers" of the binomial "staircase" c.d.f. The probability $P(Y_n \le k)$ is the height up to the upper stair tread at k, whereas the normal c.d.f. used in approximating it has a value that is appreciably less. The value of the normal c.d.f. at $k + 1/2$ is closer to the desired probability. Thus

$$P(Y_n \le k) \doteq \Phi\left[\frac{k + 1/2 - np}{\sqrt{npq}}\right].$$

When n is very large, the correction is not significant. However, by using the continuity correction, one can get by with a normal approximation for values of n as small as 10, or even 5, if p is not extreme. (See Problem 10.)

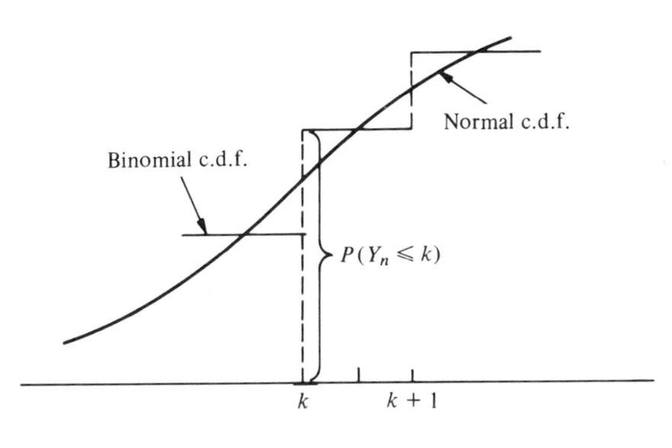

Figure 8-4

The continuity correction may always be used in approximating binomial by normal probabilities, but it will not have much effect when n is large. Moreover, when calculating a difference, as in Example 8-e, the corrections to the individual terms would tend to cancel out to some extent. Only experience with specific numbers will show the kind of improvement possible.

Example 8-f Consider 8 independent trials with $p = .4$. The normal approximation *without* a continuity correction for $P(Y \leq 4)$ is

$$\sum_0^4 \binom{8}{4}(.4)^k(.6)^{8-k} \doteq \Phi\left(\frac{4 - 3.2}{\sqrt{1.92}}\right) = .718,$$

and with it,

$$\Phi\left[\frac{4 + 1/2 - 3.2}{\sqrt{1.92}}\right] = .826.$$

The value of this probability from Table X is .8263. Clearly, the correction is effective. On the other hand, for a particular value, say, $Y = 4$, the approximation without correction is

$$P(Y = 4) \doteq \Phi\left(\frac{4 - 3.2}{\sqrt{1.92}}\right) - \Phi\left(\frac{3 - 3.2}{\sqrt{1.92}}\right) = .275$$

and with it,

$$P(Y = 4) \doteq \Phi\left(\frac{4.5 - 3.2}{\sqrt{1.92}}\right) - \Phi\left(\frac{3.5 - 3.2}{\sqrt{1.92}}\right) = .240,$$

as compared with the true value .232. ◄

It will be recalled that one approximation of binomial probabilities for large n was given in Section 7.1, namely, for the case of "small p." When p is so small that the Poisson parameter np is still in the range of Table XII (less than 15), the Poisson approximation is preferred, since the normal approximation to such a skewed distribution does not work well. If $np > 15$ and $nq > 15$, the normal approximation may be used; and, indeed, the Poisson variable itself is nearly normal, as will be explained next. (If $nq < 15$, then interchanging the 1 and 0, or the roles of p and q, makes the problem one in which the new p is small.)

★ 8.5 Approximating Poisson Probabilities

It may have been noticed that the Poisson table (Table XII of Appendix B) only extends to $m = 15$. The reason for this is that if $m > 15$, the Poisson distribution is nearly normal. This is not surprising when it is realized that a Poisson variable Y with mean m, an integer, can be expressed as

$$Y = X_1 + X_2 + \cdots + X_m,$$

where the X_i's are independent Poisson variables with mean 1. The Central Limit Theorem would then say that

$$P(Y \leq y) \doteq \Phi\left(\frac{y - m}{\sqrt{m}}\right),$$

the denominator $\sqrt{m}$ being the standard deviation of Y. (It is true in general that the variance of a Poisson variable is equal to its mean.)

Example 8-g To get some idea of the kind of success achieved by this approximation, consider the case $m = 15$. The probability $P(Y \leq 12)$, for example, is found in Table XII to be .268. The normal approximation yields

$$\Phi\left(\frac{12 - 15}{\sqrt{15}}\right) = .219.$$

A continuity correction improves the result:

$$\Phi\left(\frac{12.5 - 15}{\sqrt{15}}\right) = .260. \blacktriangleleft$$

★ 8.6 Errors

A name sometimes used in connection with the normal or Gaussian c.d.f. and normal tail probabilities is "error function," stemming from the applicability of the normal distribution† in describing various kinds of random errors—errors of measurement and errors in trying to achieve specified dimensions or amounts.

Tail probabilities are related, in general, to the variance of a distribution by *Chebyshev's inequality*:

$$P(|X - \mu_X| \geq k\sigma_X) \leq \frac{1}{k^2}$$

This bound on the amount of probability farther from the mean than a certain number of standard deviations will not be derived here; its importance is mainly theoretical, the bound being usually too weak to be of practical value. A more precise bound, indeed, a *value* of the tail probability, can be calculated for a specific distribution. When X is *normally* distributed, the area in the tails farther than k σ's from the mean can be calculated by means of Table I. Some probabilities corresponding to particular k's are shown in the following table, along with the Chebyshev bound $1/k^2$.

	$P(\|X - \mu\| > k\sigma)$	
k	Normal X	Bound, general X
1	.3174	1.00
1.5	.1336	.444
2	.0456	.25
2.5	.0124	.16
3	.0026	.111

It is evident that the probability that a normal variable takes a value farther from its mean than $\pm 3\sigma$ is very small indeed—about 1 chance in 400.

† The name of K. F. Gauss is associated with the normal distribution because of his use of it in his theory of random errors, although Laplace, a contemporary of Gauss, used the normal distribution more extensively, and Jacob Bernoulli, Gauss' predecessor, proved that binomial probabilities could be approximated by using the normal curve. (It is ironic that Bernoulli, who first proved the most beautiful and important theorem in all probability, should have his name associated with the second most trivial random variable.)

The values $\mu \pm 3\sigma$ are sometimes taken as *tolerance limits*; the value of X is almost always within these limits when X is normal. Because the relationship between probability and standard deviation or variance is well known and understood for normal variables, an analysis of errors usually involves a study of variances.

If variables combine additively, with normally distributed errors in the summands, the variance of the sum is easily calculated when the errors are *independent*:

$$\text{var}\left(\sum X_i\right) = \sum \text{var}\, X_i.$$

The standard deviation of the error in the sum is thus calculated by a Pythagorean-type relation:

$$\sigma_{\sum X_i} = \sqrt{\sum \sigma_{X_i}^2}$$

And if the tolerance is defined to be 3σ (or *any* multiple of σ), then tolerances combine in the same way. Moreover, it can be shown that if the summands are normally distributed about their means:

$$X_i = \mu_i + \varepsilon_i, \qquad \text{where } \varepsilon_i \text{ is } \mathcal{N}(0, \sigma_i^2),$$

then the *sum* is also normally distributed, with mean $\Sigma\, \mu_i$.

More generally, if a system variable which has an error ε that is, at least approximately, a linear combination of normally distributed component errors ε_i, then the system error is normal. And in any case, if the component errors are independent, they combine in root-sum-square fashion:

> If independent errors ε_i with tolerances T_i combine linearly,
>
> $$\varepsilon = a_1\varepsilon_1 + a_2\varepsilon_2 + \cdots + a_m\varepsilon_m,$$
>
> the overall tolerance is
>
> $$T = \sqrt{a_1^2 T_1^2 + \cdots + a_m^2 T_m^2}.$$
>
> If ε_i's are normal, so is ε.

Example 8-h The combined resistance of two resistors connected in "series" is the sum of their resistances: $R = R_1 + R_2$. Suppose that both R_1 and R_2 are nominally

8 ohms but are actually 8 ohms plus a random error with standard deviation .08 (or 1 per cent). Then

$$\sigma_R = \sqrt{.08^2 + .08^2} = .08\sqrt{2},$$

which is .707 per cent of the nominal resistance (16 ohms) of the series combination. The combination is more precise, relative to the nominal resistance, because of the good chance that the errors in R_1 and R_2 will tend to offset each other.

If the same resistors are connected in parallel, the combined resistance R is determined as follows:

$$\frac{1}{R} = \frac{1}{R_1} + \frac{1}{R_2}.$$

An error dR_1 in R_1 and dR_2 in R_2 produces an error dR in the combination, given approximately by the differential dR, where

$$\frac{dR}{R^2} = \frac{dR_1}{R_1^2} + \frac{dR_2}{R_2^2}.$$

Thus with $R_1 = R_2 = 8$ and $R = 4$,

$$\sigma_R^2 \doteq \left(\frac{R}{R_1}\right)^4 \sigma_{R_1}^2 + \left(\frac{R}{R_2}\right)^4 \sigma_{R_2}^2 = (.02\sqrt{2})^2.$$

The relative error σ_R/R is again about .707 per cent. ◄

Problems

1. A certain type of light bulb is found to have a life that is approximately normal with mean 700 hours and standard deviation 50 hours. Determine:

(a) The probability that a bulb will last more than 800 hours.
(b) The probability that a bulb burns out before 650 hours.
(c) A life length such that 40 per cent of the bulbs last at least that long.

2. The weight of a certain 15-cent candy bar is approximately normally distributed with mean 1.1 ounces and standard deviation .05 ounce. What is the probability that a bar is at least as heavy as the "net weight" 1.0 ounce, printed on the wrapper?

3. A certain aptitude test has been found, after being given to hundreds of thousands of students, to have scores that are normal with mean 450 and standard deviation 50.

(a) What is the score of a student who is told that he is at the 37th percentile?
(b) Determine the 1st and 3rd quartiles. (What fraction of the scores fall between these quartiles?)

4. Let Z denote a standard normal variable. Show the following:

(a) The c.d.f. is symmetric: $\Phi(-z) = 1 - \Phi(z)$. (An examination of Table I shows this, so that half the table would have been sufficient.)
(b) $P(|Z| > k) = 2\Phi(-k)$, for $k \geq 0$.

5. In Example 8-c, what would the mean weight have to be (assuming the same standard deviation) in order for the probability that a capacity load of 17 persons overloads the elevator to be .01?

6. Boxes of a certain detergent are filled by machine, and the net weight of detergent in a box is a random variable with mean 24 ounces and standard deviation .5 ounce. Determine the mean and standard deviation of the total net weight of detergent in a carton of 24 boxes, and the probability that the total exceeds 36.5 pounds.

7. Referring to Example 8-d, determine the probability of finding an average IQ of less than 99 in a group of 100 students from the population.

8. Approximate the probability that a fair coin turns up heads more than 15 times in 25 tosses.

9. If 35 per cent of the people think the President is doing a good job, what is the probability that in a random selection of 1000 people fewer than 30 per cent think so?

10. Calculate $\binom{8}{3}\left(\frac{3}{4}\right)^5\left(\frac{1}{4}\right)^3$ and compare with an approximate value obtained using the normal approximation (with continuity correction).

11. Calculate approximately the probability that there are fewer than 30 arrivals in an hour, if arrivals follow a Poisson law with mean 50 per hour.

★ **12.** Use a calculator to compute $e^{-20}(20)^{12}/12!$ and compare the result with a normal approximation to this probability.

★ **13.** Let Z be standard normal, with density

$$f(z) = \frac{1}{\sqrt{2\pi}} \exp(-z^2/2).$$

Determine $P(0 < Z < .1)$ approximately, without Table I, using the idea of differential, or "probability element." Check the result with Table I.

★ **14.** In a Poisson process with mean λ, determine the probability that the time to the 100th arrival, if $\lambda = 1$ per minute, does not exceed $1\frac{1}{2}$ hour.
(HINT: T is approximately normal, by the Central Limit Theorem.)

★ **15.** In calculating binomial probabilities, for the following cases, which method is to be preferred (given the usual tables, but no calculator)—Poisson approximation, normal approximation, or direct calculation?

(a) $n = 500, p = .15.$
(b) $n = 40, \quad p = .05.$
(c) $n = 6, \quad p = .4.$
(d) $n = 10, \quad p = .15.$

16. Compute binomial probabilities for $n = 5$, and the corresponding normal approximations using the continuity correction, given $p = 1/2$.

17. Many thousands of dollars have been lost by unsuspecting players of a popular carnival game frequently called Razzle-Dazzle. The player rolls 8 fair dice and the payoff is based on the total number of points. While there are complicated variations of the game, the essence of them is that the player wins if the sum is less than 18 or greater than 38 and loses otherwise. Approximate the probability that the player wins. (In the games as actually played, these unfavorable odds are exploited in a complicated way to trap the unwary—and even the wary!)

★ **18.** A poultry farmer has an Easter order for 30 baby chicks with black-tipped wings. He knows that black-tippedness is a recessive trait, so he will have to produce many more chicks than 30. In particular, assume that the probability of a chick with black-tipped wings is 1/4 and that he wants to be at least 99.87 per cent sure of ending up with 30 chicks with black-tipped wings. How many chicks should he produce?

(HINT: An equation of the form $a + bn = c\sqrt{n}$ can be thought of as a quadratic in $x \equiv \sqrt{n}$.)

★ **19.** Three gears are to be assembled side by side on a single shaft. The nominal widths are .500, .300, and .700 inch, with tolerances .004, .001, and .002, respectively. What is the probability that the combined width exceeds 1.505, if "tolerance" means 3σ, and the widths are normally distributed?

★ **20.** Determine the standard deviation of the error in computing the area of a lot by multiplying its dimensions x, with standard deviation σ_{dx}, and y, with standard deviation σ_{dy}. Suppose that the measured dimensions are 150 by 100 feet, with $\sigma_{dx} = \sigma_{dy} = .5$ feet.

9

Samples

The process of performing an experiment to obtain observations from a population or sample space is called *sampling*. The observations obtained constitute a *sample*. The object of the sampling is to learn something about the nature of the population. Statistical inference is the drawing of conclusions about the population from the sample.

Although such inference involves risks, this is often unavoidable because of the inaccessibility of the whole population. The population may be infinite, so that no finite process would reveal the whole population. And even if it is finite, looking at the whole population may be too costly. Sometimes the population just will not stand still long enough to be completely surveyed. And sometimes testing individual members of a population is destructive.

The question is: How should a sample be taken so that inferences about the population are reasonably reliable, and so that their reliability can be assessed? It is sometimes said that a sample should be "representative," having the same or close to the same characteristics as the population. But since population characteristics are not known (if they *were*, sampling would be pointless), it would never be known whether a sample is representative or not. About all that can be done is to sample in a way that is describable by a probability model, so that one can talk about and compute such things as expected errors or probabilities of errors in inference.

9.1 Random Sampling

Simple random sampling of a finite population means selecting population members one at a time, at random—with all available individuals having the same chance at each selection, and observing or measuring the characteristic of interest for each individual drawn. (The term *simple* here is to contrast this with more complicated schemes, not considered here, such as sampling from strata after partitioning a population according to age, or geography, or income.) The sampling can be done without replacement of the individuals, or with replacement and mixing after each draw. The latter is usually mathematically simpler.

When individuals, drawn one at a time at random from a finite population, are replaced, and the population is thoroughly shuffled or mixed before the next drawing, the population is restored to the same set of possible outcomes and same probabilities as before. The probability model for each observation is the population model; and further, no outcome or set of outcomes has any influence on the probabilities for other selections. So the observations are independent.

Experiments with infinite or conceptual sample spaces, such as measurement processes, can also be carried out (usually) so that the same probability model—the population model—applies at each trial, and so that the trials are independent. A sample of n observations $X_1, X_2, \ldots, X_n$ will be said to be a *random sample* if obtained by such a procedure.

> A *random sample* from a population is a sequence of independent observations on it (each having the distribution of the population).

The observations may be numerical, or they may be categorical—the results of classifying outcomes into one of several possible categories. When they are numerical, they are of course random variables, as defined in Chapter 3.

> A random sample from a numerical population X is a sequence of independent random variables $(X_1, \ldots, X_n)$, each having the distribution of the population variable X.

To define a sample as a sequence of random *variables* may seem odd to one who is looking at a particular set of numbers he has just obtained. Indeed, that set of numbers is not random—not even variable, just as the particular side of a die that has turned up after a toss is not random. Saying that the outcome of the toss of a die is a random variable refers to the experiment, and to what *can* happen and with what odds. Even so, saying that a sample is a random sample refers to the method by which it was obtained—to what could have happened and with what odds. One cannot tell by looking at a particular, actual sample, however hard or intelligently, whether it was obtained as a random sample.

In evaluating some procedures of statistical inference, it will be important, even when examining particular results, to know what could have happened and the appropriate probability distribution that describes the sampling process.

9.2 Random Numbers

Some calculators and most computers have a random number key or program; and the Rand Corporation has published† a book of 1 million random digits. The programs generate, and the book lists, samples of digits from the population consisting of 10 equally likely digits $\{0, 1, 2, \ldots, 9\}$. The Rand list could have been (and perhaps should have been) generated by selecting numbers from a hat, one at a time, with replacement, or by spinning a wheel of fortune with 10 possible stopping positions. They were, in fact, generated by a computer, by a program perhaps more sophisticated than that found wired into hand calculators; however, any such program is necessarily deterministic, and the results only *simulate* random digits, albeit rather well.

It is not hard to see that three successive random digits, for example, can be put together to form a three-digit number that is a random selection from the list $\{000, 001, \ldots, 999\}$. The Rand list, part of which is reproduced in Table XIII of Appendix B, gives digits in groups of five. These can be used as five random digits, or as a five-digit random decimal (you supply the decimal point), or in groups of k as k-digit random numbers. They can be read in sequence across in rows, or down in columns. A sequence should be started at a random point in the table. This can be done crudely by putting a pencil on the page blindfolded to get a starting point—or to get a random number to be used to locate a starting place, after pages and columns and rows are numbered for reference.

A random number sequence can be used as an aid in drawing a random sample from a finite population. If the population size is less than 10,000, each member can be assigned a four-digit identifying code. The random number table is then used to obtain a sequence of four-digit random

† Rand Corporation, *A Million Random Digits* (New York: Free Press, 1955).

numbers corresponding to the desired sample size, and these determine which population numbers are drawn. (The random number table does permit repetitions, so if sampling *without* replacement is desired, numbers that have appeared already in the sampling sequence should be ignored although some modification of this procedure might be more efficient if the sample is a substantial portion of the population.)

Example 9-a A campus newspaper reported results of a survey of residents of a particular suburb in which a shooting had occurred, taken to determine public reaction. The sampling was done by means of four-digit random numbers, with residents identified according to the last four digits of their telephone numbers. (Just two prefixes were involved, and the sampling alternated between them.) The population sampled, of course, was the population of households with telephones. It was conceded that more women than men responded, which is not unexpected if a woman is more likely to be at home than a man. This element of bias could have been serious with regard to the question asked ("Do you feel that the policeman was justified in shooting the suspect?"), since men and women may not agree in their responses. ◄

9.3 Statistics

In Chapters 3 and 8, certain quantities were introduced for the purpose of describing characteristics of populations. These are often referred to as *parameters* of the population.

The word *parameter* has different connotations in different contexts. As used above, it is a numerical characteristic of a population or model. Inference about such a parameter is often the statistical problem with which one is faced. Sometimes, too, a parameter may not be of primary interest, but not knowing it may complicate inference about some other parameter.

In mathematical usage, a parameter is a quantity that indexes a family of functions. This usage is common in probability, when it is clear that there is a class of similar, appropriate models. For instance, in the case of a Bernoulli experiment, the probability function is

$$P(X = x) = p^x(1 - p)^{1-x}, \qquad x = 0, 1$$

The probability p of success is usually not known, and so one considers (as possible models) the class of probability functions indexed by p, the *parameter*:

$$f(x; p) \equiv p^x(1 - p)^{1-x}, \qquad x = 0, 1.$$

Normal distributions with $\sigma^2 = 1$ constitute another parametric family of models, indexed by μ, whose specification would single out one model from the class. In both of these and in many other cases, the parameter indexing the class defines an easily interpretable characteristic of the family, a point that will be made again in Chapter 10.

Descriptive quantities can also be defined for sample characteristics, and might be called, with some justification, parameters of the sample. However, to avoid confusion, the term *statistic* is used for a quantity computed from and for the observations in a sample.

> DEFINITION. A *statistic* is a function of the observations in a sample.

In particular, the computation of a statistic does not require knowledge of any unknown population characteristics.

Some examples of functions of a sample $(X_1, \ldots, X_n)$ are $X_1 + X_4$, $1/X_2$, $X_1^2 + 7X_3$, $\max(X_1, \ldots, X_n)$, # (positive X_i's), and so on. Each of these is a statistic, although several are of little use.

A population parameter is usually unknown, and to learn something about its value is often the goal of a statistical experiment. A statistic, on the other hand, is calculable from the sample at hand. Sometimes, it seems rather clear that whatever the sample has to say about a given parameter lies in the value of a corresponding particular statistic. That is to say, not surprisingly, that the nature of the problem may suggest a relevant and useful statistic. It will be convenient, however, to have at hand a vocabulary of commonly used statistics.

The statistics to be introduced at this point are for numerical data and are of two types, descriptive of location and of variability, respectively. The *location* of a distribution of sample observations, as in the case of a population distribution, is usually given in terms of its "center" or "middle," described by some kind of "average" or "typical" value. *Variability* refers to the degree to which data are dispersed or spread about their middle.

The most familiar measure of "center" is the arithmetic *mean* or average. Given sample observations $X_1, X_2, \ldots, X_n$, this is defined to be

$$\bar{X} \equiv \frac{1}{n}(X_1 + \cdots + X_n) = \frac{1}{n}\sum_1^n X_i.$$

Mathematically, this is again a "first moment," a moment of the distribution of the whole into relative amounts $1/n$ at each observation. Thus, like the population mean, the sample mean has the property of being a balance point—the first moment about this point is zero:

$$\frac{1}{n}\sum_{1}^{n}(X_i - \bar{X}) = 0.$$

Various location statistics are defined in terms of the *ordered* observations $X_{(1)}, X_{(2)}, \ldots, X_{(n)}$, where $X_{(1)}$ is the smallest of the observations, $X_{(2)}$ is the second smallest, and so on. The *median* is the middle of this ordered sequence—the middle value if n is odd, or (arbitrarily) the average of the two in the middle if n is even. The idea is that there are as many observations greater than the median as there are smaller than the median.

Example 9-b A United States senator from New York, intending to plead that New York City needed special aid because of the unusual financial status of its residents, said in a national TV news program that "Half the people in New York City earn less than the median income!" Although this may sound bad, it simply says that New York City's median income is the same as the national median. Moreover, the logically equivalent, "Half the people earn more than the median income," does not sound bad at all! ◄

The *midrange* is defined as the center of the interval over which the observations in a sample are spread:

$$\text{midrange} \equiv \frac{1}{2}[X_{(n)} + X_{(1)}].$$

The *midmean* is defined as the mean or average of the middle half of the ordered observations—of what remains after discarding the top one fourth and the bottom one fourth of the observations.

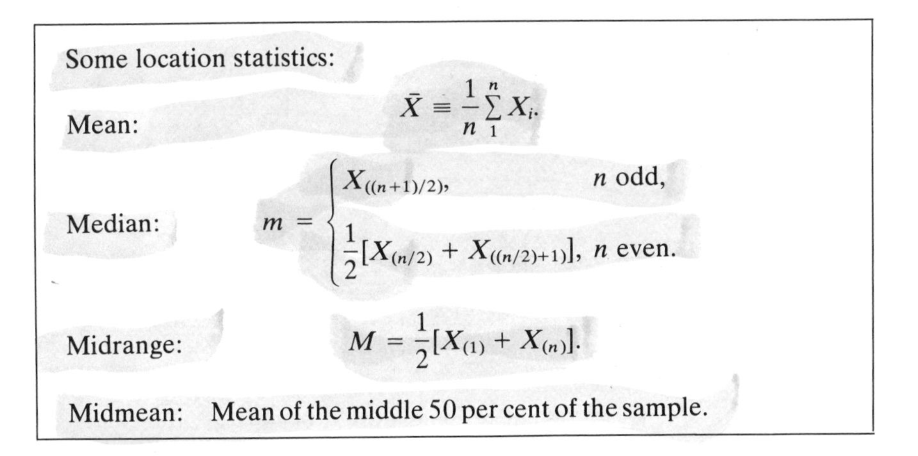

Some location statistics:

Mean: $$\bar{X} \equiv \frac{1}{n}\sum_{1}^{n} X_i.$$

Median: $$m = \begin{cases} X_{((n+1)/2)}, & n \text{ odd}, \\ \frac{1}{2}[X_{(n/2)} + X_{((n/2)+1)}], & n \text{ even}. \end{cases}$$

Midrange: $$M = \frac{1}{2}[X_{(1)} + X_{(n)}].$$

Midmean: Mean of the middle 50 per cent of the sample.

A simple measure of spread or dispersion is the *width* of the interval over which the observations are distributed; this is called the *range*:

$$R \equiv X_{(n)} - X_{(1)}$$
$$= \text{largest observation–smallest observation.}$$

The *interquartile range* is the width of the interval over which the middle half of the observations are distributed:

$$\text{interquartile range} \equiv Q_3 - Q_1,$$

where Q_1 and Q_3 are the first and third *quartiles*—numbers such that 25 per cent of the observations are smaller and 25 per cent larger, respectively.

Perhaps the most popular measure of sample dispersion is the *standard deviation*. It is defined, for the distribution $1/n$ at each sample observation, just as it is for populations—as a root-mean-square type of average deviation from the mean:

$$S \equiv \sqrt{\frac{1}{n}\sum (X_i - \bar{X})^2}.$$

The square of this is the sample *variance*, computable also as the difference between the average square and the square of the average:

$$S^2 = \frac{1}{n}\sum_1^n (X_i - \bar{X})^2 = \frac{1}{n}\sum_1^n X_i^2 - (\bar{X})^2.$$

This last relation does not need special proof, since it is a manifestation of a general property of distributions encountered earlier for population distributions. [Of course, it can be proved here, directly, by expanding the square $(X_i - \bar{X})^2$ and summing term by term.]

Some statistics of dispersion:	
Range:	$R = X_{(n)} - X_{(1)}.$
Interquartile range:	$Q_3 - Q_1.$
Standard deviation:	$\sqrt{\frac{1}{n}\sum_1^n X_i^2 - (\bar{X})^2}.$

Example 9-c In an anthropological study the following were reported as widths (in mm) of human skulls found in a certain digging:

141, 140, 145, 135, 147, 141, 154.

The same observations, given in numerical order, are

135, 140, 141, 141, 145, 147, 154.

From this, the following statistics can be computed:

$$\text{Range} = 154 - 135 = 19.$$

$$\text{Midrange} = \frac{1}{2}(135 + 154) = 144.5.$$

$$\text{Median} = 141.$$

$$\text{Mean} = 143.3.$$

$$\text{Standard deviation} = 5.62. \blacktriangleleft$$

It needs to be mentioned that many people prefer to define sample variance with a divisor $n - 1$:

$$\tilde{S}^2 \equiv \frac{1}{n-1}\sum_1^n (X_i - \bar{X})^2 = \frac{n}{n-1}S^2.$$

The reason usually given is of debatable merit, and because of simplicity, we prefer the divisor n. [One commercial hand calculator is wired with $1/n$ in its computation of variance but with $1/(n - 1)$ in its computation of standard deviation. This is curious, since the usual "justification" for $n - 1$ applies to variances, not to standard deviations.] The problem of estimating σ by using a sample standard deviation will be taken up in Section 10.4, but in other problems that require a measure of sample dispersion it is simply a matter of keeping in mind which definition is being used, and adjusting formulas accordingly.

The computation of mean and variance for data that are given in a *frequency distribution* can proceed a little more efficiently if formulas are used that take advantage of repeated observations. Thus, if the value 13 occurs 5 times in a sample, the contribution to the sample sum is

$$13 + 13 + 13 + 13 + 13 = 5 \times 13.$$

So if $x_1, \ldots, x_k$ denote the k *distinct* numbers that can occur, and x_i occurs with frequency f_i among the n sample observations $X_1, \ldots, X_n$, there follows:

$$\sum_1^n X_j = f_1x_1 + \cdots + f_kx_k = \sum_1^k f_ix_i.$$

Similarly,

$$\sum_1^n X_j^2 = f_1 x_1^2 + \cdots + f_k x_k^2 = \sum_1^k f_i x_i^2.$$

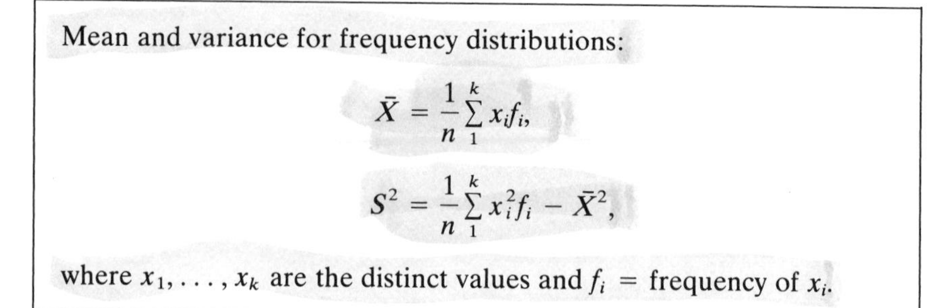

Mean and variance for frequency distributions:

$$\bar{X} = \frac{1}{n} \sum_1^k x_i f_i,$$

$$S^2 = \frac{1}{n} \sum_1^k x_i^2 f_i - \bar{X}^2,$$

where $x_1, \ldots, x_k$ are the distinct values and f_i = frequency of x_i.

Example 9-d The frequency distribution of cholesterol levels in Table 6.2 is repeated here, with columns added for computing the mean and variance.

x	f	xf	x^2f
134	1	134	17,956
165	4	660	108,900
196	24	4,704	921,984
227	17	3,859	875,993
258	27	6,966	1,797,228
289	16	4,624	1,336,336
320	9	2,880	921,600
351	0	0	0
382	2	764	291,848
Sum	100	24,591	6,271,845

The mean and variance are computed as follows:

$$\bar{X} = \frac{24{,}591}{100} = 245.91, \qquad S^2 = \frac{6{,}271{,}845}{100} - (245.91)^2 = 2246.72.$$

Of course, these have been computed from rounded data; the mean and variance computed from the original data are $\bar{X} = 245.69$ and $S^2 = 2117.89$. (Sometimes Sheppard's correction is applied to make up for the tendency of grouping to inflate the variance. It consists of subtracting $h^2/12$ from the grouped data variance, where h is the width of the class intervals.) ◄

9.4 Linear Transformations

Not everyone uses the same scale of measurement, but different scales are often *linearly* related. Thus temperature in degrees Fahrenheit (F) and temperature in degrees Celsius (C) are linearly related:

$$F = 32 + \frac{9}{5}C \qquad \text{or} \qquad C = \frac{5}{9}(F - 32).$$

As here, so in any linear transformation $Y = aX + b$, there is a shift in the origin or reference point, determined by the added constant, and also a change in the size of a "unit," or a change of scale. A question that arises naturally is how the statistics calculated using one scheme of measurement are related to those that use the other.

The sample mean and sample variance are special cases of moments of a discrete distribution, and so the rules developed in Chapter 3 apply:

> If $Y = aX + b$, then
>
> $$\bar{Y} = a\bar{X} + b,$$
> $$S_Y = |a|S_X.$$

For statistics based on the ordered observations it is merely necessary to observe that the linear transformation preserves order if $a > 0$:

$$Y_{(k)} = aX_{(k)} + b.$$

That is, the kth smallest X is transformed into the kth smallest Y. But then the ranges are related thus:

$$R_Y = Y_{(n)} - Y_{(1)} = aX_{(n)} - aX_{(1)} = aR_X.$$

Similar reasoning with $a < 0$ shows that, in general, $R_Y = |a|R_X$.

The midranges M are related in the same way as the means:

$$M_Y = \frac{1}{2}[Y_{(1)} + Y_{(n)}] = \frac{1}{2}[aX_{(1)} + b + aX_{(n)} + b] = aM_X + b,$$

a relation that is true whether a is positive or negative, as the reader should verify.

In summary, the measures that describe location are transformed just as are the individual observations in a linear transformation; and measures of dispersion, if in the same units as the measurements themselves, are multiplied by the scale factor.

A linear transformation is often referred to as a "coding," especially when used to simplify the scale so that hand computations are easier. With the ready availability of hand calculators, this technique is not so important, but is illustrated in the following.

Example 9-e Since the class intervals in the preceding example are of equal width, a linear transformation will code the class marks as integers. Let

$$y = \frac{x - 258}{31}.$$

This puts 0 at 258 (chosen arbitrarily) and makes the class marks one unit apart. (The number 31 is the original class-interval width.)

x	f	y	yf	y^2f
134	1	−4	−4	16
165	4	−3	−12	36
196	24	−2	−48	96
227	17	−1	−17	17
258	27	0	0	0
289	16	1	16	16
320	9	2	18	36
351	0	3	0	0
382	2	4	8	32
Sum	100		−39	249

Then

$$\bar{Y} = \frac{-39}{100} = -.39, \qquad S_Y^2 = \frac{249}{100} - (.39)^2 = 2.3379,$$

and from these the mean and standard deviation on the x-scale can be computed by using $x = 258 + 31y$:

$$\bar{X} = 258 + 31\bar{Y} = 245.91, \qquad S_X = 31S_Y = 47.40. \blacktriangleleft$$

9.5 Sampling Distributions

A statistic, being computed from a sample, is a *random variable*. That is, it varies from sample to sample, because its value depends on the particular observations that happen to have been obtained in a given case. In order to make any intelligent inference from the value of a statistic, it is necessary to understand the nature of its pattern of variation from sample to sample. This pattern of variation, that is, the probability distribution of the statistic, is called its *sampling distribution*.

Some sampling distributions have been encountered already, although they were not identified as such, in the study of particular random variables. For instance, the sum $X_1 + \cdots + X_n$ of the n observations on a Bernoulli population is a statistic; its distribution is binomial if the observations are independent, and hypergeometric if the observations are obtained at random without replacement.

9.6 Sampling Distribution of the Mean

The sample mean is an important statistic, and although its exact sampling distribution is generally more complicated, quite a bit can be said about it on the basis of what has been developed so far. Its first two moments are readily computed if the sample consists of *independent* observations on a population with mean μ and variance σ^2:

$$E(\bar{X}) = E\left(\frac{1}{n}\sum_1^n X_i\right) = \frac{1}{n}\sum_1^n E(X_i)$$

$$= \frac{1}{n}(\mu + \mu + \cdots + \mu) = \mu,$$

$$\operatorname{var} \bar{X} = \operatorname{var}\left(\frac{1}{n}\sum_1^n X_i\right) = \frac{1}{n^2}\sum_1^n \operatorname{var} X_i$$

$$= \frac{1}{n^2}(\sigma^2 + \sigma^2 + \cdots + \sigma^2) = \frac{\sigma^2}{n}.$$

[The formula $E(\bar{X}) = \mu$ would hold even without independence, and the formula for variance is valid as long as the X's are pairwise uncorrelated. But they *do* hold for random samples.]

Moments of the mean of a random sample:

$$E(\bar{X}) = \mu,$$

$$\operatorname{var} \bar{X} = \frac{\sigma^2}{n}.$$

The formula $E(\bar{X}) = \mu$ could be read: "The mean of the mean is the mean," but it would be much clearer to put it this way: The center of the pattern of variation of sample means is at the same point (μ) as the center of the pattern of variation of individual observations. And as to variability: The standard deviation of the pattern of variation of sample means is $1/\sqrt{n}$ times the standard deviation of the pattern of variation of individual observations. Thus averaging observations has the effect of reducing the variability without changing the location.

Example 9-f The following are artificially generated samples of size 5 from a population of values that are uniformly distributed on $0 < x < 1$ (the spinning pointer). Also shown are the sample means:

Sample number	Observations					Mean
1	.954	.954	.699	.315	.355	.655
2	.447	.717	.721	.536	.807	.646
3	.760	.387	.090	.704	.411	.470
4	.896	.562	.105	.091	.162	.363
5	.074	.398	.306	.240	.393	.282
6	.642	.621	.834	.497	.427	.604
7	.094	.800	.131	.999	.824	.570
8	.817	.871	.204	.617	.412	.584
9	.173	.862	.077	.540	.148	.360
10	.610	.845	.475	.279	.810	.604

Even from this limited number of observations of $\bar{X}$-values, it is apparent that their center is near 1/2 (the mean of these means is .514). It is also evident that there is more variation from observation to observation than there is from mean to mean. The means are less variable. These 10 samples do not give the ideal picture, in which $\sigma_{\bar{X}} = \sigma_X/\sqrt{5}$, but the tendency is clear. ◄

If $X_1, \ldots, X_n$ is a sample obtained by random sampling *without* replacement from a finite population, then the formula for $E(\bar{X})$ still applies, since the average of a sum is the sum of the averages even if the terms are dependent. However, the variance of a sum is now computed as in Section 5.3:

$$\text{var}\,(X_1 + \cdots + X_n) = n\sigma^2 \frac{N - n}{N - 1},$$

where N is the population size.

Moments of the mean, simple random sample from a finite population of size N:

$$E(\bar{X}) = \mu,$$

$$\operatorname{var} \bar{X} = \frac{\sigma^2}{n} \frac{N - n}{N - 1}.$$

The next important general fact about the sampling distribution of the sample mean to be taken up here has to do with large samples. This says that in the case of a large random sample, the distribution of $\bar{X}$ is approximately *normal*, no matter what population is being sampled (provided second moments exist). That statement is simply a rephrasing of the Central Limit Theorem (see Section 8.3):

If $(X_1, \ldots, X_n)$ is a random sample from a population with mean μ and variance σ^2, then for large n, $\bar{X}$ is approximately normal with mean μ and variance σ^2/n:

$$\bar{X} \sim \mathcal{N}\left(\mu, \frac{\sigma^2}{n}\right).$$

This normality permits computation of probabilities of events concerning $\bar{X}$, as illustrated in Example 8-d, and again in the following example.

Example 9-g Suppose X is uniform on $0 < x < 1$ so that $\mu = .5$ and $\sigma^2 = 1/12$, $\sigma = .289$. Then, for instance, in a sample of size 5, so that $\sigma_{\bar{X}} = \sigma/\sqrt{5} = .129$, it follows that

$$\begin{aligned} P(.25 < \bar{X} < .75) &= P(\bar{X} < .75) - P(\bar{X} < .25) \\ &\doteq \Phi\left(\frac{.75 - .50}{.129}\right) - \Phi\left(\frac{.25 - .50}{.129}\right) \\ &= .9737 - .0263 \doteq .95. \end{aligned}$$

Thus only about 1 sample in 20 will have a mean value outside the interval (.25, .75). Observe that in the preceding example none of the 10 sample means was outside that interval. (Even though the sample size is only 5, the normal approximation used in the calculation of probability for $\bar{X}$ is quite good. The uniform population being sampled is not skewed and is sufficiently like the normal for the approximation to be useful even for small n. See also Figure 6-4.) ◄

Problems

1. Given the following exam scores, compute the mean, median, midrange, range, and standard deviation:

42, 46, 54, 56, 57, 59, 59, 60, 66,

68, 72, 73, 81, 82, 82, 82, 85, 87.

2. A certain clinic reported the following data on ages of women who had abortions during a given period. Calculate the (approximate) mean and standard deviation of ages from the given frequency distribution. (Use 14 as the class mark for the interval ≤ 15; and interpret 16–17 as meaning age 16 up to 18, and so on.)

Age	Frequency
≤15	3
16–17	17
18–19	25
20–21	20
22–23	8
24–25	6
26–30	12
31–35	4
36	1
37	1
38	1
41	1

3. Calculate means and standard deviations for the frequency distributions in Table 6.3. Do they agree? (Would you expect them to agree?)

4. Determine the median and range of the cholesterol levels in Table 6.1.

5. Determine approximately the median age of the women included in the data of Problem 2, by reasoning in terms of the histogram—locating that point that divides the area in half. (It is not necessary to draw the histogram. Just find the class interval that contains the median and look at the rectangle over that interval.)

6. Let $X_1, X_2, \ldots$ be a sequence of random digits, and define S to be the sum of three successive digits.

(a) What is the sample space for S?

(b) Use Table XIII to obtain 100 observations on S, by using 300 random digits and summing in groups of three. Draw a histogram with class intervals 0–1, 2–3, . . . , 26–27. (Use a bar even though the only possible values within a bar are integers.) Comment on the shape and what it might be like ideally (as sample size becomes infinite).

7. If the mean and standard deviation of January 1 average temperatures for 30 years are $\bar{X} = 19°F$ and $S = 7°F$, determine the mean and standard deviation in degrees Celsius.

8. Repeat the steps of Problem 6 for the statistic M defined as the maximum of five random digits. (Read the maximum digit in each of 100 successive groups of five digits.)

9. If Y is the number of heads in 25 tosses of a coin.

(a) What is the sampling distribution of Y?
(b) How might the distribution in part (a) be approximated for easier computation of probabilities (with tables but no computer)?

10. Compute the mean and variance of the number of points in a bridge hand. (See Problem 4 in Chapter 3.)

11. Let $\bar{X}$ denote the mean of a sample of size 50 from a population with $\mu = 50$ and $\sigma = 2$. Find

(a) $P(\bar{X} > 51)$.
(b) $P(|\bar{X} - 50| < 0.5)$.
(c) K, such that $P(\bar{X} < K) = .05$.
(d) K, such that $P(|\bar{X} - 50| < K) = .90$.

12. In taking samples of size 50 from a population of size 500, what is the reduction in standard deviation of $\bar{X}$ achieved by sampling *without* replacement, as opposed to sampling *with* replacement?

13. If $n = 100$, what is the probability that the mean of a random sample is farther from the population mean than one fifth of a population standard deviation?

Estimation

The idea of estimating an unknown constant is common. One estimates the height of a building, or the speed of light, or the number of beans in a jar, or a person's weight, by some means or other but without being able to make actual counts or exact measurements. The problem in statistics is that of estimating a *parameter* of a distribution—an unknown constant that identifies a particular one of a class or family of models admitted as possible representations for a given experiment, a constant not available for direct measurement. The estimation is statistical (and, one hopes, better than a sheer guess) if it is based on the observations in a sample.

Several parametric families have been encountered in earlier chapters, families identified by densities or probability functions $f(x; \theta)$, where θ is a parameter. Some examples are given in Table 10.1.

Observe that in most of these families the population mean is either the parameter of interest or is closely related to it. Clearly, the estimation of the mean is an important special estimation problem.

More generally, it is often desirable to estimate the mean, or some other measure of population center, even when the population distribution is not known to be a member of a parametric family such as those listed. One may want to know average income, or average concentration of a chemical, or average achievement, or average growth, when little is known about the shape of the population distribution. And, at least in the case of large samples, it is possible to estimate such averages with reasonable success.

Table 10.1

Name	$f(x;\theta)$	Interpretation of θ
Binomial	$\binom{n}{x}\theta^x(1-\theta)^{n-x}$ $(x = 0, 1, \ldots, n)$	$\theta = \frac{1}{n}E(X)$ $= \frac{\text{mean number of successes}}{\text{number of trials}}$
Exponential	$\theta e^{-\theta x} \quad (x > 0)$	$\theta = \frac{1}{E(X)}$ $= (\text{mean time to failure})^{-1}$
Poisson	$\frac{e^{-\theta t}}{x!}(\theta t)^x$ $(x = 0, 1, \ldots)$	$\theta =$ mean number of failures per unit time
Geometric	$\theta(1-\theta)^{x-1}$ $(x = 1, 2, \ldots)$	$\theta = \frac{1}{E(X)} = (\text{mean number of trials to get success})^{-1}$
Hypergeometric	$\frac{\binom{\theta}{x}\binom{N-\theta}{n-x}}{\binom{N}{n}}$	$\theta = \frac{N}{n}E(X)$
Normal	$\frac{1}{\sigma\sqrt{2\pi}}\exp\left[-\frac{1}{2\sigma^2}(x-\mu)^2\right]$	$\theta = (\mu, \sigma^2)$, where $\mu = E(X)$

10.1 Mean-Squared Error

To estimate a parameter statistically means to come up with some number, as the estimate, by a judicious use of the data in a sample. An *estimator*, then, is some function of the observations in the sample and so is a statistic. Various statistics might be tried as estimators. For instance, in estimating a population mean one might think to use any of several measures of center of the sample—the sample mean, the sample median, and so on.

Some estimators that might be proposed are better than others. To consider this notion of "goodness" in a general setting, let T be used to denote some particular estimator of a parameter θ. (In a particular case this parameter might be μ, or σ, etc.) The *error* of estimation is defined to be

$T - \theta$, which locates the estimated value with respect to the true value. This error is inaccessible because θ is not known. Moreover, the estimator T varies from sample to sample, as described by its sampling distribution, so that the error is sometimes small, sometimes large. About all one can expect is to know how well the estimator performs on the average. But to avoid cancellation of positive and negative errors, in averaging, it is customary to square first, to obtain a criterion for successful estimation:

Mean-squared error:

$$\text{m.s.e.} \equiv E[(T - \theta)^2].$$

Measuring the degree of success in terms of mean-squared error is a reasonable approach in light of the fact that T will be close to θ with high probability if its m.s.e. is small. This follows from the following fact (a form of the Chebyshev inequality), stated without proof:

$$P(|T - \theta| > \varepsilon) \leq \frac{E(T - \theta)^2}{\varepsilon^2}.$$

In terms of the estimation problem, this becomes

$$P(|\text{error}| > \varepsilon) \leq \frac{\text{m.s.e.}}{\varepsilon^2}.$$

It is evident from this that if the m.s.e. is small, the probability of a large error is small. So it may be asserted that an estimator of θ with small m.s.e. has a good chance of being close to θ, whatever value θ may actually have.

10.2 Estimating the Mean—Large Samples

The *law of large numbers* (L.L.N.) asserts that if one averages successively more and more independent observations from a given population, the sample mean will "converge" to the population mean: $\bar{X}_n \to \mu$. The convergence here can be taken to mean that there is no chance of obtaining a sequence of observations for which the sample mean does not approach μ (strong law of large numbers); or it can mean that for any given small interval about μ, the probability that $\bar{X}_n$ falls in that interval is close to 1 when the sample size is large (weak law of large numbers).

Laws of large numbers:

If $X_1, \ldots, X_n$ are independent observations from a population with mean $E(X) = \mu$,

(1) $P\left(\frac{X_1 + \cdots + X_n}{n} \to \mu\right) = 1.$

(2) $P(|\bar{X} - \mu| > \varepsilon) \to 0,$ as $n \to \infty$ for any $\varepsilon > 0$.

Both statements are true, if $E(X)$ exists as a finite number, although it happens that the type of convergence in (1) implies the type used in (2). [The Chebyshev inequality, given above, easily yields (2) if σ^2 is finite.]

However one chooses to think of the law of large numbers intuitively, it is surely reasonable, in large samples, to take $\bar{X}$ as an estimate of μ, since chances are that it will be close to μ. (Nevertheless, there may actually be better estimates of μ, especially when the sample is small. See Problem 17 and Example 13-g.)

The mean-squared error in estimating μ by $\bar{X}$ is given by $E(\bar{X} - \mu)^2$, which (because $E\bar{X} = \mu$) is just the variance of $\bar{X}$:

$$\text{mean-squared error} = \frac{\sigma^2}{n}.$$

The square root of this can be more readily interpreted, since it has the dimension of X (or of μ):

$$\text{r.m.s. error} = \frac{\sigma}{\sqrt{n}}.$$

However, σ^2 is not usually known, so this may seem to be of no value. But (as can be shown using the L.L.N.) it is also true that the sample standard deviation approaches the population standard deviation as $n \to \infty$, so that for large n it is close to σ. Thus a good approximation to the r.m.s. error is obtained by replacing σ by S:

Standard error of the mean:

$$\text{s.e.} = \frac{S}{\sqrt{n}}.$$

More precisely, this is the standard error in the sample mean $\bar{X}$ when used as an estimate of the population mean μ.

The importance of the standard error lies, first, in the fact that, unlike the m.s.e., it is calculable from the sample observations—from what is known; and second, in the fact that it is an approximate measure of reliability of the estimator being used. In particular, for the case of $\bar{X}$, which is approximately normal for large n, the standard error can be interpreted as describing the width of the distribution of $\bar{X}$, in the same way that the σ of any normal distribution describes its width.

It is common practice to give, along with the *point estimate* $\bar{X}$, its standard error $S/\sqrt{n}$. Indeed, it is usual to give along with the point estimate T of any parameter θ its standard error (similarly defined), if T is approximately normal. (The term *point estimate* is in contrast to *interval estimate*, to be introduced in Section 10.5.)

Example 10-a In a dental research study, a sample of 294 children showed an average of 10.88 new cavities per child over a 30-month period. The standard deviation of the sample was 6.36. The average $\bar{X} = 10.88$ is a point estimate of the population mean μ, with standard error $6.36/\sqrt{294} = .371$. ◄

10.3 Estimating a Proportion

The proportion p of a population members having a certain characteristic is just the population mean, when that characteristic is coded 1, and its absence is coded 0:

	x	Probability
Population:	1	p
	0	$1 - p$

$$\mu = p, \quad \sigma^2 = pq.$$

Moreover, if one obtains a random sample of size n from a Bernoulli population, the sample mean is just the sample proportion:

	x	Frequency	Relative frequency
Sample	1	Y	Y/n
	0	$n - Y$	$1 - Y/n$

$$\bar{X} = \frac{Y}{n}, \quad S^2 = \frac{Y}{n}\left(1 - \frac{Y}{n}\right).$$

[The computation of S^2 resulting in $(Y/n)(1 - Y/n)$ is exactly the same as the computation of σ^2, with Y/n playing the role of p.] So it is reasonable, in line with the more general discussion of means, to take the sample proportion or relative frequency Y/n as an estimate of the population proportion p. Indeed, as in the general case $\bar{X} \to \mu$, so in this special case $Y/n \to p$ (as $n \to \infty$), in the same sense.

It will be convenient here to adopt a notation sometimes used in estimation: the circumflex over a parameter will denote an estimate thereof. Thus $\hat{\theta}$ means an estimate of θ, and in the present instance $\hat{p}$ will mean an estimate of p. Of course, there are many estimators for a given parameter, but in what follows, $\hat{p}$ will mean the sample proportion:

Estimate of a population proportion p:

$$\hat{p} \equiv \bar{X} = \frac{\#(\text{successes})}{\#(\text{trials})}.$$

The estimator $\hat{p}$ has mean p and variance pq/n.

The mean-squared error involved in estimating p by $\hat{p} = \bar{X}$ is (as for any $\bar{X}$) the variance of $\bar{X}$:

$$\text{m.s.e.} = \frac{\sigma^2}{n} = \frac{p(1-p)}{n},$$

since the variance of the Bernoulli population is $pq = p(1-p)$. And as before, the standard error is obtained by approximating σ^2 by the sample variance:

Standard error of the sample proportion $\hat{p}$:

$$\text{s.e.} = \sqrt{\frac{\hat{p}(1-\hat{p})}{n}}.$$

Example 10-b A student newspaper (*Minnesota Daily*, January 13, 1976) reported that 28 per cent of 433 students polled were smokers. This proportion, $\hat{p} = .28$, would be a point estimate of the proportion of smokers in the student population. The standard error is

$$\sqrt{\frac{.28 \times .72}{433}} = .0216,$$

or just over 2 percentage points. The paper also reported that in general population the proportion of smokers is 37 per cent. The figure 28 per cent is perhaps too many standard errors away from 37 to permit believing that students are like the general population—a notion that will be discussed in more detail in Chapter 11. ◀

★ 10.4 Estimating a Standard Deviation

The fact that the sample variance S^2 is close to the population variance σ^2 in large samples was used earlier (in Section 10.2) as a reason to replace σ by S in approximating the mean-squared error in estimating μ by $\bar{X}$. One may well have wondered about the degree of success in using S as an estimator of σ. Using mean-squared error to measure success, one would need to know

$$\text{m.s.e.} = E[(S - \sigma)^2].$$

In general this is rather complicated, so only the special case of a *normal* population will be considered here.

If the population is normal, the mean-squared error in estimating σ by S, for large samples, is given by the approximate formula

$$\text{m.s.e.}(S) \doteq \frac{\sigma^2}{2n}.$$

A standard error could be defined by using S in place of σ in the square root of the m.s.e.:

Standard error of S (assuming a normal population):

$$\text{s.e.} = \frac{S}{\sqrt{2n}}.$$

If n is large, the S that appears here could be computed using either n or $n - 1$ as the denominator in the averaging process. If n is not large, it would make a difference; but it turns out that a smaller mean-squared error can be achieved by using yet another constant factor. Among estimators of the form γS, the choice of multiplier γ that minimizes the mean-squared error is given in Table 10.2, as is the corresponding minimum r.m.s. error. Observe that for $n = 20$, the asymptotic formula $1/\sqrt{2n}$ gives .158, which is close to the r.m.s. error in the third column for $n = 20$.

Table 10.2 Estimation of σ, small sample from normal population

Sample size	Multiple of S to minimize m.s.e.	Root-mean-square error as a multiple of σ	Multiple of R to minimize m.s.e.	Root-mean-square error as a multiple of σ
2	1.1284	.603	.564	.603
3	1.0854	.463	.463	.465
4	1.0638	.389	.411	.393
5	1.0509	.341	.378	.348
6	1.0424	.308	.355	.318
7	1.0362	.282	.338	.295
8	1.0317	.262	.324	.277
9	1.0281	.246	.313	.260
10	1.0253	.232	.304	.251
11	1.0230	.221	.297	.241
12	1.0210	.211	.290	.232
13	1.0194	.202	.285	.225
14	1.0180	.194	.280	.219
15	1.0168	.187	.275	.215
16	1.0157	.181		
17	1.0148	.175		
18	1.0140	.170		
19	1.0132	.165		
20	1.0126	.161	.258	.191

If the population is normal, its standard deviation can be estimated, with some degree of success, as a multiple of the sample range R. The multiple that gives the smallest mean squared error is given in Table 10.2, along with the corresponding r.m.s. error as a multiple of σ. It is to be noted that the optimal multiple of R does almost as well as the optimal multiple of S for small samples. (Indeed, for $n = 2$, R is itself a multiple of S, so the minimum mean-squared errors agree in that case.) The estimate based on range, then, might be preferred on the basis of computational simplicity if n is not large.

If n is large, the performance of the sample range can be improved by dividing the sample into, say, k subsamples of size m, where $mk = n$. The quantity $\bar{R}$, the average of the ranges of the k subsamples, can be used as the basis for an estimate of σ. The quantity $a\bar{R}/(1-b^2 + b^2/k)$, where a and b are the entries opposite m in the last two columns of Table 10.2, gives an estimate of σ that is better than that based on the range of the whole sample—but still not as good as that based on S. (Actually, $\bar{R}/a_m$, where a_m comes from Table V of Appendix B, is not a bad estimate and is commonly used.)

Example 10-c A sample of 20 observations on a standard normal population, obtained from Table XIII, is as follows:

−.802	−.716	−.097	−.415
.603	2.154	.648	−2.356
−.247	−.406	1.004	−.222
−.373	1.620	.628	−1.108
.280	−1.793	.564	−.209

The standard deviation is $S = 1.012$, and the sample range is $R = 2.154 - (-2.356) = 4.510$. The minimum mean-squared-error estimates of σ based on S and R are then

$$1.0126S = 1.024,$$

$$.258R = 1.16.$$

If the sample is divided into four subsamples of size 5, as listed above, the ranges of the subsamples are 1.405, 3.947, 1.101, and 2.147, with mean $\bar{R} = 2.15$. The quantity

$$\frac{.378\bar{R}}{1 - (.348)^2 + (.348)^2/4} = .894$$

is a point estimate of σ, obtained by a process that has a mean-squared error between those for the estimates based on S and on R. ◀

10.5 Interval Estimates for μ

With the realization that a particular value of an estimator T, called a point estimate, is almost surely wrong, it is natural to want to indicate the degree of fuzziness or anticipated error by giving an *interval* of parameter values that, with some assurance or confidence, will contain the true value of the parameter. This interval would have to be determined from the data available for inference.

Two approaches suggest themselves. One is to view an interval computed from data as a particular realization of an experiment of chance and to construct the interval in a way that has a specified probability of success in covering or including the true parameter value. The other is to treat the parameter as a random variable and to construct an interval so that the probability that the parameter lies in the interval, *given* the sample, is a specified amount. The latter approach is that of Bayesian inference (see

Chapter 13); the former will be considered here, for the particular problem of estimating a mean.

In constructing an interval that is expected to contain a population mean μ with some assurance, it seems reasonable to center it at or near $\bar{X}$, the point estimate, the "fuzziness" extending on either side thereof. Now, the interval with limits $\bar{X} \pm A$ will include or cover μ if and only if $\bar{X}$ is sufficiently close to μ, namely, within A units of μ: $|\bar{X} - \mu| < A$. Thus for large n,

$$P[(\bar{X} - A, \bar{X} + A) \text{ covers } \mu] = P(|\bar{X} - \mu| < A)$$
$$\doteq \Phi\left(\frac{A}{\sigma/\sqrt{n}}\right) - \Phi\left(\frac{-A}{\sigma/\sqrt{n}}\right).$$

This will have the value 95 per cent, say, if A is chosen so that

$$\frac{A}{\sigma/\sqrt{n}} = z_{.975} \quad \text{or} \quad A = z_{.975}\frac{\sigma}{\sqrt{n}}.$$

The resulting interval of μ-values is called a 95 per cent *confidence interval* for μ.

> A 95 per cent confidence interval for μ (large n, σ *known*):
>
> $$\left(\bar{X} - z_{.975}\frac{\sigma}{\sqrt{n}},\ \bar{X} + z_{.975}\frac{\sigma}{\sqrt{n}}\right).$$

The probability 95 per cent used here is called the *confidence coefficient.*

If σ is not known, these confidence limits, or endpoints of the confidence interval, cannot be computed from the sample. But if n is large, the unknown σ can be replaced, with impunity, by the sample version S:

> A 95 per cent confidence interval for μ (large n, σ unknown):
>
> $$\left(\bar{X} - z_{.975}\frac{S}{\sqrt{n}},\ \bar{X} + z_{.975}\frac{S}{\sqrt{n}}\right).$$

Observe that the distance from $\bar{X}$, the center of the interval, out to the endpoints is just a multiple of the standard error, a multiple determined (from Table I) by the specified confidence coefficient. Multipliers for various confidence coefficients are as follows (and repeated in Table Ic):

Confidence coefficient	Multiple of standard error
.80	$z_{.90} = 1.28$
.90	$z_{.95} = 1.645$
.95	$z_{.975} = 1.96$
.99	$z_{.995} = 2.58$

Note that the higher the "confidence," the larger must be the confidence interval in order to have that degree of assurance that μ is trapped within the interval.

Example 10-d A sample of 200 measurements of viscosity has mean $\bar{X} = 31.7$ and standard deviation $S = 1.81$. A 90 per cent confidence interval for the population mean viscosity then has limits

$$31.7 \pm 1.645 \times \frac{1.81}{\sqrt{200}} \quad \text{or} \quad (31.49, 31.91).$$

For a 99 per cent confidence interval, the multiplier 2.58 would be used (in place of 1.645), yielding the wider interval (31.37, 32.03). ◄

Whether an interval, such as (31.49, 31.91) in the preceding example, actually *does* include μ on its interior cannot be known as long as μ is unknown. Either it does or it does not include μ. The degree of assurance given by the confidence coefficient is a relative frequency of success over many such experiments; that is, the procedure outlined for obtaining a 95 per cent confidence interval succeeds 95 per cent of the time in producing an interval that includes μ.

If the sample size is not large, the distribution of $(\bar{X} - \mu)/S$ is more complicated. It depends on the population distribution, which was not the case—thanks to the Central Limit Theorem—for large samples. One important population for which available tables permit construction of a confidence interval is the ***normal***. If $(X_1, \ldots, X_n)$ is a random sample from a normal population, then $\sqrt{n-1}(\bar{X} - \mu)/S$ has what is called a *t-distribution with* $n - 1$ *degrees of freedom*, given in Table III. The confidence interval for μ in this case is just about the same as in the large-sample case except for the multiplier, and the use of $\sqrt{n-1}$ in place of $\sqrt{n}$:

A 95 per cent confidence interval for μ (small samples from a normal population, σ unknown):

$$\left(\bar{X} - k\frac{S}{\sqrt{n-1}},\quad \bar{X} + k\frac{S}{\sqrt{n-1}}\right),$$

where $k = t_{.975}$, the 97.5 percentile of t with $n - 1$ degrees of freedom (Table III).

The change from $\sqrt{n}$ to $\sqrt{n-1}$ would not be necessary if one used the version of sample variance with $n - 1$ in the denominator. For, if $\tilde{S}^2 = \Sigma (X_i - \bar{X})^2/(n - 1)$, then

$$\frac{S}{\sqrt{n-1}} = \left[\frac{1}{n(n-1)}\Sigma (X_i - \bar{X})^2\right]^{1/2} = \frac{\tilde{S}}{\sqrt{n}}.$$

Example 10-e Readings of oil temperature in a certain application were obtained as follows:

93, 97, 99, 99, 90, 96, 93, 88, 89,

with mean $\bar{X} = 93.78$ and variance $S^2 = 15.73$: For $n - 1 = 8$ degrees of freedom, the 95th percentile of the t-distribution is 1.86, and 90 per cent confidence limits for μ are then

$$93.78 \pm 1.86\sqrt{\frac{15.73}{8}} \qquad \text{or} \qquad 91.25 \text{ to } 96.31,$$

assuming a normal population. ◄

10.6 Interval Estimates for p

The population proportion p in a Bernoulli population is the population mean, and a large-sample confidence interval for p is therefore given in terms of the sample mean—which is the sample proportion—as a special case of the interval given in the preceding section. Given a sample proportion $\hat{p} = \bar{X}$, the standard error of $\hat{p}$ in estimating p is

$$\text{standard error} = \sqrt{\frac{\hat{p}(1-\hat{p})}{n}}.$$

The desired confidence limits are then constructed as the mean plus and minus a multiple of the standard error. For a confidence coefficient of 95 per cent:

A 95 per cent confidence limits for p (large n):

$$\hat{p} \pm z_{.975}\sqrt{\frac{\hat{p}(1-\hat{p})}{n}}.$$

The replacement of $\sigma^2 = p(1-p)$ by $\hat{p}(1-\hat{p})$ works fairly well for $n \geq 100$ (if p is not too extreme); but for $20 < n < 100$, where the normal approximation for $\hat{p}$ is still valid, a more precise interval may be given as follows:

Confidence limits for p (moderate n):

$$\frac{1}{1+k^2/n}\left[\hat{p} + \frac{k^2}{2n} \pm k\sqrt{\frac{\hat{p}(1-\hat{p})}{n} + \frac{k^2}{4n^2}}\right].$$

The multiple k is a percentile of the standard normal distribution, as before. For instance, for a confidence coefficient of 95 per cent, $k = z_{.975} = 1.96$.

The more refined formula is obtained as follows. Starting with the normal approximation

$$P\left(|\hat{p} - p| < k\sqrt{\frac{p(1-p)}{n}}\right) = 1 - 2\Phi(-k),$$

one inverts the inequality, solves it for p, by first squaring:

$$\hat{p}^2 - 2p\hat{p} + p^2 < \frac{k^2}{n}p(1-p),$$

collecting terms of this quadratic in p:

$$\left(1 + \frac{k^2}{n}\right)p^2 - 2\left(\hat{p} + \frac{k^2}{2n}\right)p + \hat{p}^2 < 0,$$

and then locating the zeros (by the quadratic formula) of the quadratic function on the left. The inequality holds when p is between those zeros, which are then the confidence limits.

Example 10-f Ten of 100 articles are found to be defective. The point estimate of the population proportion defective is $\hat{p} = .10$, and a 95 per cent confidence interval for p is given, in the simpler form, by the limits

$$.10 \pm 1.96\sqrt{\frac{.10 \times .90}{100}}$$

or $.04 < p < .16$. Using the more refined formula, one obtains

$$\frac{1}{1 + (1.96)^2/100}\left[.1 + \frac{(1.96)^2}{200} \pm 1.96\sqrt{\frac{.1 \times .9}{100} + \frac{(1.96)^2}{4(100)^2}}\right]$$

or $.055 < p < .176$. ◄

★ 10.7 Interval Estimates for σ

Confidence intervals for σ and σ^2 will be constructed here only for the case of a *normal population.*

If a statistic T can be found such that the distribution of T/σ does not depend on any population parameters, and is determined by just the sample size n, then a confidence interval can be obtained as follows. First, according to the definition of a percentile,

$$P\left(U_{.025} < \frac{T}{\sigma} < U_{.975}\right) = .95,$$

where $U_{.025}$ and $U_{.975}$ are the 2.5 and 97.5 percentiles of $U \equiv T/\sigma$. The inequality within parentheses is equivalent to

$$\frac{1}{U_{.025}} > \frac{\sigma}{T} > \frac{1}{U_{.975}}.$$

Hence

$$P\left(\frac{T}{U_{.025}} > \sigma > \frac{T}{U_{.975}}\right) = .95.$$

The inequality within the parentheses defines a 95 per cent confidence interval for σ.

Now, for normal populations, the distribution of $W \equiv R/\sigma$, where R is the sample *range*, depends only on n; and so W can be used as the T in the method just explained. The distribution of W is given (by percentiles) in Table V of the Appendix. From that table one can read $W_{.025}$ and $W_{.975}$ to use in calculating the confidence interval.

A 95 per cent confidence interval for σ:

$$\frac{R}{W_{.975}} < \sigma < \frac{R}{W_{.025}}.$$

Clearly, for a 90 per cent interval one would use $W_{.05}$ and $W_{.95}$, and so on.

The distribution of $\chi^2 \equiv nS^2/\sigma^2$ depends only on n, so χ^2 is another T that can be used to obtain a confidence interval. The distribution of χ^2 is given (by percentiles) in Table II of Appendix B. This distribution is called the "chi-square distribution with $n - 1$ degrees of freedom." A 95 per cent confidence interval for σ^2 is then given as follows:

A 95 per cent confidence interval for σ^2:

$$\frac{nS^2}{\chi^2_{.975}(n-1)} < \sigma^2 < \frac{nS^2}{\chi^2_{.025}(n-1)},$$

where $\chi^2_p(n - 1)$ is the $100p$ percentile of χ^2 with $n - 1$ degrees of freedom. The corresponding interval for σ is obtained by taking square roots. Typically, it is narrower than an interval with the same confidence coefficient but based on R (if $n > 2$) and is therefore preferred when S is available.

Example 10-g The nine oil-temperature readings given in Example 11-e have variance $S = 15.728$ and range $R = 11$. If it can be assumed that they come from a normal population, confidence intervals can be constructed for σ as follows. Using R, with $W_{.05} = 1.03$ and $W_{.95} = 3.86$ from Table V, the 90 per cent interval is given by

$$\frac{11}{3.86} < \sigma < \frac{11}{1.03},$$

or (2.85, 10.68). Using S^2, with $\chi^2_{.05}(8) = 2.73$ and $\chi^2_{.95}(8) = 15.5$ from Table II, one obtains

$$\frac{9 \times 15.728}{15.5} < \sigma^2 < \frac{9 \times 15.728}{2.73}.$$

The confidence limits for σ are the square roots (3.02, 7.20), defining an interval that is narrower than the one based on range, as is usually the case. (The point estimates, as discussed in Section 10.4, would be $.313R = 3.44$ and $1.028S = 4.08$, using R and S, respectively.) ◄

Example 10-h The sample of 200 viscosity measurements referred to in Example 10-d had $S = 1.81$. For a 95 per cent confidence interval one would need the 2.5 and 97.5 percentiles of χ^2 with 199 degrees of freedom. Table II does not extend that far but gives a formula to use for large n:

$$\chi_p^2 = \frac{1}{2}(z_p + \sqrt{2k - 1})^2.$$

With $p = .025$ and $.975$, this yields (using $z_{.025} = -z_{.975} = -1.96$):

$$\chi^2_{.025}(199) = 161.4, \qquad \chi^2_{.975}(199) = 239.5,$$

and the confidence interval is (2.736, 4.06) for σ^2, or (1.646, 2.015) for σ.

[Incidentally, since n is large, one might try the formula $S \pm 2$(standard error), or $S \pm 2S/\sqrt{2n}$; this gives the interval (1.63, 1.99), which is not too different from the interval obtained above using S^2.] ◄

★ 10.8 Maximum Likelihood Estimation

A common technique for obtaining estimators is based on the idea that among the competing models for an experiment, the most plausible one is the one in which an observed result has the greatest probability.

Example 10-i The family of models for two independent trials of a Bernoulli experiment is given by $\binom{2}{x}p^x(1 - p)^{2-x}$, where x is the number of successes. If one observes one success and one failure, the probability $2p(1 - p)$ of this result is 1/2 for $p = 1/2$ and no larger than 1/2 for any other value of p. Thus $p = 1/2$ is taken to be the most plausible explanation of what happened. ◄

Given an outcome of a process of sampling from a discrete sample space, this outcome has different probabilities in the various possible models as indexed by θ; it is a *function* of θ and is called the *likelihood function* corresponding to an observed result.

> Likelihood function for the data $Z = z$:
>
> $$L(\theta) \equiv P_\theta(Z = z).$$

(The subscript θ here means that the probability is calculated in the model indexed by θ.) A value of θ that maximizes the likelihood function is called a *maximum likelihood estimate* of θ.

> Maximum likelihood estimate:
>
> $$\text{m.l.e.} \equiv \hat{\theta}, \qquad \text{such that } L(\hat{\theta}) \geq L(\theta) \text{ for all } \theta.$$

Example 10-j Two beads are picked at random from a container with five beads that are some black and the rest white, in unknown proportion. Suppose that the beads picked are both black. This result has probability

$$P(\text{both black}) = \frac{\binom{M}{2}\binom{5-M}{0}}{\binom{5}{2}},$$

where M is the (unknown) number of blacks in the container. For various values of M, this probability is as follows:

M	0	1	2	3	4	5
P_M(both black)	0	0	.1	.3	.6	1

The greatest of these probabilities is 1, for $M = 5$. This value 5 is the maximum likelihood estimate of M. ◄

If $(X_1, \ldots, X_n)$ is a random sample from a population with density function or probability function, as the case may be, given by $f(x; \theta)$, then the joint density or probability function (see Section 10.1) is given by the product

$$\prod f(x_i: \theta) \equiv f(x_1; \theta) f(x_2; \theta) \cdots f(x_n; \theta).$$

For the observed result $X_1 = x_1, \ldots, X_n = x_n$, this probability or density depends on θ. As such, it is the *likelihood function* corresponding to the given observations:

> Likelihood function for the random sample $(X_1, \ldots, X_n)$:
>
> $$L(\theta) = \prod f(x_i; \theta).$$

Again, a value $\hat{\theta}$ that maximizes this function is called a maximum likelihood estimate of θ.

Example 10-k If X is distributed continuously on $0 < x < \infty$ with density $\lambda e^{-\lambda x}$ for $\lambda > 0$, the likelihood function is given by

$$L(\lambda) = \prod(\lambda e^{-\lambda x_i}) = \lambda^n e^{-\lambda \sum x_i}.$$

This function of λ can be maximized by differentiating and setting the derivative equal to zero. Alternatively, it is maximized for the same value of λ as is its logarithm:

$$\log L(\lambda) = n \log \lambda - \lambda \sum x_i.$$

Differentiation yields

$$\frac{L'(\lambda)}{L(\lambda)} = \frac{n}{\lambda} - \sum x_i,$$

which vanishes at the value $\hat{\lambda} = 1/\bar{x}$. Since $L'(\lambda)$ changes sign from + to −, upon passing through 0 at $\hat{\lambda}$, the critical value $1/\bar{x}$ yields a maximum. So $1/\bar{X}$ is the maximum likelihood estimate of $\lambda = 1/E(X)$. ◄

The method of maximum likelihood is most successful in producing good estimators when based on samples of moderate to large size. (Example 13-g deals with the case of a small sample, pointing out what can go wrong in maximum likelihood estimation and offering an alternative approach.)

Problems

1. The cholesterol data in Table 6.1 have a mean and a standard deviation $\bar{X} = 245.69$ and $S = 46.02$.

(a) Determine the standard error of $\bar{X}$.
(b) Determine the standard error of S.
(c) Estimate the probability that an individual selected from this population will have a cholesterol level less than 200, by referring to Table 6.1 and counting.
(d) Estimate the probability in part (c) by replacing parameters by statistics in $\Phi((200 - \mu)/\sigma)$, which gives the desired probability if the population is normal.

2. In Example 10-b a standard error in estimating a proportion was found to be .0216, based on a sample of size 433. How large a sample would be needed to obtain an estimate with a standard error of .01?

3. In planning a survey by random sampling, it is desired that the standard error should not exceed .02. What sample size should be used?

[HINT: Although p is not known, $p(1 - p)$ is bounded above; find its maximum and use it in your calculations.]

4. In a sample 1000 individuals from a certain population, 10 are found to be color-blind. Compute the *relative* standard error of .01 as an estimate of the population proportion.

★ **5.** A multiple-choice examination question reads as follows: Compute the standard deviation of the following set of data:

34.4,	32.1,	33.1,	32.7,	31.5	A. 1.78
33.3,	34.2,	32.2,	32.1,	30.8	B. 0.90
31.6,	31.6,	31.3,	34.6,	32.6	C. 0.33
					D. 1.27

Pick the correct answer *without* actually computing S and give a rationale for your choice.

★ **6.** Estimate σ for the population from which the cholesterol data in Table 6.1 were taken. Use the mean of the ranges in groups of five (in columns).

7. Suppose that you are in the business of constructing confidence intervals, and you regularly use a confidence coefficient of 95 per cent.

(a) Among your first 200 clients, how many would you expect to have confidence intervals that cover the true parameter value?
(b) Would you expect complaints from those whose confidence intervals do *not* cover the true parameter value?

8. In Example 10-a, statistics were given for the number of new dental cavities in a sample of 294 children over a 30-month period: $\bar{X} = 10.88$ and $S = 6.36$.

(a) Construct a 95 per cent confidence interval for the mean number of cavities in the population of children from which this sample was taken.
(b) Would it be proper to estimate the proportion of children with fewer than two cavities as $\Phi((2 - 10.88)/6.36)$? Explain.

9. Give 95 and 99 per cent confidence intervals for μ, the mean cholesterol level in the population, based on the data of Problem 1.

10. Through careful examination of sound and film records it is possible to measure the distance to an insect when a bat first detects it. Given these eight observations (in centimeters):

28, 79, 62, 53, 40, 31, 55, 63,

construct a 90 per cent confidence interval for μ, the population mean distance, assuming normality of the distribution.

11. A sample of 10 rats averages 38 seconds to solve a certain maze, with a standard deviation of 6.5 seconds. Construct a 90 per cent confidence interval for the mean time required, assuming a normal population.

12. Determine the probability that the mean of a sample of size 10 from a normal population differs from the population mean by more than 3/4 of the sample standard deviation.

13. Check the simple versus the refined formula for a confidence interval for p when $\bar{X} = .5$ and $n = 200$. (Use $k = 2$, corresponding to 95.44 per cent confidence.)

★ **14.** Construct a 90 per cent confidence interval for σ based on the data in Problem 10, assuming (as in that problem) a normal population, for the following cases:

(a) Using the sample standard deviation.
(b) Using the sample range.

★ **15.** Construct a confidence interval for the population standard deviation based on the S of the cholesterol data as given in Problem 1.

16. In a survey with questions about a certain behavior pattern that people may not want to reply to truthfully, it is possible to reduce the potential embarassment, while retaining half of the information, by the following kind of question: "I will ask you to toss this coin, but do not show or tell me how it lands. If it lands heads, answer (yes or no): Is your first child a male? If it lands tails, answer (yes or no): Have you had an extramarital affair?" Suppose that you got 175 yes's from a random sample of 300 respondents. What would you estimate to be the proportion of the population who have had affairs?

17. We are interested in μ, the average number of children per family attending a certain elementary school, in the population of families that have at least one child in attendance. We take a sample of 10 students, selected randomly while the students are at lunch, and ask each one how many children in his family attend the school. Their answers are 3, 3, 2, 4, 1, 3, 2, 4, 1, and 2.

(a) Explain what would be wrong with using $\bar{X}$ to estimate μ.
(b) Can you see a reasonable way to use the data given to estimate μ? (If so, do it.)
(c) One approach to part (b) is to think of a child who comes from an i-child family as representing $(1/i)$th of a family. The average number of families per child is then estimated as $\overline{1/X}$, and the average number of children per family as the reciprocal of this. Is $\overline{1/X}$, the average of the reciprocal, the same as the reciprocal of the average, $1/\bar{X}$?

18. In a certain war the enemy has N tanks. The number N is unknown to us, but we do know that the tanks are numbered consecutively from 1 to N. The numbers of seven of their tanks have been observed:

$$21, \quad 459, \quad 4, \quad 12, \quad 27, \quad 53, \quad 124.$$

Assume that these were obtained as independent observations on the uniform distribution with probability function

$$f(x; N) = \frac{1}{N}, \qquad x = 1, 2, \ldots, N,$$

which has mean $\mu = (N + 1)/2$. Does the sample mean $\bar{X}$ give a reasonable estimate of μ? The function $2\bar{X} - 1$ has expectation N; does it give a reasonable estimate of N? What is wrong with estimates based on $\bar{X}$?

★ **19.** Obtain $Y/n = \hat{p}$ as the maximum likelihood estimate of the population proportion p in a Bernoulli population, where Y is the number of successes in n independent trials.

(HINT: It is easier to work with $\log L(p)$.)

★ **20.** Let $(X_1, X_2, \ldots, X_n)$ be a random sample from a Poisson distribution:

$$P(X = x) = \frac{m^x}{x!}e^{-m}.$$

(a) Write out the likelihood function $L(m)$.
(b) Find out the maximum likelihood estimate of m.

21. Two techniques sometimes used in dealing with "outliers" (spurious observations) in estimation are as follows. Comment on the validity of each, from an intuitive basis:

(a) In some laboratories, measurements are made in triplicate, and the value used is the average of the two closest together. (The third is discarded.)
(b) Individual athletic performances are sometimes judged by a panel of several judges. The highest and lowest scores are discarded, and the contestant is given the average of the remaining scores.

11

Testing Hypotheses

Questions such as "Are these performance figures up to standard?" and "Is the response increased by this treatment?" suggest, first, a recognition of the fact that scores and responses—even on the average—are random; and second, a realization that a sample average that is quite different from usual may be so simply because of sampling fluctuations and not because the population being sampled is really unusual. In both situations there may be various models or "hypotheses" that need to be checked against the data, and the process of doing so is referred to as *testing* the hypotheses.

Example 11-a A problem of testing for ESP was introduced in Example 5-a. The skeptic's hypothesis would be that there is no such thing as ESP, or that $p = 1/4$, where p denotes the probability that a subject can correctly guess the suit of a card without seeing it. The subject himself might claim that $p > 1/4$, and the point of a test is to see which of these hypotheses is the more plausible. (*To be continued.*) ◄

11.1 Hypothesis Testing—General

The usual situation is that one has a particular model to be tested, or checked for accuracy in representing a certain experiment. The hypothesis that this model is the right one is called the *null hypothesis*, often denoted by H_0. This hypothesis may have emerged from long experience with an experiment, and it is desired to see whether the hypothesis is still correct, when there has perhaps been some change in circumstances that call it into question. Or, the hypothesis may be the result of a theoretical analysis, or a logical argument, and the theory needs to be verified. In comparison problems, the hypothesis H_0 is usually that there is "no difference" between using a treatment and not using it, or between two different treatments, or between "before" and "after" some treatment. In these latter problems, where "null" means "no treatment effect," the experimenter is usually expecting that there *is* a treatment effect, and he is conducting the experiment with the express purpose of being able to reject H_0 on reasonably solid grounds.

A null hypothesis is ordinarily taken to be quite specific—specific enough to permit probability calculations. The models that are considered possible if the null hypothesis is not true, taken together, define what is called the *alternative hypothesis*, called H_A, or sometimes H_1.

Example 11-a (continued) The hypothesis of *no* ESP in card guessing, or $p = 1/4$ in the Bernoulli model, would be taken to be the null hypothesis. It defines the experiment completely so that probabilities can be calculated. The hypothesis that ESP *is* present is more general: $p > 1/4$; this hypothesis about p includes any model in which the probability of a correct guess exceeds 1/4. This hypothesis is the *alternative* hypothesis. It is typical that alternative hypotheses, like this one, are *composite*, including many possible models. (*To be continued.*) ◄

The usual approach in testing is something like this: A sample statistic is chosen to serve as a *test statistic*; for the present let it be denoted by T. If the hypothesis H_0 to be tested specifies values of a parameter θ of a class of models defined by densities or probability functions $f(x; \theta)$, then T will often be some natural estimator of θ. (For instance, if θ is the population mean, then T could well be $\bar{X}$, the sample mean.) In any case, T is devised, often according to intuition, so that its value will be intimately related with the truth or falsity of H_0 (as opposed to the alternative H_A).

Example 11-a (continued) If, as in the earlier discussion in Example 5-a, the test for ESP is to be based on the outcomes of 10 independent trials of the suit-guessing experiment, it is natural to work with the summary statistic Y, the number of correct guesses in the 10 trials. This test statistic is a random variable with possible values $0, 1, \ldots, 10$. The probability distribution of Y when $p = 1/4$ (no ESP) was given in Example 5-a as an example of a binomial distribution. It is repeated here:

y	0	1	2	3	4	5	6	7	8	9	10
$f(y)$	.056	.188	.282	.250	.146	.058	.016	.003	.000	.000	.000

Intuitively, it would seem that values such as $Y = 9$ or 10 suggest the presence of ESP; that is, $p > 1/4$ would appear to be more plausible as an explanation of 9 or 10 correct guesses in 10 tries. Of course, $Y = 0$ is also an unlikely result if $p = 1/4$, and one might wonder if $p = 1/4$ is believable when there are *no* correct guesses. But the subject would claim that his ESP makes $p > 1/4$, and he would be the first to concede that 10 incorrect guesses would do little to make his claim credible. (*To be continued.*) ◄

A test statistic T is usually devised so that if H_0 is not true, the value of T will be "extreme," and that if H_0 is true, the value of T will not be extreme, but more toward the center of its distribution. "Extreme," here, may mean exceptionally large, or it may mean exceptionally small, or in some cases, either too large or too small—all depending on the way T is constructed and on the models alternative to H_0.

When H_0 is true (and sufficiently specific), the test statistic T will have a corresponding sampling distribution, a pattern of variation from sample to sample. This will be described by a density or probability function $f_T(t)$, often like those shown in Figure 11-1. In such distributions, the values of T that commonly occur are those near the middle of its distribution. On the other hand, values of T in the tails, the extreme values, will occur only rarely.

When an extreme value, such as t' in Figure 11-1, *is* observed to occur, it is natural to ask whether that value t' really came from the null distribution of T, or whether some hypothesis other than H_0 would be a more plausible explanation of what has happened. In particular, if t' is the sort of value that would be usual under some alternative to H_0, then perhaps that alternative is more plausible than H_0.

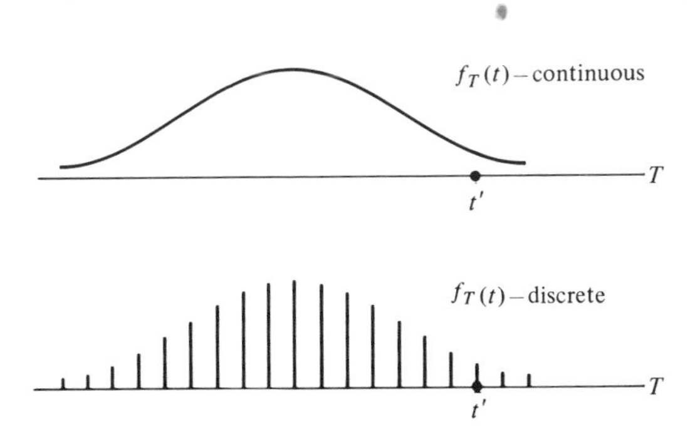

Figure 11-1 Typical null distributions of a test statistic.

So, when data are gathered and summarized in the value $T = t'$, what conclusion can be drawn? Is a value t' out in the tail of the null distribution reason to discard H_0?—to reject the null hypothesis? If it is, then it would also seem reasonable to insist that any value of T still *farther* out would be all the more reason to reject H_0. Pressure to choose between accepting and rejecting H_0 might then lead one to adopt a policy of the form: Reject H_0 if T (as computed from a given set of data) turns out to have a value as extreme as t^*, or more so. Different choices of the *critical boundary* t^* correspond to how easily one is swayed from accepting H_0. Those values of T whose occurrence causes rejection of H_0, for a particular t^*, constitute the *critical region* of the test. If it is large T-values that make the possible alternatives to H_0 more plausible, the critical region is of the form $T \geq t^*$ (or $T > t^*$).

The T-values in the critical region are thought of as signifying that H_0 is not the right model, and such a value is said to be *significant* (or sometimes *statistically significant*).

Example 11-a (continued) Suppose, in testing for ESP using the number Y of correct guesses in 10 trials of the suit-guessing experiment, one follows his intuition and decides that H_0 is not plausible—or rather, that $p > 1/4$ is *more* plausible—if it should turn out that $Y = 7, 8, 9$, or 10. This set of four Y-values constitutes the critical region of the test, and any one of the four values would be deemed to signify that H_0 is not correct. Following this rule, however, can lead one astray—in two ways. He might get a value in the critical region just by chance, and so reject H_0, even when there is no ESP. And he might *not* get such an extreme value even if ESP *is* present, especially if the degree of ESP is only moderate. (*To be continued.*) ◄

Whenever inferences are based on data—on a less than complete look at a population, they can be wrong. This was noted in the estimation problems of Chapter 10 and again in the case of hypothesis testing just cited. In testing generally, just owing to sampling variability, a significant value t' *can* occur when H_0 is really true; and rejection of H_0 on the basis of this value of T would be an error. On the other hand, one may have a sample that yields a moderate or nonsignificant value of T, and so be led to accept H_0, even when H_0 is not the correct model. This is a different kind of error, and will usually have different consequences.

Type I error:	Rejecting H_0 when H_0 is true.
Type II error:	Accepting H_0 when H_0 is false.

The hypothesis H_0 is often a specific model in which probabilities for T can be calculated. The probability of the critical region, computed under such an hypothesis H_0, is called the *significance level* of the test, or also, the *size* of the type I error; it is almost universally denoted by α.

Significance level of a test (size of type I error):

$$\alpha \equiv P(\text{test rejects } H_0|H_0) = P(\text{critical region}|H_0).$$

(The vertical bar in the notation above does not mean that this is a conditional probability in the earlier sense, although it will be so interpreted in Chapter 13. All that is meant here is that what follows the bar is the model in which probability is to be computed.)

Example 11-a (continued) Rejecting the null hypothesis of no ESP ($p = 1/4$) when the subject makes 7, 8, 9, or 10 correct identifications of card suit in 10 trials can lead to erroneous conclusions. Let C denote this critical region: $\{7, 8, 9, 10\}$. Using C, one rejects H_0 when it is true with probability

$$\alpha_C = P(7, 8, 9, \textit{or } 10|p = .25) = .003,$$

from the table given earlier. This very small probability for the type I error when H_0 is *true* is achieved at the expense of insensitivity to the presence of ESP ($H_A: p > 1/4$). In particular, if $p = .5$, the probability of (correctly) rejecting H_0 would be only

$$P(7, 8, 9, \text{or } 10|p = .5) = \left\{\binom{10}{7} + \binom{10}{8} + \binom{10}{9} + \binom{10}{10}\right\}(.5)^{10} \doteq .17.$$

(The corresponding type II error size at this value of p would be $1 - .17 = .83$.) Enlarging the critical region to $C' \equiv \{5, 6, 7, 8, 9, 10\}$ would increase the size of the type I error:

$$\alpha_{C'} = P(Y \geq 5|p = .25) = .077,$$

but there would be a greater chance of detecting $p = .5$:

$$P(Y \geq 5|p = .5) = \sum_{5}^{10} \binom{10}{k}(.5)^{10} \doteq .62.$$

The type II error size at this p is .38. (*To be continued.*) ◄

If T has a continuous distribution with density $f_T(t)$ under H_0, then α is the area under the density curve above the critical region. Figure 11-2 shows

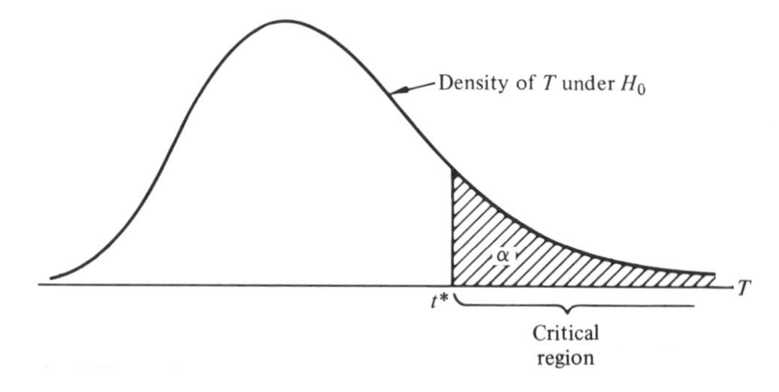

Figure 11-2 Significance level as an area.

such a density, a critical region $T > t^*$, and the area defining the significance level α. If T is discrete, then α is a sum of probabilities of those T-values that are deemed significant.

A type I error can usually be made small by insisting that a T-value be far out in the tail before it is deemed significant; and a type I error can be avoided completely by simply accepting H_0 no matter what the data have to say! But of course this increases the tendency to accept H_0 when it is not true—a type II error. So some kind of compromise has to be struck, by adopting a policy that has a good chance of uncovering falsity of H_0 when this is the case, without being so easily swayed by an extreme T-value as to run much risk of rejecting H_0 when it is true.

Where should the line be drawn? Sometimes one is forced to take action, as for instance in an industrial situation, where a lot is being sampled for acceptability. In such a case, the only really rational thing to do is to study the costs of wrong decisions and, according to the probabilities or errors involved, to choose the test that minimizes the expected cost. (See Chapter 13.) If a decision is not forced, then it is usually a matter of simply *reporting* results—although someone else may make a decision on the basis of them.

Results can be reported in various ways. One way is just to present all the data, perhaps summarized in a frequency distribution. Or one can give the value of the summary statistic T together with some information as to its distribution. Along this line, it is the practice in some fields to report a *P-value*—the probability, under H_0, that T would have a value at least as extreme as the value computed from the data. This would be the boundary of a critical region that would call for rejection with significance level P. The crudest report is one of the form "H_0 is rejected at the 5 per cent level of significance." Such a report does not say whether the result is such as to suggest *strongly* that H_0 should be rejected or whether it leads to a value of T

so close to the critical boundary that the proper conclusion is not at all clear. A modified version of the crude report is to follow a tradition used by some: Report that a value of T is "significant" if the corresponding P-value is between 5 per cent and 1 per cent and "highly significant" if it is less than 1 per cent. The 5 per cent and 1 per cent figures, although traditional, are quite arbitrary.

Example 11-a (continued) To conclude the ESP example, suppose that the experiment of 10 trials yields $Y = 6$. That is, the subject correctly identifies the suit in 6 of 10 trials. Using the critical region $C = \{7, 8, 9, 10\}$, with $\alpha = .003$, one would *not* reject H_0, since 6 is not a value in C. If he were using the critical region $C' = \{5, 6, 7, 8, 9, 10\}$, with $\alpha = .077$, he *would* reject H_0, since 6 *is* in C. The P-value is the probability, under H_0, of 6 or more successes:

$$P\text{-value} = P(6, 7, 8, 9, 10 \,|\, p = .25) = .019.$$

This is the α of the test ($Y \geq 6$) that just barely rejects H_0; the value 6 is the least extreme value in this critical region. ◄

11.2 Large-Sample Tests for μ

The sample mean $\bar{X}$, or some function of it, seems natural to use in drawing inferences about the population mean μ. In particular, for testing

$$H_0\colon \mu = \mu_0,$$

it would be natural to judge its plausibility by measuring the distance from $\bar{X}$ to μ_0; for, if μ_0 *is* the true mean, then $\bar{X}$ tends to close to μ_0 in large samples. But there is the problem of scaling if the distance $\bar{X} - \mu_0$ is used as it stands, because changing units of measurement would change its numerical value. Moreover, the kind of discrepancies $\bar{X} - \mu_0$ that would be encountered would depend also on the population variance, σ^2.

When σ^2 is *known*, the test statistic used is a standardized version of $\bar{X}$ (a "Z-score"):

Test statistic for $\mu = \mu_0$ (σ known):

$$Z = \frac{\bar{X} - \mu_0}{\sigma/\sqrt{n}}.$$

This is insensitive to change of scale. For large samples, Z is approximately normal with unit variance:

> Sampling distribution of Z (large n):
>
> $$Z \sim \mathcal{N}\left(\frac{\mu - \mu_0}{\sigma/\sqrt{n}}, 1\right)$$

Thus, *if* H_0 *is true*, $\mu - \mu_0 = 0$ and Z is *standard* normal; but if H_0 is false, then the center of the sampling distribution of Z shifts according to how far the actual mean is from the value μ_0 being tested.

It will be recalled (from Chapter 8) that what is a "large" sample depends both on how accurate a result if desired and on how close the population is to normal. If the population is itself normal, then Z is exactly normal for any sample size. (This fact can be established using moment generating functions, studied in Appendix A.) If the population is unimodal and reasonably symmetric, then $n = 10$ or even $n = 5$ may be large enough for an adequate approximation.

Example 11-b The diameter of a certain type of ball bearing is a random variable whose mean is designed to be $\mu_0 = .610$ centimeter and standard deviation .004 centimeter. A sample of 25 taken from the production line yields an average diameter of $\bar{X} = .612$. Is the process still "in control"—still producing bearings with diameters as specified?

Perhaps one can assume (as might well be the case in a real situation) that σ has not changed. Then

$$\sigma_{\bar{X}} = \frac{\sigma}{\sqrt{n}} = \frac{.004}{\sqrt{25}} = .0008.$$

test statistic Z has the value

$$Z = \frac{\bar{X} - \mu_0}{\sigma_{\bar{X}}} = \frac{.612 - .610}{.0008} = 2.5.$$

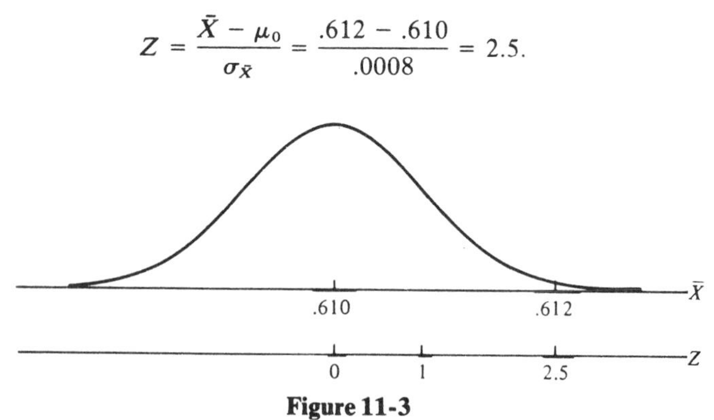

Figure 11-3

That is, the observed $\bar{X}$ is 2.5 standard deviations away from $\mu_0 = .610$, in the right tail of the sampling distribution of $\bar{X}$ under H_0. (See Figure 11-3.) To protect against a two-sided alternative to H_0, namely, $\mu \neq \mu_0$, a two-sided critical region would be appropriate, rejecting H_0 if Z is *either* very large or very small. For $\alpha = .05$, any Z-value such that $|Z| > 1.96$ would be deemed significant; and so the observed $Z = 2.5$ would call for rejection of H_0 at this level. Perhaps the production process should be halted and readjusted. ◄

When σ is *not* known, the sample standard deviation can be used in its place, in defining the test statistic, if n is large:

Test statistic for $\mu = \mu_0$ (σ unknown, n large):

$$Z' \equiv \frac{\bar{X} - \mu_0}{S/\sqrt{n}}.$$

The sampling distribution of Z' is approximately $Z \sim \mathcal{N}\left(\frac{\mu - \mu_0}{\sigma/\sqrt{n}}, 1\right)$. It is standard normal if $\mu = \mu_0$.

11.3 Small-Sample Tests for μ—Normal Case

One might think to use the test statistic Z' defined at the end of the preceding section even when n is not large, but then the distribution of Z' may not be close enough to normal for Table I to give good approximations of probabilities. What makes the approximation poor is the variability of the S in the denominator of Z'. Even though $\bar{X}$ may still be close to normal, Z' will have too much probability in the tails of its distribution.

The exact sampling distribution of Z' depends on the population being sampled. When the population is *normal*, the distribution of Z' is known. A slight modification of Z', to be called T, has the "Student's t-distribution† with $n - 1$ degrees of freedom." (This was introduced in Section 10.5 in the discussion of small-sample confidence limits for μ.) For degrees of freedom not exceeding 30, various percentiles of the distribution are given in Table III of Appendix B. For $n > 30$, it suffices to use the normal percentiles, shown for convenience in the last line of the t-table opposite "∞" degrees of freedom. This is because, as would be expected in view of the earlier claim

† "Student" is the pseudonym of W. S. Gosset, who was not permitted to publish under his own name while employed by the Guinness brewery.

about Z', the statistic T is asymptotically normal as the number of degrees of freedom becomes infinite.

> Test statistic for $\mu = \mu_0$ (σ unknown, n small, normal population):
> $$T \equiv \frac{\bar{X} - \mu_0}{S/\sqrt{n - 1}}.$$
> The sampling distribution of T is $t(n - 1)$ when $\mu = \mu_0$.

The expression "$t(n - 1)$" will be used to denote the t-distribution with $n - 1$ degrees of freedom.

In this last box, as in several that follow, we have given only the null distribution of the test statistic. This is sufficient for determining the critical region for a specified α. But the distribution of the test statistic when H_0 is *not* true is important for considering type II errors, as will be done in Section 11.6. As is the case here, these "nonnull" distributions of the test statistic are often complicated, especially when the alternative is highly composite.

A t-table is not usually given in as much detail as a normal table. This is in part because there is not a simple relationship among the t-distributions, whereas the distributions of the normal family are all generated by the standard normal distribution. Table III of Appendix B gives, in each line, several useful percentiles for a particular number of degrees of freedom, lower percentiles being obtained from those given by symmetry: $t_p = -t_{1-p}$.

As mentioned in Section 10.5, if one is using $\tilde{S}$ (the sample standard deviation with divisor $n - 1$), then $\sqrt{n - 1}\tilde{S} = \sqrt{n}S$, and

$$T = \frac{\bar{X} - \mu_0}{\tilde{S}/\sqrt{n}},$$

the same value for T as given by the earlier formula using S.

Example 11-c In Example 9-c were given widths of 7 skulls found in a certain digging. Might these be a sample from a race previously found nearby, known to average 146 millimeters in width? For the newly found skulls, $\bar{X} = 143.3$ and $S = 5.62$, so

$$T = \frac{143.3 - 146}{5.62/\sqrt{6}} = -1.18.$$

If the $t(6)$-distribution applies here, as it would if the population could be considered normal, this value of T is not sufficiently extreme to warrant rejection of $\mu = 146$ at usual levels [$t_{.95}(6) = 1.94$]. ◄

It is perhaps worth repeating a basic assumption in the t-test, namely, that the population being sampled is normal. This assumption is used in showing that the statistic T has a t-distribution under H_0. Because distributions occurring in practice are never exactly normal, one might well wonder if the t-test is *ever* valid. Experience, theory, and artificial sampling experiments have shown that for moderate departures from normality, the t-distribution for T is usually applicable for the kind of accuracy that can be appreciated.

11.4 Tests and Confidence Intervals

In testing a particular value θ_0 of parameter θ, it might occur to one quite naturally to use the following rule: Construct a confidence interval for θ and reject θ_0 if it does not fall in the confidence interval. The significance level of such a test is easily determined.

For instance, suppose that $I_{.95}$ is a confidence interval of θ-values, a random interval based on the value of the estimator T, with the property that it has a 95 per cent chance of covering the true value of θ. But then, if θ_0 *is* the true value, there is a 5 per cent chance that it will not be covered by the confidence interval:

$$\alpha = P(\text{reject } \theta_0 | \theta = \theta_0) = P(I_{.95} \text{ does not cover } \theta_0 | \theta_0) = .05.$$

More generally, the significance level α would be 1 minus the confidence coefficient.

α-level test for $\theta = \theta_0$:
Reject θ_0 unless the $1 - \alpha$ confidence interval for θ includes θ_0.

If the confidence interval is a finite interval defined by an upper confidence limit and a lower confidence limit, the corresponding test is two-sided and will reject θ_0 for T-values in either tail of its null distribution.

The correspondence between test and confidence interval actually goes both ways. For, if an α-level critical region C_θ is defined for testing θ, for each θ, then the set of θ such that a given sample result T is in C_θ defines a $1 - \alpha$ confidence region for θ.

Giving a confidence interval for θ is a way of reporting experimental results that is more informative than a report that merely declares that H_0 is rejected because the value of T was found to be significant. The confidence interval can, as seen above, be used as the basis of a test; but it also gives an estimate of the true θ along with a notion of the possible error of estimation.

Example 11-d In the preceding example, the mean width of seven skulls was given as $\bar{X} = 143.3$, and the standard deviation $S = 5.62$. A 90 per cent confidence interval for the true mean has limits

$$143.3 \pm 1.94 \frac{5.62}{\sqrt{6}},$$

where $1.94 \equiv t_{.95}(6)$, assuming a normal population. This interval, $138.85 < \mu < 147.75$, *does* include the value 146, so this value is accepted in a two-sided test with $\alpha = 1 - .90 = .10$. ◄

11.5 A Rank Test for Location

In some cases it may not be realistic to assume that the population is close enough to normal for the t-test to give valid results for small samples. Various "distribution-free" statistics have been devised—statistics whose sampling distribution under H_0 do not depend on any assumptions about the nature of the population. In particular, it is no longer assumed that the population is one of a "parametric family" [such as $\mathcal{N}(\mu, \sigma^2)$, or exponential (λ)], and the methods used are often referred to as *nonparametric.*

Consider first the testing of a hypothesis about the median of a continuous population:

$$H_0: \quad \text{median} = m_0.$$

If this is true, the observations are as likely to fall on one side as the other of m_0, with probability 1/2 for each side. One could simply count the number of observations to the right, say, and use this as a statistic for detecting a shift in location. It has a binomial distribution, with $p = 1/2$ under H_0. The idea that extreme values of this statistic suggest that H_0 is false leads to what is called a "sign test." However, a more powerful test—one more sensitive to a shift in the median—can be devised if one takes into account how *far* to the right or left of m_0 each observation falls.

Consider then the differences obtained by subtracting m_0 from each observation:

$$X_1 - m_0, \quad X_2 - m_0, \quad \ldots, \quad X_n - m_0,$$

and *order* these according to the magnitude of the difference—the absolute distances from m_0: $|X_i - m_0|$. When so arranged, the differences present a pattern of plus and minus signs, and these *signs* are assigned ranks according to their positions in the sequence.

For example, one might have this sequence of signs and corresponding ranks:

Signs:	+	+	−	+	−	−	−	+	−	−.
Ranks:	1	2	3	4	5	6	7	8	9	10.

A plus sign with a low rank denotes an observation close to m_0 on its right; a plus sign with high rank comes from an observation far to the right of m_0.

The test statistic R_+ to be used is the sum of the ranks of the +'s. An extremely large value of R_+ would mean that there are many observations to the right of m_0 and that if there are any to the left, they are not very far to the left. An extremely small value of R_+ would mean very few observations to the right, and those (if any) only a little bit to the right. [The sum of the ranks of the −'s would be a complementary quantity, equal to $n(n + 1)/2$ minus the sum of the + ranks; if one rank sum is large, the other is small.]

The distribution of R_+ needed to determine a significance level is derivable from the fact (not proved here, but intuitively reasonable) that under H_0 the 2^n possible sequences of +'s and −'s are equally likely, provided the population is symmetric about its median. Table VII gives cumulative probabilities for the left-hand tail of this distribution.

Wilcoxon signed-rank test for H_0: median $= m_0$ (for population symmetric about its median):

1. Arrange differences $X_i - m_0$ according to magnitude.
2. Sum the ranks of the positive differences, R_+.
3. Take large or small values R_+ as "extreme," according to the alternatives to H_0.
4. Determine critical value from Table VII.

Although, ideally, ties will not occur if the population is really continuous, they can occur with actual, recorded data. When they do, tied observations are assigned the average rank of the positions they occupy. Again, ideally, the probability of the event $X_i - m_0 = 0$ is zero, but it can occur in practice. If it does, drop that observation from the sample.

Example 11-e The seven skull widths, used in Example 12-c to illustrate the t-test of the hypothesis that $\mu = 146$, were given in Example 10-c as follows:

$$135, \quad 140, \quad 141, \quad 141, \quad 145, \quad 147, \quad 154.$$

If the population is symmetric, then the median is also equal to 146 under H_0, and the differences $X_i - 146$ are

$$-11, \quad -6, \quad -5, \quad -5, \quad -1, \quad 1, \quad 8.$$

In order of *magnitude*, they are (with some "ties")

$$\begin{Bmatrix} -1 \\ 1 \end{Bmatrix}, \quad \begin{Bmatrix} -5 \\ -5 \end{Bmatrix}, \quad -6, \quad 8, \quad -11,$$

with ranks

$$1.5, \quad 3.5, \quad 5, \quad 6, \quad 7.$$

The sum of the ranks of the positive differences is

$$R_+ = 1.5 + 6 = 7.5.$$

Suppose that a two-sided test with an α of about 10 per cent is in order. This would mean putting values of R_+ from each end of its range of values into the critical region so that the probability in each tail is about .05. Going down the $n = 7$ column in Table VII, one finds the cumulative probability .055 opposite $c = 4$. That is,

$$P(R_+ \leq 4) = P(R_+ \geq 28 - 4) = .055,$$

where $28 = n(n + 1)/2$ is the right-hand end of the range of R_+. So a two-sided test at $\alpha = .11$ would accept H_0, since $4 < 7.5 < 24$. ◄

When $n > 15$, the asymptotic distribution of R_+ may be used to approximate critical values. This distribution is normal with mean and variance given by

$$E(R_+) = \frac{1}{4}n(n + 1), \qquad \text{var } R_+ = \frac{1}{24}n(n + 1)(2n + 1).$$

The "Z-score," $(R_+ - \mu_{R_+})/\sigma_{R_+}$, can then be assessed in terms of Table I, as usual.

11.6 Power

The significance level of a test tells how the test performs—but only when H_0 is true. How does the test perform when H_0 is *not* true? Has it been constructed so as to be sensitive to the possible falsity of H_0? This is an important consideration, because it would usually be the case that if H_0 specifies a particular value θ_0, it is *not* true. Indeed, an investigator often conducts an experiment for the expressed purpose of rejecting a postulated value θ_0, having reason to expect or to hope that θ_0 is not correct. He should then be concerned about the *size* of the type II error—the probability (if there be one) that a test accepts H_0 when the alternative H_A is true.

If H_A is of the form $\theta > \theta_0$ or $\theta \neq \theta_0$, for example, then to say that H_A is true (or H_0 false) does not specify a particular model, and probabilities are not uniquely defined. However, for a *particular* θ in H_A, one *can* usually calculate probabilities for the test statistic, and the probability that a test rejects H_0 is uniquely defined for that θ. This probability is called the *power* of that test.

Power function of a test with critical region C:

$$\pi(\theta) \equiv P_\theta(C) = P_\theta(\text{test statistic falls in } C),$$

as a function of θ.

[The notation P_θ means that probability is to be calculated in the model defined by θ. The model depends on what value θ is assumed to have, and so $P_\theta(C)$ is a function of θ.]

Example 11-f It is desired to test $\mu = 0$ using the critical region $\bar{X} > K$, based on a random sample of size 25. For simplicity suppose that $\sigma^2 = 25$, so that $\sigma_{\bar{X}} = 1$. Then the power function of the test is

$$\pi(\mu) = P_\mu(\bar{X} > K) = 1 - \Phi\left(\frac{K - \mu}{1}\right) = \Phi(\mu - K).$$

If K is chosen to make $\alpha = .10$, then

$$\pi(0) = \Phi(-K) = .10,$$

from which it follows that $K = 1.28$. And so

$$\pi(\mu) = \Phi(\mu - 1.28).$$

Some values of power are given as follows, computed with the aid of Table I:

μ	0	.5	1.28	2	2.5
$\pi(\mu)$	.10	.218	.500	.764	.889

Figure 11-4 shows the distribution of $\bar{X}$ for these particular μ's and in each case the shaded area is the probability of the critical region, or power. It is evident that the power of this test (to discover that μ is not 0) increases as μ increases. The power function $\pi(\mu)$ is shown in Figure 11.5 along with the (equivalent) OC-function to be defined next. ◀

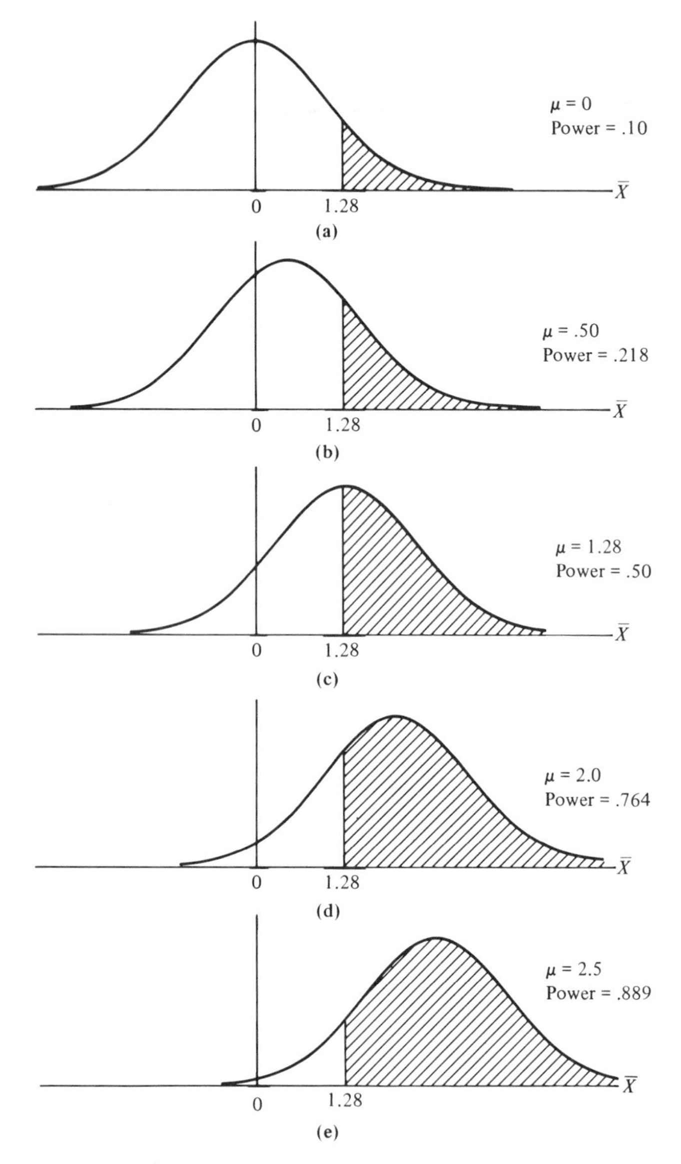

Figure 11-4 Rejection probabilities for Example 11-f.

In some applications it is customary to work in terms of the *operating characteristic function*, which is simply the "complement" of the power function:

> Operating characteristic function of a test:
>
> $$OC(\theta) \equiv 1 - \pi(\theta) = P_\theta(\text{test accepts } H_0).$$

Clearly, if you know one, you automatically know the other, and which to use is simply a matter of taste. Figure 11-5(b) shows the *OC* function for the test of Example 11-f.

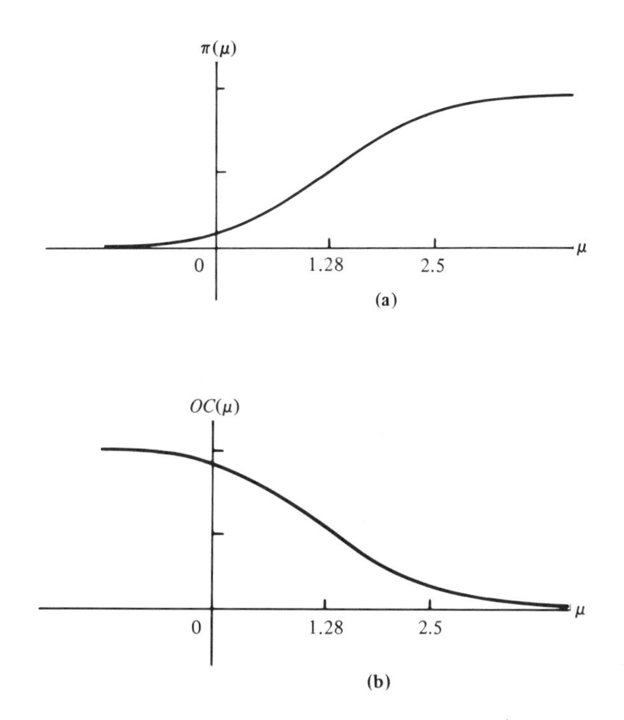

Figure 11-5 Power function and *OC*-function for Example 11-f.

If θ is in the alternative H_A, then accepting H_0 is an error of type II; so for θ in H_A, the acceptance probabilities $1 - \pi(\theta)$ [or $OC(\theta)$] are type II error sizes. Moreover, even though the term *power* derives from the notion that $\pi(\theta)$ measures the ability of a test to detect that H_0 is *not* true, the power function can be evaluated at $\theta = \theta_0$ and gives there the size of the type I error:

> Significance level from the power function:
>
> $$\alpha = \pi(\theta_0) = P_{\theta_0}(\text{test rejects } \theta_0).$$

Thus the power function of a test [or the OC-function] incorporates both the type I and type II error sizes and so gives a complete picture of the test's performance.

The size of the type I error and the various type II error sizes should all be small. Ideally, they would be zero, but this ideal is ordinarily unattainable when the test is based on a finite amount of data—on just a partial look at the population. So one settles for a "good" test, one with small error sizes.

> *Ideal* test of $\theta = \theta_0$:
>
> $$\text{Power should be } \begin{cases} 0 & \text{for } \theta = \theta_0 \\ 1 & \text{for } \theta \text{ in } H_A \end{cases}$$
>
> *Good* test of $\theta = \theta_0$:
>
> $$\text{Power should be } \begin{cases} \text{close to 0,} & \text{for } \theta = \theta_0 \\ \text{close to 1,} & \text{for } \theta \text{ in } H_A. \end{cases}$$

Because the power function describes the way a test performs in the various situations that might be encountered, it is of considerable practical importance in decision making.

The notion of power also serves as the basis for comparing tests. If two tests have the same α, but one of them is more powerful over the class of alternatives that are of particular concern, that one is naturally to be preferred. As in the preceding sections, critical regions and test statistics are often proposed on the basis of intuition; but they are then used if they turn out to have good power characteristics.

The power function of a test with critical region of the form $\bar{X} > K$, appropriate for testing μ_0 against μ_1 (where $\mu_1 > \mu_0$), is determined by the sample size n and the constant K defining the critical region. On the other hand, if one is free to choose any sample size, he can make the power function pass through *two* arbitrarily chosen points by his choice of n and K. This is done as follows.

The conditions

$$\begin{cases} \pi(\mu_0) = \alpha \\ 1 - \pi(\mu_1) = \beta, \end{cases}$$

where α and β are specified for testing μ_0 against μ_1, are two constraints that can be satisfied by proper choice of the two design parameters, sample size and critical boundary. The general one-sided problem, say $\mu \leq \mu^*$ against $\mu > \mu^*$, can be handled by considering parameter values μ_0 in H_0 and μ_1 in H_1 for which error sizes α and β can be specified, and choosing an n and K for μ_0 against μ_1 to meet these specifications. The critical region $\bar{X} > K$, designed for the simpler problem but applied to the more general one, defines a test whose power function passes through the two points (μ_0, α) and $(\mu_1, 1 - \beta)$.

Example 11-g Consider a test of the form $\bar{X} < K$, where $(X_1, \ldots, X_n)$ is a random sample of lifetimes of a certain type of fluorescent lamp. How large a sample is needed, and what critical boundary should be used, if it is decided that there should be only a 5 per cent chance of rejection when the mean life is $\mu = 2000$ and a 10 per cent chance of acceptance when $\mu = 1900$? Of course, the answer depends on the variability in lifetimes, and this must be known or estimated. Suppose that $\sigma = 200$ hours is given. The conditions are then

$$.05 = P(\bar{X} < K | \mu = 2000) = \Phi\left(\frac{K - 2000}{200/\sqrt{n}}\right)$$

$$.10 = P(\bar{X} > K | \mu = 1900) = 1 - \Phi\left(\frac{K - 1900}{200/\sqrt{n}}\right)$$

from which (using Table I) one obtains

$$K = 1900 + 1.28 \times \frac{200}{\sqrt{n}},$$

$$K = 2000 - 1.645 \times \frac{200}{\sqrt{n}}.$$

Subtracting yields

$$\sqrt{n} = \frac{2.925 \times 200}{100},$$

or an n of about 34. The corresponding value of K is 1944 hours. ◄

The power functions for large-sample tests involving the population mean μ can be easily determined (as in Example 11-f), when σ^2 is known, from the fact that the test statistic $Z = \sqrt{n}(\bar{X} - \mu_0)/\sigma$ is approximately normal with mean $\sqrt{n}(\mu - \mu_0)/\sigma$ and variance 1. The power function for a one-sided test ($Z >$ constant or $Z <$ constant) is a monotonic function of μ, as in Example 11-f, low on H_0 and high on H_A. In the case of a two-sided test, the power function is symmetrical, dipping to a minimum of α at $\mu = \mu_0$.

Example 11-h Consider the test with critical region $|\bar{X}| > 1$, appropriate for testing $\mu = 0$ against $\mu \neq 0$. Suppose that $n = 100$ and $\sigma = 5$. Then $\sigma_{\bar{X}} = 5/\sqrt{100}$, and

$$\pi(\mu) = P(|\bar{X}| > 1) = P(\bar{X} > 1) + P(\bar{X} < -1)$$

$$\doteq 1 - \Phi\left(\frac{1 - \mu}{1/2}\right) + \Phi\left(\frac{-1 - \mu}{1/2}\right).$$

The graph of this power function is shown in Figure 11-6. It has a minimum value $\alpha = .0456$ at $\mu = 0$. It shows, for example, that a mean of $\mu = 1$ is detected with probability about .5, and a mean of $\mu = 2$ with probability .977. ◄

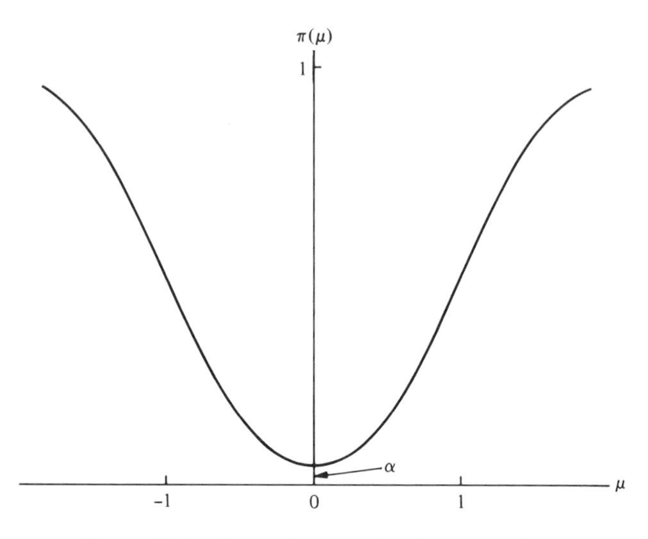

Figure 11-6 Power function for Example 11-h.

The power function for a t-test involves what are called "noncentral t-distributions." Tables and charts are available,† but will not be given here.

† Operating characteristic functions $[1 - \pi(\theta)]$ for this and various other tests in later chapters are given in Mary G. Natrella, *Experimental Statistics*, Handbook 91 (Washington, D.C.: U.S. Department of Commerce, National Bureau of Standards).

The signed-rank test, like most nonparametric tests and tests based on statistics whose *null* distribution do not depend on population shapes, are not easy to analyze in terms of power. The term power "function" assumes implicitly that the admissible models are indexed by a parameter, and power is then a function of that parameter. But when the alternatives are as general as they are when signed-rank tests are used, the power is not expressible as a function of anything simple. What is often done is to assess the power empirically, by sampling experiments, using various types of alternative distributions that might include or come close to those ordinarily encountered.

In summary, power is important but not always easy to study. In the tests that will be discussed in later chapters for problems involving parametric classes of models, there will be power functions; these usually involve extensive tables and/or graphs—needed by, available to, and easily used by those who make decisions. We do not present them in this text.

★ 11.7 Goodness of Fit

Testing a particular model against the very general alternative that the model is wrong is referred to as testing *goodness of fit.* The tests for $\mu = \mu_0$ considered earlier in the chapter did test "fit," in a sense, but the alternatives admitted were models of the same type as those in H_0, differing only in location. And the tests presented there were designed with such alternatives in mind and are powerful against those alternatives. A test will be presented here (for the continuous case, and in Section 15.1 for the discrete case), designed to pick up various kinds of deviations from a null model. Being more general in this sense, the test would not be as good as a t-test, say, when the alternatives are, specifically, shifts in location.

The test to be given here is one of several that have been proposed and used. It is based on the idea of a "sample distribution function," a statistic that is the sample version (and good large-sample estimate) of the population distribution function.

Given a sample $(X_1, \ldots, X_n)$, the *sample distribution function* is the cumulative distribution function of a mass distribution which assigns weight $1/n$ to each X_i. It increases in *jumps*, of amount $1/n$ at each sample value, rising from 0 to the left of smallest X_i to 1 at the largest X_i.

Example 11-i Consider a sample consisting of these five observations:

$$2.22, \quad -.83, \quad .18, \quad 1.18, \quad 2.05$$

[which were obtained from $\mathcal{N}(.5, 4)$—a secret to which you would not be privy in practice]. The sample distribution function is easily constructed after the sample values are marked on the x-axis: Starting at the far left at height 0, move the pencil to the right, changing the level to 1/5 unit higher as you pass each sample value. The result is shown in Figure 11-7. ◀

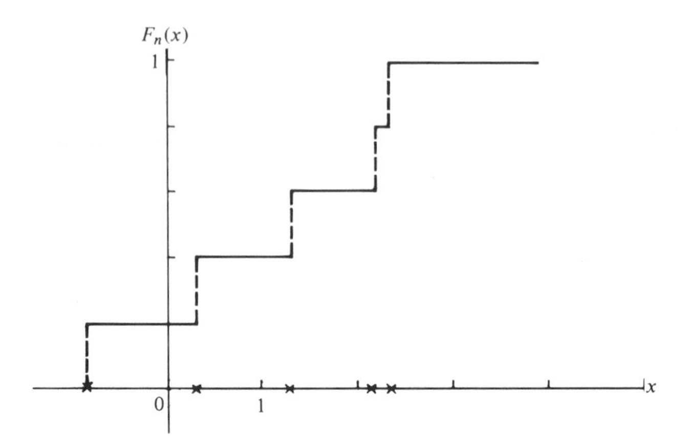

Figure 11-7 Sample distribution function for Example 11-j.

The Kolmogorov–Smirnov test statistic is defined as the maximum vertical absolute deviation of the sample d.f. from the particular population c.d.f. (cumulative distribution function) given by H_0. If the fit is good, this statistic will have a small value; and if the fit is poor, it will be large, so very large values are taken to be reason to reject H_0. The distribution of this statistic, fortuitously, depends only on the sample size—not on the shape of the distribution being tested. Percentiles are given in Table VI.

Kolmogorov–Smirnov statistic for goodness of fit:

$$D_n \equiv \sup_x |F_n(x) - F_0(x)|,$$

where $F_n(x)$ is the sample d.f. and $F_0(x)$ the c.d.f. being tested. Percentiles of D_n under H_0 are given in Table VI. Reject $F_0(x)$ if D_n is too large.

(The term *sup* here means supremum or *least upper bound*. It is equal to the maximum when there is a maximum. For example, the set of numbers in $0 < x < 1$ has no maximum, but the supremum is 1.)

Example 11-j The sample d.f. of the preceding example (with data 2.22, −.83, .18, 1.18, and 2.05) is repeated in Figure 11-8, which shows on the same axes the c.d.f. of $\mathcal{N}(0, 1)$, the standard normal distribution. The supremum D_n occurs at a sample value (as it always will), an amount $.88 - .40 = .48$ at the value 1.18, as shown in Figure 11-8. Reference to Table VI shows that this corresponds to a P-value of about 14 per cent; at $\alpha = .10$ one would accept H_0, even though (as we know, from Example 11-i) H_0 was not the model used to generate the data. A sample of only five observations has limited potential in detecting that H_0 is false. ◄

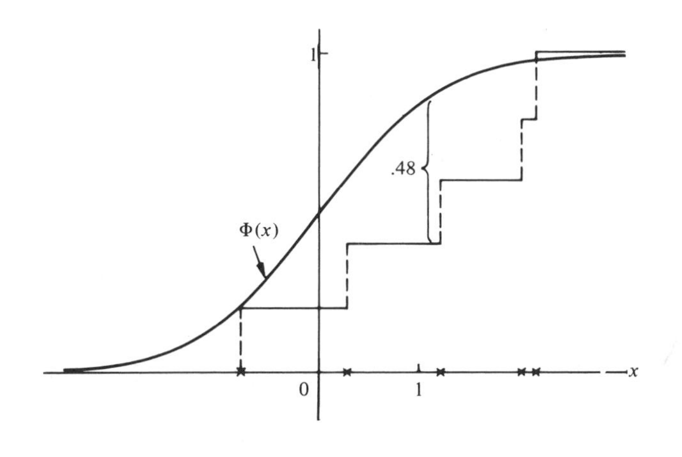

Figure 11-8

It is often of interest, in view of the assumption of normality underlying many test procedures, to test whether a population is normal—not that it is a particular normal population, but that it belongs to the normal family. This null hypothesis of normality is not specific enough to draw a c.d.f. for comparison with the sample d.f. However, a Kolmogorov–Smirnov type of statistic can be computed, if in place of a given $F_0(x)$ one uses $\tilde{F}_0(x)$, the normal c.d.f. defined in terms of the sample, by taking μ to be $\bar{X}$ and σ^2 to be S^2. This statistic, $\sup|F_n(x) - \tilde{F}_0(x)|$, will have a tendency to be small simply because the sample is being compared with an aspect of itself; so it might be expected that the critical value for a given α is smaller than it would be for the Kolmogorov–Smirnov statistic. Lillefors† has used artificial sampling experiments to determine percentiles of the modified statistic under the null hypothesis of normality:

n	$\alpha = .10$	$\alpha = .05$
5	.315	.337
6	.294	.319
7	.276	.300
8	.261	.285
9	.249	.271
10	.239	.258
11	.230	.249
12	.223	.242
13	.214	.234

n	$\alpha = .10$	$\alpha = .05$
14	.207	.227
15	.201	.220
16	.195	.213
17	.189	.206
18	.184	.200
19	.179	.195
20	.174	.190
25	.165	.173
30	.144	
$n > 30$	$.805/\sqrt{n}$	$.886/\sqrt{n}$

Example 11-k The five observations of the preceding examples have mean $\bar{X} = .96$ and standard deviation $S = 1.152$. Since the supremum occurs at a jump point, it suffices to give the ordinates at these points:

x	$\dfrac{x - .96}{1.152}$	$\tilde{F}_0(x) = \Phi(z)$	$F_n(x)$
−.83	−1.554	.060	.2
.18	−.677	.250	.4
1.18	.191	.576	.6
2.05	.946	.828	.8
2.22	1.094	.863	1.0

The supremum occurs at 2.05, namely, $.828 - .6 = .228$. [A sketch shows that $\tilde{F}_0(x)$ must be compared with the $F_n(x)$ at that and at the preceding x.] This value .228 is not as large as .315, so normality is not rejected, even at $\alpha = .10$. ◄

11.8 Statistical and Practical Significance

The phrase "statistically significant" has been defined to mean that some statistic has turned out to have a value which, if the null hypothesis is true, is out in the tail of its distribution. But the reader should be warned that there

† *Journal of the American Statistical Association*, **64**, 399–402 (1967).

is a common tendency to reinterpret this kind of significance as practical significance. The following examples illustrate this deplorable tendency.

1. From an evaluation of a teacher training program:

   "The training system had measurable and significant influences on the instructors' actual performances."

   "It was found that (a) student performances and (b) instructor ratings both increased significantly."

2. From a TV commercial for headache remedies:

   "Two studies at major hospitals on pain other than headache [sic!] showed Excedrin to be significantly more effective than aspirin."

3. From remarks by a professor of bacterial physiology at a meeting of the AAAS, regarding genetic potentials of races:

   ". . . there may be significant statistical differences in potentials between groups."

The first two of these quotations clearly imply not only that there is a treatment effect, but also that this effect is of practical significance. This is not what "statistically significant" means. The third quotation is using something like the language of statistical significance, presumably to lend authority to the remark, but it is not at all clear what is meant. The statement refers to groups or populations, whereas the phrase statistically significant refers to samples.

To illustrate the problems associated with the notion of statistical significance, consider testing the value $\mu = 100$ on the basis of the sample mean. Suppose that the sample mean is $\bar{X} = 101$. Then, depending on what size sample this was computed from, various conclusions might be drawn. The following table gives some pertinent quantities computed for various sample sizes and assuming $\sigma = 10$:

n	$\bar{X}$	σ	s.e.	Z	One-sided critical value for $\bar{X}$	Reject H_0?	P-value	95% C.I. (two-sided)
25	101	10	2	.50	103.29	No	.31	101 ± 4
100	101	10	1	1.00	101.645	No	.16	101 ± 2
400	101	10	.5	2.00	100.82	Yes	.023	101 ± 1
1600	101	10	.25	4.00	100.41	Yes	.000	101 ± .5

Now, suppose that μ is really 101. Then $\bar{X}$ will not necessarily be 101, but as n increases, $\bar{X}$ will tend toward 101. And at the same time the confidence-interval width will shrink toward to zero. So for a large enough n, the value

$\mu = 100$ can be rejected at any preassigned significance level. And the same thing would be true if μ were really 100.1, or 100.0001, and so on.

> Any actual departure from a specified model, *however slight,* and however practically insignificant, will be detected as "statistically *significant*" (by a good test) if the sample is large enough.

A confidence-interval estimate would then appear to have the advantage not only of telling us if $\mu = 100$ is to be rejected, but also of giving an idea or estimate as to what μ really is so that an assessment of practical significance can be made. For instance, if a confidence interval turns out to be $100.02 < \mu < 100.04$, the value $\mu = 100$ is rejected—because the sample is large; and the value 100.03 for μ, which is the center of the confidence interval, may or may not be different enough from 100 to be practically significant.

Example 11-l A zealous consumer weighs 900 "1-lb" bags of peanuts and finds a mean weight of 15.99 ounces and a standard deviation of .12 ounce. To test the hypothesis that $E(W) = 16$ ounces, he computes

$$Z = \frac{15.99 - 16}{.12/\sqrt{900}} = 2.5$$

and so concludes that the average weight "is significantly different" from the 16 ounces printed on the package. Should he bring suit for deceptive labeling?

A 95 per cent confidence interval for $E(W)$ would have limits

$$15.99 \pm 1.96\frac{.12}{\sqrt{900}},$$

or about 15.982 to 15.998. Even if the population mean were 15.98, the consumer might have a hard time convincing the court that this departure from 16 has any practical significance. ◄

Problems

1. Power consumption, although random, increases steadily over the years. After adjusting for this trend, a power company estimates that the average domestic use in a certain month will be 490 kilowatt-hours (kWh). To help in the present attempt to reduce power consumption, the company puts on an advertising campaign prior to the given month to try to influence customers to use less power. A sample of 400 bills for the month in question later shows an average consumption of 485 kWh, with a standard deviation of 60 kWh. Was the campaign successful? (Test the hypothesis that the mean consumption for the month is 490 kWh against the alternative that it is less than 490 kWh.)

2. A certain type of cable is supposed to have a mean breaking strength of at least 1500 lb.

(a) To test $\mu = 1500$ (against $\mu < 1500$), 50 sample cables are checked, with the results $\bar{X} = 1490$ lb and $S = 31$ lb. Should one accept $\mu = 1500$?

(b) Suppose that the specifications which were to be met were, instead, that there is to be at most a 1 per cent chance that a cable would break under a 1400-lb load. Is this specification met?

3. A filling machine is set so that the expected weight of a "5-lb" bag of flour is 5.1 lb. The standard deviation is assumed to be .05 lb.

(a) A sample of 25 bags has a mean weight of 5.08 lb. Should the machine be reset? Assume that there is to be at most a 5 per cent chance of stopping production for resetting when this is unwarranted.

(b) When the expected weight is 5.1, what is the chance of getting a bag that weighs less than the nominal weight of 5 lb?

(c) What is the probability that the needed adjustment will *not* be made if the expected weight is really 5.09 and the machine is shut down when $\bar{X} < 5.08$? (How serious would a deterioration to 5.09 be?—What would the chance then be of getting a bag weighing less than the nominal weight?)

4. A high school principal, who knows the national average in a certain achievement test to be 424, feels that his senior class of 144 students is rather special, having an average score of 434, with a standard deviation of 60.

(a) Is he justified in this?

(b) Is his class a random sample from the national population of students in that age group?

5. A certain production process is "in control" if a particular dimension is distributed with mean $\mu = 2.7$ and standard deviation $\sigma = .1$. A sample of five parts is checked and the average dimension is found to be $\bar{X} = 2.8$.

(a) Assuming that σ has not changed from its specified value, has the process mean shifted?

(b) If it is not assumed that σ is .1 but that the standard deviation of the five dimensions is $S = .12$, would you accept $\mu = 2.7$?

6. An extruding machine for polyethylene tubing is rated at 45 minutes per 6000 feet. On successive runs the following times were recorded (in minutes): 45, 93, 45, 50, 51, 49, 45, 48, 40, 45, 45, 65, 55, and 40. The 93 is suspiciously large (a possible outlier), and a check revealed that it was caused by a screen pack change during the run. Omit this observation and use the remaining ones to carry out a t-test of the hypothesis that the mean time is 45 minutes. Use a two-sided test and a 10 per cent significance level. (Is there anything about the data that might make a t-test questionable?)

7. The increase X in a subject's reaction time in milliseconds after he is given a certain drug is assumed to be approximately normal. The increases for 10 subjects are found to have mean $\bar{X} = 10.0$ and variance $S^2 = 112$.

(a) Give a 95 per cent confidence interval for μ.
(b) Test the hypothesis that the mean increase is 0 at $\alpha = .05$.

8. The following data were obtained to test $\mu = 0$:

$$-6, \quad 0, \quad 12, \quad 18, \quad -2, \quad 5, \quad 1, \quad 18, \quad 60.$$

Calculate the signed-rank statistic and the T-statistic that might be used, and discuss the appropriateness of each, in light of the assumptions underlying the corresponding tests, and the nature of the data.

9. Suppose the raw data for the 10 subjects in Problem 7 are as follows:

$$11, \quad -5, \quad 8, \quad 20, \quad 3, \quad 16, \quad 14, \quad 30, \quad -7, \quad 10.$$

(a) Use the signed-rank test of the hypothesis that $\mu = 0$ (μ is the median when the population is normal).
(b) Observe that there are two $-$'s (two observations below $\mu = 0$). The number of $-$'s is binomial with $p = 1/2$ if 0 is the median. Compute the P-value. (This is the probability of 2 or fewer or 8 or more $-$'s if the alternative is two-sided.)

10. Referring to the report in Problem 3, Chapter 5, that 54 of 100 drivers preferred car M, test the hypothesis that $p = 1/2$ by using a 90 per cent confidence interval for p. What is the α of your test?

11. The computation in Problem 3(c) is essentially one of power. Choosing a few convenient values of μ, calculate the corresponding power

$$\pi(\mu) = P(\bar{X} < 5.08|\mu)$$

and sketch the power function.

12. To distinguish between the means $\mu = .50$ and $\mu = .52$ in a certain population with 5 per cent accuracy (i.e., with $\alpha = \beta = .05$), what sample size is required if $\sigma = .1$?

13. A manufactured part is to fit other parts, and a certain critical dimension is 2 centimeters. Owing to variations in the production process, the actual dimension is random with a standard deviation $\sigma = .05$. Determine a sample size and critical region for testing $\mu = 2$ against $\mu \neq 2$ such that there is a 5 per cent chance of rejecting $\mu = 2$ when really $\mu = 2$, and a 10 per cent chance of accepting $\mu = 2$ when μ differs from 2 by .02 centimeter either way.

★ **14.** Are the data in Problem 8 from a normal population? (Estimate μ and σ^2 from the sample and use the Lillifors table.)

★ **15.** Testing goodness of fit of "random numbers" is one aspect of accepting them as such. (Besides being observations from the specified population, they should also exhibit independence, a matter we have not covered in testing.)

(a) Taking 10 successive observations from Table XIII, test the hypothesis that the population is uniform by using the Kolmogorov–Smirnov statistic.
(b) Taking 10 successive observations from Table XIV, test the hypothesis that the population is standard normal.
(c) Use the Lillifors table to test the observations in part (b) for normality. (One would expect it to be easier to satisfy "normality" than "*standard* normality." Is it?)
(d) A serious challenge of a specified distribution could of course be checked by taking a much longer record. What maximum departure from the null c.d.f. would call for rejection at $\alpha = .01$ if $n = 10{,}000$?

16. A basketball coach will choose 3 players by lot from 25 nonregular seniors to start the last game, along with two key regulars. The lottery yields players with heights 73, 74, and 78 inches.

(a) In sampling without replacement,

$$\sigma_{\bar{X}}^2 = \frac{N - n}{N - 1}\sigma^2/n.$$

Evaluate the finite population factor in this case to see that it is not much different from 1.
(b) If the given sample of three were used to test the hypothesis that the population mean height is $\mu = 72$ inches, would you say the observed deviation of the sample mean $\bar{X}$ from 72 is significant? (Use a t-statistic, ignoring the finite population difficulty, and assuming that t is robust enough to get you by.)
(c) Would the coach think that the observed difference is of any practical significance?

12

Comparing Locations

The comparison of two populations on the basis of incomplete information in samples is a common problem. Often, one sample is termed the *control* group, usually untreated or given a standard treatment. The other is then called the *treatment* group, and its members are given some experimental treatment. A measure of response or performance is obtained for each member of each sample, and these comprise the data upon which inference is to be based.

12.1 Tests Based on Means

The question of whether the treatment has an effect, or whether there is a difference in treatments, is usually interpreted in terms of average performance or response:

Null hypothesis—No treatment effect

$$H_0: \quad \mu_C = \mu_T,$$

where the subscripts C and T refer to the control and treatment populations. In terms of the parameter $\Delta\mu = \mu_T - \mu_C$, the null hypothesis asserts that $\Delta\mu = 0$.

Given samples from each population, treatment and control; the obvious estimate of $\Delta\mu$ is $\bar{X}_T - \bar{X}_C \equiv \Delta\bar{X}$, the difference of the sample means. Whether or not to reject H_0 depends on how far the sample mean difference $\Delta\bar{X}$ is from $\Delta\mu = 0$, but in order to interpret this distance properly, one should use a standardized version:

$$Z = \frac{\Delta\bar{X} - 0}{\sigma_{\Delta\bar{X}}}.$$

Now, if the samples are independent,

$$\text{var}(\Delta\bar{X}) = \text{var}\,\bar{X}_T + \text{var}\,\bar{X}_C = \frac{\sigma_T^2}{n_T} + \frac{\sigma_C^2}{n_C},$$

where n_C and n_T are the sample sizes. But when, as is usually the case, the population variances are unknown, sample variances S_T^2 and S_C^2 may be used instead, for large n's:

Statistic for testing $\mu_C = \mu_T$ (large, independent samples):

$$Z = \frac{\bar{X}_T - \bar{X}_C}{\sqrt{\dfrac{S_T^2}{n_T} + \dfrac{S_C^2}{n_C}}}.$$

Z is approximately standard normal if $\mu_C = \mu_T$.

Example 12-a Students in a certain economics course in the fall quarter were compared as to grade-point average (G.P.A.) with those in the same course in the winter quarter (as part of a study of possible inhomogeneities between the two groups, which were then to be used in a study of pedagogical techniques). The data are as follows:

	Number of students	Mean G.P.A.	S.D.
Fall	323	2.73	.51
Winter	438	2.77	.48

The Z-statistic for comparing means is

$$Z = \frac{2.77 - 2.73}{\sqrt{\frac{.51^2}{323} + \frac{.48^2}{438}}} \doteq 1.1.$$

This is not significant at $\alpha = .05$, since $|Z| < 1.96$. The hypothesis that fall- and winter-quarter students in the course have the same average G.P.A. was not rejected. ◀

The problem of comparing means of two normal populations with *small* samples is simpler when the population *variances are equal*. This will be assumed:

$$\sigma_T^2 = \sigma_C^2 \equiv \sigma^2.$$

The common value σ^2 could be estimated from either sample, but a better estimator is obtained by *pooling* the sample variances, with weights proportional to sample sizes:

> Pooled variance:
>
> $$S^2 \equiv \frac{n_C S_C^2 + n_T S_T^2}{n_C + n_T - 2}.$$

The variance of $\bar{X}_T - \bar{X}_C$ is then estimated in terms of S^2 as follows:

$$\sigma_{\bar{X}_T - \bar{X}_C}^2 = \frac{\sigma^2}{n_C} + \frac{\sigma^2}{n_T} \doteq S^2\left(\frac{1}{n_C} + \frac{1}{n_T}\right),$$

and the test statistic used has a $t(n_C + n_T - 2)$ distribution:

> Statistic for testing $\mu_C = \mu_T$ (small samples, normal populations, equal variances):
>
> $$T = \frac{\bar{X}_T - \bar{X}_C}{S\sqrt{\frac{1}{n_C} + \frac{1}{n_T}}}, \qquad \text{where } S^2 = \text{pooled variance}.$$
>
> The null distribution of T is $t(n_C + n_T - 2)$.

When the populations are normal but the variances are *unequal*, the t-test just given is not appropriate. In this case (called the Behrens–Fisher problem) there seems to be no best procedure. One procedure is to use the Z-score given for the large-sample case, and to define a critical region for given α by using the t-distribution with a number of degrees of freedom equal to the minimum of $n_C - 1$ and $n_T - 1$. The size of the type I error, although not uniquely defined (depending on σ_C^2/σ_T^2), will not exceed α.

12.2 Confidence Interval for $\Delta\mu$

Reporting that the null hypothesis of no difference in means is rejected or accepted is a very crude summary of results. It may be desirable to report, instead, an interval estimate of the true difference, $\mu_T - \mu_C \equiv \Delta\mu$.

A large-sample confidence interval for $\Delta\mu$ is based on the estimator $\bar{X}_T - \bar{X}_C \equiv \Delta\bar{X}$ and, as in the case of estimating a mean, extends a multiple of the standard error on either side. The multiple depends on the confidence coefficient and is obtained from Table Ic as in the one-sample case.

Large-sample confidence limits for the difference of two means:

$$\Delta\bar{X} \pm k\sqrt{\frac{S_T^2}{n_T} + \frac{S_C^2}{n_C}},$$

where k is a percentile of the standard normal distribution (Table Ic in Appendix B), and $\Delta\bar{X}$ is the difference in sample means.

Example 12-b Consider again the situation of Example 12-a. The mean difference in G.P.A.'s of the students in the fall and winter quarters was $\Delta\bar{X} = .04$, and the standard error

$$\text{s.e.} = \sqrt{\frac{.51^2}{323} + \frac{.48^2}{438}} = .037.$$

A 95 per cent confidence interval for the population difference is then

$$.04 - 1.96 \times .037 < \Delta\mu < .04 + 1.96 \times .037,$$

or $-.033$ to $+.113$. The fact that this interval *includes* $\Delta\mu = 0$ corresponds to the test result (in Example 12-a) that the observed $\Delta\bar{X}$ is not statistically significant at $\alpha = .05$. That is, one can account for the observed $\Delta\bar{X}$ in terms of sampling fluctuation, even if the populations involved are identical. (Whether a difference of .04, if that were the true value of $\Delta\mu$, is of practical significance is another matter.) ◄

A small-sample confidence interval for the difference $\Delta\mu$ in means of two normal populations with equal variances can be constructed using the fact that the quantity

$$T = \frac{\Delta\bar{X}}{S\sqrt{\frac{1}{n_C} + \frac{1}{n_T}}}$$

has a $t(n_T + n_C - 2)$ distribution.

Small-sample confidence limits for $\Delta\mu$ (normal populations with $\sigma_T^2 = \sigma_C^2$):

$$\Delta\bar{X} \pm t_\alpha S\sqrt{\frac{1}{n_C} + \frac{1}{n_T}},$$

where S^2 is the pooled variance and t_α is the 100α percentile of the $t(n_C + n_T - 2)$ distribution. The confidence coefficient is $2\alpha - 1$.

12.3 A Two-Sample Rank Test

A simple statistic based on ranks can be constructed for the problem of comparing locations, to test the null hypothesis of no difference. In terms of the names X and Y for the population variables to be compared, the null hypothesis is this:

H_0: X and Y have the same distribution.

The kind of alternative against which the test to be presented affords protection is a "shift" alternative—the hypothesis that the distribution of one variable is shifted in location, to one side or the other of the second variable.

The test statistic is to be based on independent random samples from the two populations. To provide an effective test, it should tend to be extreme if there *is* a difference in location, and not extreme if H_0 is true. Such a statistic is the *Wilcoxon rank-sum* statistic, defined as the sum of the X-ranks when the samples are combined and ordered according to numerical value. For instance, if there are 4 X's and 6 Y's, these 10 observations, when arranged in numerical order, might yield the pattern

$$X \quad X \quad Y \quad X \quad Y \quad X \quad Y \quad Y \quad Y \quad Y.$$

That is, the smallest of the 10 observations is an X, the next smallest also an X, the third smallest a Y, and so on. The X-ranks are 1, 2, 4, and 6, with sum 13; and this small a value suggests that perhaps the X-distribution is really to the left of the Y-distribution. (The sum $1 + 2 + 3 + 4 = 10$ is the smallest *possible* value.)

The distribution of the sum of the X-ranks under the null hypothesis of no difference is *independent of the nature of the common distribution.* Tail probabilities for various combinations of small-sample sizes can therefore be given in one table (Table VIII of Appendix B).

> Wilcoxon two-sample rank-sum statistic:
>
> $R_X \equiv$ sum of X-ranks in the ordered sequence of X's and Y's from independent samples combined into one. Extremely large or small values are significant (Table VIII), depending on the alternative.

Table VIII assumes continuous populations, with zero probability of a tie when ordering the combined sample. As in the case of the one-sample rank test of Section 12.5, tied observations can occur in practice and are assigned their average rank.

Example 12-c Five male and five female drivers drove a car over a given course, with gasoline consumption figures as follows:

Male (X)	Female (Y)
0.94	1.40
1.20	0.98
1.00	1.22
1.06	1.16
1.02	1.34

The combined sample, in numerical order and written with the X's in italics, is

0.94, 0.98, *1.00*, *1.02*, *1.06*, 1.16, *1.20*, 1.22, 1.34, 1.40

and the corresponding pattern of X's and Y's:

$$X, \quad Y, \quad X, \quad X, \quad X, \quad Y, \quad X, \quad Y, \quad Y, \quad Y.$$

The X-ranks are 1, 3, 4, 5, 7 with sum $R_X = 20$. From Table VIII it is seen that, if there is no difference between males and females, $P(R_X \leq 20) = .075$. Thus $R_X = 20$ would reject the hypothesis of no difference in a one-sided test at $\alpha = .10$ but accept it at $\alpha = .05$. (The P-value is .075.)

The reader who gives any thought to the experiment described here may well wonder how the drivers were chosen. A basic assumption of the test is that the observations constitute random samples from the populations to be compared. If the 10 drivers in the test happened to be the workers in an office among whom an argument had arisen, to be settled by the test, then perhaps the only tenable conclusion—and a nonstatistical one at that—might be that the women in that office use more gasoline than the men in that office. To extend the conclusion of statistical significance to any larger population, one would have to pick drivers at random from that population. ◄

Table VIII gives rejection limits for sample sizes up to $m = n = 10$. It stops there because the large-sample distribution provides an adequate approximation for $n > 10$. It can be shown that the test statistic R_X is approximately normally distributed, with mean

$$E(R_X) = \frac{1}{2}n_X(n_X + n_Y + 1)$$

and variance

$$\text{var } R_X = \frac{1}{12}n_X n_Y(n_X + n_Y + 1),$$

where n_X and n_Y are the numbers of X's and Y's, respectively.

> Large-sample rank statistic for comparing locations:
>
> $$Z = \frac{R_X + 1/2 - n_X(n_X + n_Y + 1)/2}{\sqrt{n_X n_Y(n_X + n_Y + 1)/12}}.$$
>
> Z is approximately standard normal under H_0 (n_X, n_Y large).

(The 1/2 used in defining the Z-score is a "continuity correction" that improves the approximation when the sample sizes are only moderate.)

Example 12-d   Final composite scores for 42 students in an elementary statistics course are given in Table 12.1, according to sex, in numerical order. Also given are the ranks, and the ranks of the F-scores.

The sum of the F-ranks is 399, with corresponding Z-score

$$Z = \frac{399 + 1/2 - 18 \times 42/2}{\sqrt{\dfrac{18 \times 23 \times 42}{12}}} = .55.$$

Table 12.1

M	F	Rank	F rank sum
92		1	
	87	2	2
85, 85	85	3, 4, 5	4
84		6	
83		7	
82, 82	82, 82, 82	8, 9, 10, 11, 12	30
	81	13	13
80		14	
79		15	
77		16	
	73	17	17
72	72	18, 19	18.5
71		20	
70		21	
	68	22	22
66	66	23, 24	23.5
65		25	
63		26	
61		27	
	60	28	28
	59, 59	29, 30	59
57	57	31, 32	31.5
56, 56	56	33, 34, 35	34
55		36	
	54	37	37
51		38	
	46	39	39
42	42	40, 41	40.5

This slightly higher than average Z is not usually thought of as significant. ("Higher" here would mean, since ranks were arbitrarily numbered from the top down, that the F-scores are lower than the M-scores.) On the basis of this sample that there is no reason to reject the null hypothesis that male and female students do equally well, among students taking that particular course at that school in that time of year. ◄

12.4 Paired Data

Data in the form of n pairs $(X_1, Y_1), \ldots, (X_n, Y_n)$ will often arise in such a way that the X's and Y's are not independent. For instance, if X_i is the "before" measurement and Y_i the "after" measurement on the ith individual in a random sample of individuals, there is a natural pairing of X_i with Y_i, and these values will usually be correlated. Similarly, the dimensions of a person's left hand and right hand are related, and naturally paired, by virtue of their common source and support.

If the individuals which provide the pairing are randomly drawn from a population, then the X's constitute a random sample and the Y's constitute a random sample; but these samples are not necessarily independent, and so $\bar{X}$ and $\bar{Y}$ are not apt to be independent. The variance of $\bar{X} - \bar{Y}$ could only be given in terms of σ_X^2 and σ_Y^2 if the correlation between X_i and Y_i were also known. On the other hand, the differences $d_i \equiv X_i - Y_i$ themselves constitute a random sample from a population of differences. Moreover,

$$\bar{d} = \bar{X} - \bar{Y},$$

so that the mean of the sample of differences ought to have something to say about the difference in population means. Indeed,

$$E(\bar{d}) = E(\bar{X}) - E(\bar{Y}) = \mu_X - \mu_Y,$$

so if $\Delta\mu = \mu_X - \mu_Y = 0$, the value of $\bar{d}$ is distributed around 0. The standardized deviation of $\bar{d}$ about 0, or "Z-score,"

$$Z = \frac{\bar{d} - 0}{\sigma_{\bar{d}}},$$

would be approximately standard normal for large n. If $\sigma_{\bar{d}} = \sigma_d/\sqrt{n}$ is not known, it can be approximated by $S_d/\sqrt{n}$, where S_d is the standard deviation of the sample of differences.

> One sample Z-statistic for $\Delta\mu = 0$ (large n):
>
> $$Z \equiv \frac{\bar{d}}{S_d/\sqrt{n}}.$$
>
> Z is approximately standard normal if $\Delta\mu = 0$.

Whether a two-sided or one-sided critical region is used depends on the alternative. Thus the alternative $\Delta\mu \neq 0$ would call for a two-sided critical region $|Z| > k$.

In the case of small samples, the Z-test given above is not appropriate and should be replaced—by a t-test if the difference d can be assumed normal, otherwise by a rank test.

Example 12-e Experimentation suggests that it may be possible for a person to control certain body phenomena previously thought to be beyond his control if he is trained to do so in a program of "biofeedback" exercises. The following data and blood-pressure measurements (in millimeters of mercury) before and after the training or conditioning, for each of six subjects:

Subject	Before	After	Difference
1	136.9	130.2	−6.7
2	201.4	180.7	−20.7
3	166.8	149.6	−17.2
4	150.0	153.2	3.2
5	173.2	162.6	−10.6
6	169.3	160.1	−9.2

The mean and standard deviation are −10.2 and 7.662, so that

$$T = \sqrt{n-1}\,\frac{\bar{d}}{S_d} = -2.98.$$

This suggests $[t_{.05}(5) = -2.01]$ a real reduction in blood pressure achieved by self-control as learned in the training program.

It should be observed that the variation in blood pressure reading from person to person, either before or after, is considerable, and the difference in means of two independent samples would have much greater variability if the X's and Y's were not paired.

The sequence of differences *ordered by magnitude* is

$$(3.2, -6.7, -9.2, -10.6, -17.2, -20.7),$$

the only positive term having rank 1. The signed rank statistic is $T = 1$, so at $\alpha = .05$ one would reject the hypothesis of no difference (see Table VII, Appendix B) and infer that the special biofeedback training is effective. ◄

Paired data are sometimes obtained by design. For example, to study the effect of a new diet on pig growth one could select a random sample of young pigs for the "treatment" group and another random sample for a "control" group. The treatment group would get the new diet and the control group a standard diet. However, the variability in the difference of mean weight gains can be reduced significantly by choosing, instead, a pair of pigs from each of n litters, randomly assigning one of the pair to the new diet and one to the standard diet. Pigs in the same litter tend to be more nearly alike in their response to diet than are pigs generally, and this means that the difference between the weight gains in a pair is less variable than the difference between weight gains of two randomly chosen pigs. Even when there is nothing like a litter to use in pairing, it can often be effectively done by selecting pairs of subjects so that the two in a pair are nearly alike in many of the ways that may contribute to variability of response.

Example 12-f Two types of construction of automobile tires are to be compared as to wear. Randomly selected drivers would drive in widely varying ways and environments, so that data for n tires of each type obtained from their use by $2n$ randomly chosen drivers would have an unnecessarily large amount of variation. This could be reduced by pairing. Suppose that n drivers are asked to use one tire of each type, one on the left front wheel and one on the right front wheel. To avoid a possible difference between left and right, the tires should be assigned to left and right at random, or alternatively. After a given number of miles, the amount of rubber worn away can be measured for each tire. The differences in wear between tire brands on the individual cars would constitute a (single) random sample $(d_1, \ldots, d_n)$ whose mean $\bar{d}$ is an estimate of the true or population mean difference and could be used in constructing a (one-sample) test statistic. ◄

12.5 More on Significance

Example 12-g In a compilation of seven studies on fluoride toothpaste the Colgate Company reported results as follows:

	Mean number of cavities per year	
Study	Colgate with MFP	Control
I	1.50	2.09
II	6.66	7.46
III	1.53	1.86
IV	1.29	1.97
V	2.12	2.68
VI	4.35	5.36
VII	4.88	6.11

No information as to variability was presented, the presumption being, perhaps, that the consumer would not know how to use it—and could see from the means given that Colgate's is superior, without any fancy statistical test. The original sources, in dental journals, did give the standard deviations, and using them one could carry out a Z-test in each case to show statistically significant results. That is, it would be concluded that "there is a treatment effect."

The practical significance of the report, however, seems to be something entirely different. Surely one would wonder about the great discrepancy between averages on the order of 2 per year and 6 per year. Reading more closely one finds that the various studies were carried out under different conditions. The very low averages were achieved when there was supervised brushing two or three times daily (in an institution), and the high averages when brushing had little or no supervision. Moreover, there were geographical difference—and differences in water supply and diet. (In II, VI, and VII, brushing was unsupervised at home. In II the water supply had fluoride. In I, IV, and V, meals and supervised brushing were institutional.) So the most practically significant thing about the report, although fluoride in the toothpaste might be of some value, is that other factors seem to be much more important! ◄

Example 12-h In the study referred to in Example 12-a, it was concluded that the difference in average G.P.A. between the fall- and winter-quarter classes was not significant, the t-statistic having the value 1.09. This means, simply, that so far as the average G.P.A.'s are concerned, the samples could have been drawn from the same population. However, the study states that "the groups were considered adequately matched for purposes of the evaluation study." This is another matter. The study was not going to draw further samples from the *populations*, accepted as equal; it was going to use the samples already *known* to differ, and by how much (2.73 versus 2.77). Indeed, the study later indicates that, in a more sophisticated regression analysis, achievement in G.P.A. is significantly related to achievement in economic understanding, which was the variable of primary interest. ◄

Example 12-i An elementary statistic textbook gives the following data on gasoline tax rates (cents per gallon):

New England		Mideast		Far West	
Maine	8	N.Y.	7	Wash.	9
N.H.	7	N.J.	7	Ore.	7
Vt.	8	Pa.	8	Nev.	6
Mass.	6.5	Del.	7	Calif.	7
R.I.	8	Md.	7	Ala.	8
Conn.	8	D.C.	7	Hawaii	5

The reader is then asked to test for equality of means (using a method of Chapter 17, an extension of the t-test for two samples). An examination of the data shows (1) that the samples constitute the whole populations, and (2) the means *are* different—there is no question of statistical significance. Still, one might wonder if the differences among the three populations are meaningful; but this is a political or social question, not subject to statistical testing. ◄

Problems

1. A study of dentifrices [article by Frankl and Alman, *Journal of Oral Therapeutics and Pharmacology*, **4**, 443–449 (1968)] showed 19.98 new cavities per child over a three-year period in a sample of 208 children in a treatment group, and 22.39 per child in a sample of 201 children in a control group. The standard deviations are given as 10.61 and 11.96, respectively. Test the hypothesis that the dentifrice used by the treatment group is no better than that used by the control group. (Use $\alpha = .01$.)

2. Suppose that a new drug has been discovered in the treatment of leukemia. Its proponents suggest that if taken when the disease reaches a certain stage, it will increase the patient's expected life. Half of 200 patients are given the drug and half are not. The average number of months of life in the control group is 20.4, with a standard deviation of 6.0, while for the treated group these statistics are 23.5 and 15.0. Is there a significant difference? What aspect of the drug's performance do you think deserves further investigation? (Suppose that you have leukemia and want to be sure to live 12 months; would you want to be treated?)

3. A flock of turkeys was split into two groups of 50 each, and different rations were fed to the two groups for a given period of time. At the end of this period, weight gains were recorded and these statistics calculated:

$$\bar{X}_1 = 15.2, \quad S_1 = 2.2; \qquad \bar{X}_2 = 14.8, \quad S_2 = 2.0.$$

Test the hypothesis of no difference in rations (as they affect weight gains) at $\alpha = .10$.

4. Independent random samples are obtained as follows:

Control	12.9	12.4	14.7	13.8	14.2	15.4	11.8	13.3
Treatment	13.0	15.1	13.7	16.6	15.2	12.3	14.8	15.8

Test the hypothesis of no difference between treatment and control populations, using $\alpha = .10$, against the (two-sided) alternative that they differ in location, in the following ways:

(a) Using a t-test, given

$$\bar{X} = 13.56, \quad \sum (X_i - \bar{X})^2 = 10.10, \quad \bar{Y} = 14.56, \quad \sum (Y_i - \bar{Y})^2 = 14.74.$$

(b) Using a rank test.

5. Two fly sprays are tested by successive applications of each to batches of flies. The results, in per cent mortality, are as follows:

Spray A	68	68	59	72	64	67	70	74
Spray B	60	67	61	62	67	63	56	58

(a) Carry out the rank-sum test of the hypothesis that there is no difference between the sprays, at the 10 per cent significance level.
(b) Use the t-test for the same problem. (What must be assumed for this to be valid?)

6. Two samples of 10 subjects, one sober and one treated with two martinis each, were administered a psychological test involving some reasoning and some motor facility. Scores were as follows (on a scale 0 to 20):

Sober	16	13	13	17	11	16	14	16	17	12
Treated	16	16	12	8	10	14	12	9	12	13

(a) Use a t-test for the null hypothesis of no treatment effect.
(b) Use the Wilcoxon test for the null hypothesis. (In the case of ties, assign the average rank.)

7. Consider two independent samples, of size $m = 3$ from X and of size $n = 4$ from Y.

(a) Calculate R_x given that all the X's turn out to be to the left of all the Y's.
(b) Calculate R_x given that all the X's turn out to be to the right of all the Y's.
(c) Determine the probability that if X and Y have the same continuous distribution, one or the other of the extreme values in parts (a) and (b) will occur.
(d) Suppose, to test the null hypothesis that X and Y have the same distribution, one uses the critical region $R_x = 6$ or 18. What is the significance level?
(e) Suppose that X is *really* shifted to the left of Y. How would the probability of the event $\{XXXYYYY$ or $YYYYXXX\}$ compare with the probability you found in part (b)?

8. A newspaper report of a study of the effect of smoking on infant size included these data: There were 88 matched pairs of mothers in the study (one smoker and one nonsmoker in each pair), and the mean difference in birth weight was 1/2 lb. (The nonsmokers had heavier babies.)

(a) What additional information would be needed to be able to draw a conclusion as to the significance of the 1/2-lb difference?
(b) The original data would provide an estimate of σ_{X-Y} needed to measure the significance of an observed $\bar{X} - \bar{Y}$. Even without the data, perhaps something can be learned: The X and Y are not independent, and in this case σ_{X-Y} would tend to be *less* than $\sqrt{\sigma_X^2 + \sigma_Y^2}$, which would apply if they were independent. By using $S_X = 1$ from other birth-weight data to estimate σ_X and σ_Y, it would

follow that σ_{X-Y} would here not exceed $\sqrt{1^2 + 1^2} = 1.414$. Use this as σ_{X-Y} in judging the possible significance of $\bar{X} - \bar{Y} = .5$.
(c) Discuss the factors that might be taken into account when matching.

9. Each child among six pairs of twins in a children's home received a cup of milk at 9:00 o'clock, and one of the twins in each pair received in addition 1 tablespoon of honey dissolved in the milk. After six weeks, increases in hemoglobin were recorded:

Pair	Given honey	Not given honey
1	19	14
2	12	8
3	9	4
4	17	4
5	24	11
6	22	15

(The increase in each case was expected since they had all been in poor health before admission to the home.) Use an appropriate test based on means for the null hypothesis of no treatment effect against the expected alternative, that the honey increases the hemoglobin level.

10. Consider again testing for the effectiveness of honey in the diet on hemoglobin, as in Problem 9.

(a) Apply the signed-rank test, giving your result as a *P*-value.
(b) If there were no treatment effect, the differences between twins in a pair would be as likely positive as negative. In a test based on the number of positive differences, give the *P*-value for the given data.

11. Female killdeer lay four eggs each spring. A scientist claims that the egg that hatches first yields a larger bird than the one that hatches last. To test his claim, he weighs the oldest and youngest of eight families, with the following results:

Family	Oldest	Youngest
1	2.92	2.90
2	3.58	3.68
3	3.39	3.33
4	3.29	3.06
5	3.44	3.30
6	3.13	2.99
7	3.22	3.26
8	3.80	3.51

Test the scientist's claim in two ways:

(a) Using a t-test.
(b) Using a rank test.

13

Bayesian Inference

In the classical testing of hypotheses, it is disconcerting to realize that in choosing a critical region, there is no rational way to strike a balance between the size of α (the significance level) and the power of the test. The common procedures of setting α at an "acceptable" level and achieving as much power as possible with that α stems from the notion that the type I error is the more serious. And this is a matter of considering the *consequences* or costs of making one kind of error or the other. Taking these costs into account, when inferences are actually decisions, is one of the elements of Bayesian inference.

The proper interpretation of a given significance level is not at all clear, as seen in this remark of R. A. Fisher: "The calculation is based solely on a hypothesis, which, in the light of the evidence, is often not believed to be true at all, so that the actual probability of erroneous decision . . . may be . . . much less than the frequency specifying the level of significance."† The idea that one may never or seldom encounter an H_0 situation, the only setting in which α has meaning, suggests a consideration of the chances that one will encounter H_0 and the various hypotheses making up H_A. The Bayesian decision maker—the Bayesian statistician—regards the true model or "state

† *Statistical Methods and Scientific Inference*, 3rd ed. (New York: Hafner Press, 1973), p. 45.

of nature" as a random variable. That is, he represents his uncertainties and beliefs about nature by probabilities. Such representation is the major element of Bayesian inference.

In dealing with statistical problems, those involving the use of data to aid in making decisions or inferences, one finds a need for a probability model to represent his state of uncertainty about nature *prior* to the gathering and reduction of data, as well as a probability model for his view of nature *after* the data are at hand and digested. The source of the term "Bayesian" is in the use of Bayes' theorem to obtain the latter or *posterior* model from the former or prior model and the data.

13.1 Nature as a Random Variable

The class of probability models or states of nature admitted for consideration in a given problem is often indexed by a parameter θ, and this is the type of situation to be considered here†. Thus the datum X [which may actually be a vector $(X_1, \ldots, X_n)$ of sample observations] will be assumed to have a distribution defined by a parametric density function or probability function:

$$f(x; \theta) = \begin{cases} \text{density at } x \text{ for given } \theta, & \textit{or} \\ \text{probability of } x \text{ for given } \theta. \end{cases}$$

Now, if the state or model is to be considered as random, it will be denoted by Θ, and possible values by θ. The function $f(x; \theta)$ becomes a *conditional* density or probability, given $\Theta = \theta$, denoted by $f(x|\Theta = \theta)$ or, more simply, $f(x|\theta)$.

"Probability" here is *subjective* in that it represents personal beliefs, so it can vary from one individual to the next. (When one makes a wager, it is often because his probability of the event in question is different from the other person's.) Until this point, a probability has been thought of as *objective*, associated with the "object" in an experiment (even though an observer may never know its value). When an observer (or "subject") assigns a probability to an event based on the information at his disposal, that probability is subjective. For example, if a coin that is to be tossed is known to be either two-headed or two-tailed, an observer may feel that the coin is equally likely to be either and say, quite reasonably, "The probability

† The very simple case of testing one specific model against another (Section 13.4) involves a "parameter" with only two possible values.

of heads is one-half," whereas the objective probability can only be 0 or 1. Interpretations may differ, but all probabilities have the same mathematical structure.

13.2 From Prior to Posterior Distribution

The aim of statistical inference is to draw conclusions about Θ from a given observation $X = x$, but the function $f(x|\theta)$ gives probabilities for X given the model $\Theta = \theta$. The desired reversal of roles of X and Θ is provided by Bayes' theorem, which, when X and Θ are discrete, says that

$$P(\Theta = \theta_j|X = x_i) = \frac{P(X = x_i|\Theta = \theta_j)P(\Theta = \theta_j)}{\sum_k P(X = x_i|\Theta = \theta_k)P(\Theta = \theta_k)}.$$

In terms of the notation $h(\cdot)$ for the conditional probability function of Θ and $g(\cdot)$ for the probability function of Θ:

Bayes' theorem:

$$h(\theta_j|x_i) = \frac{f(x_i|\theta_j)g(\theta_j)}{\sum_k f(x_i|\theta_k)g(\theta_k)}.$$

The unconditional probabilities $g(\theta_j)$, that is, probabilities for Θ with no knowledge of X (or before X is observed) define the *prior distribution* of Θ. The distribution for Θ that is appropriate after observing $X = x$ is defined by $h(\theta_j|x_i)$ and is called the *posterior distribution* of Θ. The conditional probability of $X = x$ given $\Theta = \theta$ is called, as a function of θ, *the likelihood function*: $L(\theta) = f(x|\theta)$.

In these terms, Bayes' theorem says the following:

$$\text{posterior} = \frac{(\text{likelihood})(\text{prior})}{\text{constant}}$$

where the constant is simply the divisor that is needed (for given x_i), so that the posterior probabilities $h(\theta_j|x_i)$ add up to 1, namely, the unconditional probability $P(X = x_i)$.

If X is continuous, then $f(x|\theta)$ is to be interpreted as a density function; and if Θ is a continuous parameter, $g(\theta)$ and $h(\theta|x)$ are density functions. The basic relation giving posterior in terms of prior and likelihood is valid for both cases.

The case in which there are two possible values of Θ is of special interest. Then the posterior probabilities are

$$h(\theta_1|x) = \frac{L(\theta_1)g(\theta_1)}{L(\theta_1)g(\theta_1) + L(\theta_2)g(\theta_2)}, \quad h(\theta_2|x) = \frac{L(\theta_2)g(\theta_2)}{L(\theta_1)g(\theta_1) + L(\theta_2)g(\theta_2)}.$$

In the language of "odds," the posterior odds ratio is the product of the prior odds ratio and the *likelihood ratio*:

$$\frac{h(\theta_1|x)}{h(\theta_2|x)} = \frac{g(\theta_1)}{g(\theta_2)} \frac{L(\theta_1)}{L(\theta_2)}.$$

Example 13-a I have two pair of dice; one pair is of the usual type, but the other is specially made. Of this special pair, one die has 5's on all faces, and the other has four 2's and two 6's. I select, using a method not completely known to you, one of the pairs to be tossed some number of times. Suppose p is the probability that the sum of the faces on the dice is 7, so that p is either 1/6 or 4/6. The probability, for you, that $p = 1/6$ is $g(1/6)$ and that $p = 4/6$ is $g(4/6) = 1 - g(1/6)$.

When the dice are rolled, if ever a sum other than 7 or 11 obtains, it is evident (and the application of Bayes' theorem is trivial) that the new (posterior) probability that $p = 1/6$ is 1. Suppose that in five tosses each results in a 7; then

$$\frac{h(1/6|\text{five 7's})}{h(4/6|\text{five 7's})} = \frac{g(1/6)}{g(4/6)} \frac{(1/6)^5}{(4/6)^5} = \frac{g(1/6)}{g(4/6)}\left(\frac{1}{4}\right)^5 \doteq (.001)\frac{g(1/6)}{g(4/6)}.$$

If the original choice of a pair of dice is thought to be random, then $h(4/6|\text{five 7's}) \doteq .999$, but the probability that the special dice were selected has been significantly increased no matter what $g(4/6)$ is. ◀

Example 13-b Suppose that $\bar{X}$ is the mean of a random sample, obtained for inference as to whether the population mean μ has the value $\mu = 0$ or the value $\mu = 1$. That is, the problem is one of testing

$$H_0\colon \ \mu = 0 \qquad \text{versus} \qquad H_1\colon \ \mu = 1.$$

The sample mean is normal if the population is normal, or at least approximately normal if the sample is large. The likelihood function based on $\bar{X}$ is then

$$L(\mu) = \left(\frac{2\pi}{n}\right)^{-n/2} \exp\left[-\frac{n}{2}(\bar{X} - \mu)^2\right],$$

if the variance is assumed known, and for simplicity taken equal to 1. In particular, for the candidate values of μ:

$$H_0: \quad L(0) = \left(\frac{2\pi}{n}\right)^{-n/2} \exp\left(-\frac{n}{2}\bar{X}^2\right),$$

$$H_1: \quad L(1) = \left(\frac{2\pi}{n}\right)^{-n/2} \exp\left[-\frac{n}{2}(\bar{X}-1)^2\right].$$

Suppose that the prior probabilities are $g_0 = P(H_0)$ and $g_1 = P(H_1)$. Then, given $\bar{X}$, the posterior probabilities are

$$h_0 = \frac{g_0 L(0)}{g_0 L(0) + g_1 L(1)}, \qquad h_1 = \frac{g_1 L(1)}{g_0 L(0) + g_1 L(1)} = 1 - h_0.$$

The posterior odds ratio is

$$\frac{h_0}{h_1} = \frac{g_0}{g_1}\frac{L(0)}{L(1)} = \frac{g_0}{g_1}\exp\left[-n\left(\bar{X} - \frac{1}{2}\right)\right].$$

Notice that if $\bar{X} = 1/2$, then $L(0) = L(1)$, and the posterior odds are the same as the prior. Thus a sample mean halfway between $\mu = 0$ and $\mu = 1$ supports both values equally, so one's prior inclination is unchanged. On the other hand, if $\bar{X} = .6$ in a sample of size 25, then

$$\frac{h_0}{h_1} = \frac{g_0}{g_1}e^{-25\times .1} \doteq \frac{1}{8}\frac{g_0}{g_1}.$$

So, for instance, equal prior odds on $\mu = 0$ and $\mu = 1$ would become odds of 1 to 8 against $\mu = 0$, when modified by the 25 observations with mean .6. ◄

Example 13-c Let Y denote the number of boys in a family of n children, with $P(\text{boy}) \equiv p$; the likelihood function is given by the binomial formula:

$$L(p) = f(y|p) = \binom{n}{y} p^y (1-p)^{n-y}.$$

The probability p varies from couple to couple, and any value of p on (0, 1) is possible. Sometimes, openmindedness about p is represented by a *uniform prior* with density

$$g(p) = 1, \qquad 0 < p < 1.$$

With this uniform prior, one obtains the posterior

$$h(p|y) = (\text{constant})p^y(1-p)^{n-y}, \qquad 0 < p < 1.$$

Then if, for instance, all n children are boys, or $Y = n$, the posterior density is

$$h(p|n) = (\text{constant})p^n.$$

This density concentrates probability near $p = 1$, as might be expected in view of the data. If, on the other hand, half the children are boys (for n even), the posterior density is

$$h\left(p \middle| \frac{n}{2}\right) = (\text{constant})p^{n/2}(1 - p)^{n/2},$$

which is symmetric about its maximum at $p = 1/2$. (If n is large, the concentration is high near $p = 1/2$, representing the evidence in the data that p is near $1/2$.) ◄

Example 13-d Let $\bar{X}$ denote the mean of a random sample from a normal population with mean μ and known variance σ^2. For given $\bar{X}$, the likelihood is

$$L(\mu) = \left(\frac{2\pi\sigma^2}{n}\right)^{-n/2} \exp\left[\frac{-n}{2\sigma^2}(\bar{X} - \mu)^2\right].$$

Suppose now that a normal prior is assumed for μ, a distribution with mean ν and variance τ^2 on the values $-\infty < \mu < \infty$, defined by the density

$$g(\mu) = (2\pi\tau^2)^{-1/2} \exp\left[\frac{-1}{2\tau^2}(\mu - \nu)^2\right].$$

The posterior density $h(\mu|\bar{X})$ is the product of the likelihood $L(\mu)$ and the prior $g(\mu)$, except for a constant; this product is an exponential with an exponent that is quadratic in μ. This means, omitting details, that $h(\mu|\bar{X})$ is a *normal* density in μ; the expected value of μ is

$$E(\mu|\bar{X}) = \frac{n\tau^2\bar{X} + \nu\sigma^2}{n\tau^2 + \sigma^2}$$

(a weighted average of $\bar{X}$ and ν), and the variance is

$$\text{var}(\mu|\bar{X}) = \frac{\sigma^2\tau^2}{n\tau^2 + \sigma^2}.$$

Now, a small τ^2 represents a strong conviction that μ is close to ν; in the extreme case $\tau^2 = 0$ [interpreted as $P(\mu = \nu) = 1$], the posterior distribution is the same as the prior. On the other hand, if ν and τ^2 are fixed and n gets large, then the posterior distribution becomes concentrated around $\bar{X}$, thus imitating the likelihood function. In practice, the way in which the data will affect the distribution of μ should be kept in mind when selecting ν and τ^2. ◄

13.3 Making a Decision

It is essential to consider costs of wrong decisions if one is to make a decision wisely. However, costs are not always exclusively monetary, for they can include pain, embarrassment, the loss of good will, wasted time, and so on. Economists have developed a theory of utility showing that it is usually possible, in principle, to construct a finite scale for losses (or costs) and gains that preserves preferences, and in terms of which *expected* loss is the measure of loss for prospects that are random.

We assume that the consequences of choosing action a when nature is in state θ are measured by a "utility" such as described, and that there is defined a *loss* (or negative utility) $\ell(\theta, a)$:

Loss function:

$$\ell(\theta, a) = \text{loss incurred with action } a \text{ when nature is } \theta.$$

When the state of nature is treated as a random variable Θ, the loss incurred by taking action a is random: $\ell(\Theta, a)$, and actions are judged by the Bayesian according to expectations:

Bayes loss resulting from taking action a:

$$B(a) \equiv E[\ell(\Theta, a)].$$

The *Bayes action* is the action a_b that minimizes this Bayes loss:

Bayes action a_b:

$$B(a_b) = \min_a B(a).$$

Example 13-e Suppose that there are three actions to choose from and that nature is in one of two possible states, with losses as given in the following table:

		Actions		
		a_1	a_2	a_3
Nature	θ_1	0	1	5
	θ_2	4	3	2

If θ_1 and θ_2 are equally likely, $g(\theta_1) = g(\theta_2) = 1/2$, then the Bayes losses are

$$B(a_1) = 0 \cdot \frac{1}{2} + 4 \cdot \frac{1}{2} = 2.0,$$

$$B(a_2) = 1 \cdot \frac{1}{2} + 3 \cdot \frac{1}{2} = 2.0,$$

$$B(a_3) = 5 \cdot \frac{1}{2} + 2 \cdot \frac{1}{2} = 3.5.$$

Either a_1 or a_2 would be a "Bayes action" for the given prior probabilities, $g(\theta_i)$. On the other hand, if $g(\theta_1) = .1$, then

$$B(a_1) = 3.6, \qquad B(a_2) = 2.8, \qquad B(a_3) = 2.3,$$

so that now a_3 minimizes the Bayes loss and is the Bayes action. ◄

13.4 Testing a Simple Hypothesis Against a Simple Alternative

Suppose that Nature is in one of just two states, called H_0 and H_1, both simple hypotheses. A statistic Y is chosen as the test statistic, and a choice is to be made between H_0 and H_1, on the basis of the observed value of Y. Let the density of Y under H_i be $f_i(y)$ for $i = 0, 1$. Then for given prior probabilities (g_0, g_1), as in Example 13-b, the posterior odds are

$$\frac{h_0}{h_1} = \frac{g_0}{g_1} \frac{f_0(y)}{f_1(y)}.$$

Suppose there is no loss in making a correct decision, a loss of ℓ_{I} when a type I error is committed, and a loss of ℓ_{II} for a type II error:

		H_0 true	H_1 true
Action	Accept H_0	0	ℓ_{II}
	Reject H_0	ℓ_{I}	0

Given data and the corresponding value of Y, the appropriate probabilities for Θ are given by the posterior distribution; and the expected losses (with respect to *posterior* probabilities) are as follows:

$$B(\text{accept } H_0) = 0 + \ell_{\text{II}} h_1,$$

$$B(\text{reject } H_0) = \ell_{\text{I}} h_0 + 0.$$

The ratio of these is

$$\frac{B(\text{Rej})}{B(\text{Acc})} = \frac{\ell_{\text{I}} h_0}{\ell_{\text{II}} h_1} = \frac{\ell_{\text{I}} g_0 f_0(y)}{\ell_{\text{II}} g_1 f_1(y)}.$$

and clearly one should decide as follows:

Bayes test:

$$\text{Reject } H_0 \text{ if } \quad \frac{f_0(y)}{f_1(y)} < \frac{\ell_{\text{II}} g_1}{\ell_{\text{I}} g_0}.$$

$$\text{Accept } H_1 \text{ if } \quad \frac{f_0(y)}{f_1(y)} > \frac{\ell_{\text{II}} g_1}{\ell_{\text{I}} g_0}.$$

Example 13-f Consider again the situation considered in Example 13-b, in which $Y = \bar{X}$ is normal with mean 0 under H_0 and mean 1 under H_1, with variance $1/n$. The likelihood ratio for $\bar{X} = y$ is

$$\frac{f_0(y)}{f_1(y)} = \exp\left[-n\left(y - \frac{1}{2}\right)\right],$$

so the Bayes test to reject H_0 if

$$\exp\left[-n\left(\bar{X} - \frac{1}{2}\right)\right] < \frac{\ell_{\text{II}} g_1}{\ell_{\text{I}} g_0},$$

or if

$$\bar{X} > \frac{1}{2} + \frac{1}{n}\log\left(\frac{\ell_{\text{II}} g_1}{\ell_{\text{I}} g_0}\right).$$

For example, if H_0 is twice as likely as H_1 but rejecting it erroneously is twice as costly as a type II error: $g_0/g_1 = 2$ and $\ell_{\text{I}}/\ell_{\text{II}} = 2$, then the test rejects H_0 if

$$\bar{X} > \frac{1}{2} + \frac{1}{n}\log\frac{1}{4}.$$

It is to be observed that this *type* of test, rejecting $\mu = 0$ and accepting $\mu = 1$ if $\bar{X} >$ constant, is the kind that was proposed as intuitively reasonable in Chapter 11. ◄

13.5 Bayesian Estimation

Losses are difficult to assess in estimation problems, because consequences are seldom easy to identify, let alone measure. A commonly used loss function in estimating a parameter θ by the number t is the squared error:

Quadratic loss:

$$\ell(\theta, t) = (\theta - t)^2.$$

We define the *Bayes loss* to be the expected value of this with respect to the distribution of Θ:

Bayes loss:

$$E[\ell(\Theta, t)] = E[(\Theta - t)^2].$$

This is reminiscent of the "mean-squared error" of Chapter 10, but here it is Θ that is random. The second moment of Θ about t is minimized by choosing t to be the mean of the Θ-distribution:

Bayes estimator for Θ, assuming quadratic loss:

$$\hat{\theta}_B \equiv E(\Theta),$$

the expectation being taken with respect to whatever distribution describes one's knowledge of Θ. The minimum Bayes loss is var(Θ).

If $Y = y$ is given, the expectation is taken with respect to the *posterior* distribution as given by $h(\theta|y)$; this is the appropriate distribution for Θ, given the value $Y = y$ as calculated from a set of data.

Example 13-g In Example 13-c, it was found that the posterior density for the parameter of a Bernoulli population given a uniform prior distribution on $0 < p < 1$ is

$$h(p|y) = (\text{constant})\, p^y(1 - p)^{n-y},$$

corresponding to y successes observed in n trials. The Bayes estimate of p is the mean of this distribution:

$$E_h(p) = \frac{\int_0^1 p^{y+1}(1 - p)^{n-y}\, dp}{\int_0^1 p^y(1 - p)^{n-y}\, dp}.$$

Using the following formula from calculus,

$$\int_0^1 x^r(1 - x)^s\,dx = \frac{r!s!}{(r + s + 1)!},$$

one obtains

$$E_h(p) = \frac{y + 1}{n + 2}.$$

(This expression is called *Laplace's rule of succession* and has a notorious history.)

The class of prior densities of the form $g(p) = (\text{constant})p^r(1 - p)^s$ is useful because in the choice of r and s there is a wide variety of available distributions for modeling one's beliefs. With such a prior, the posterior becomes

$$h(p|y) = (\text{constant})p^{y+r}(1 - p)^{n-y+s}$$

with mean

$$E_h(p) = \frac{y + s + 1}{n + s + r + 2},$$

and this is the Bayes estimate. The numbers r and s can be thought of as the "effective numbers of prior successes and failures." For instance, if p, the probability of a boy, is thought a priori to be close to 1/2, say, with prior density $g(p) = (\text{constant})p^{10}(1 - p)^{10}$, and if a family has two children, both boys, then

$$E_h(p) = \frac{2 + 10 + 1}{2 + 10 + 10 + 2} = .542.$$

This number compared with the prior estimate of

$$E_g(p) = \frac{10 + 1}{10 + 10 + 2} = .50,$$

and the maximum likelihood estimate of $y/n = 2/2 = 1$. (The latter estimate, $\hat{p} = 1$, is evidently extreme and illustrates what can happen when using the maximum likelihood method with small samples.)

The posterior expected value of p has another kind of meaning. If $X_1, \ldots, X_n, \ldots$ is a sequence of independent Bernoulli random variables with parameter p, then the distribution of X_{n+1} given $X_1, \ldots, X_n$, *but not p*, is Bernoulli with probability of success equal to the posterior expected value of p. For the family of two boys, the probability that the next is a boy [assuming, as before, that $g(p) = (\text{constant})p^{10}(1 - p)^{10}$] is .542. ◄

Example 13-h In Example 13-d it was found that for a problem dealing with a normal family with given variance σ^2, the assumption of a normal prior for μ and an observed value of $\bar{X}$ results in a normal posterior with mean

$$E(\mu|\bar{X}) = \frac{n\tau^2\bar{X} + \nu\sigma^2}{n\tau^2 + \sigma^2},$$

where ν and τ^2 are the prior mean and variance, respectively. This posterior mean would be the Bayes estimate of μ, if a quadratic loss is assumed. As noted in Example 13-d, for given ν and τ^2, the estimate approaches $\bar{X}$ as $n \to \infty$; that is, the effect of the prior disappears in large samples, and the Bayes estimate is close to $\bar{X}$—is based entirely on the sample information. As $\tau^2 \to 0$, with n fixed, corresponding to increasingly *precise* prior knowledge that μ is close to ν, the estimate becomes just ν, ignoring the data. A very flat prior (corresponding to little prior information) is defined by a large prior variance; as $\tau^2 \to \infty$, the Bayes estimate of μ again approaches $\bar{X}$, the sample mean. ◄

A Bayesian interval estimate for a parameter θ can be constructed from its posterior distribution by choosing limits L_1 and L_2 such that the posterior probability between them is a specified amount:

$$P_h(L_1 < \theta < L_2) = \eta.$$

This is the sort of interpretation people would like to put on a 100η per cent confidence interval (L_1, L_2), and no doubt subconsciously do. The confidence interval and the Bayesian probability interval may coincide, usually in the case of a flat prior. In many cases, such as the following, the corresponding probability interval is somewhat smaller than a confidence interval.

Example 13-i In Example 10-d, statistics were reported for a sample of 200 viscosity measurements: $\bar{X} = 31.7, S = 1.81$. A 90 per cent confidence interval was then obtained for $\mu: 31.49 < \mu < 31.91$. Suppose now that the investigator describes his beliefs about μ by assigning it a distribution that is $\mathcal{N}(31.5, .25)$. Then the posterior is normal with parameters

$$E(\mu|\bar{X}) = \frac{n\tau^2\bar{X} + \nu\sigma^2}{n\tau^2 + \sigma^2} = 31.69,$$

$$\text{var}(\mu|\bar{X}) = \frac{\sigma^2\tau^2}{n\tau^2 + \sigma^2} \doteq \frac{(1.81)^2 \times .25}{200 \times .25 + 1.81^2} = (.124)^2,$$

where σ^2 has been replaced by S^2 to obtain the approximate variance. The shortest 90 per cent probability interval for μ is symmetric about the posterior mean, with limits

$$E(\mu|\bar{X}) \pm 1.645\sqrt{\text{var}(\mu|\bar{X})}$$

or

$$31.69 \pm .204.$$

The resulting interval (31.49, 31.89) is narrower than the confidence interval, but it is only slightly narrower, because the assumed prior is diffuse compared to the likelihood ($\tau^2 = .25$ compared with $\sigma^2/n = .0164$). The data overwhelm the prior information. ◄

13.6 How Large a Sample?

In classical statistics, of which most of this book is an example, the question of an appropriate number of observations is usually indirect and sometimes artificial. An advantage of a Bayesian approach to statistics is that the cost of making observations can be considered explicitly and sometimes weighed against the cost (or loss) of taking an incorrect action, in order to determine the optimum sample size along with the optimal procedure.

When the data have already been collected, as in Section 13.3, sampling costs play no role in the determination of the Bayes action, which is chosen according to the current (posterior) distribution of Θ. In the initial planning, however, the sample size is a design variable whose choice, along with the choice of a decision procedure, should take the cost of sampling into account.

Suppose that sampling costs an amount c for each observation, so that a sample of size n costs nc. Assume further that the sampling cost and decision cost can be combined additively, and that therefore an optimum sample size and decision rule would minimize their sum. The decision rule that is part of the optimal solution would simply be the one that minimizes the Bayes loss term; but the minimum Bayes loss will ordinarily depend on the data, since the minimizing action depends on the posterior, which in turn depends on the data. So, before minimizing over n, that term in the sum must be averaged with respect to the (unconditional) distribution in the data space.

These calculations are usually complicated, so we content ourselves with an example in which the minimum Bayes loss depends on the size of the sample but not otherwise on the data. The example has been chosen for its mathematical simplicity; more realistic loss functions may require extensive tables and numerical calculations.

Example 13-j Example 13-h treated the problem of estimating the mean μ of a normal random variable with known variance σ^2, assuming quadratic loss. Suppose, for convenience, that the prior variance is infinite ($\tau^2 = \infty$), that the Bayes estimate based on a sample of n observations is $\bar{X}$, the sample mean. According to Example 13-d, the Bayes loss using this estimate is

$$\text{var}\,(\mu|\bar{X}) = \frac{\sigma^2}{n},$$

a quantity which (curiously) does not depend on $\bar{X}$.

If each observation costs c, then the total loss plus sample cost for a sample of size n is

$$\frac{\sigma^2}{n} + cn.$$

This quantity is minimized by

$$n = \frac{\sigma}{\sqrt{c}},$$

which suggests that, for example, if the cost of observation increases by a factor of 4, the sample size should be halved. ◄

Problems

1. I take a coin from my pocket that is to be tossed 10 times. Suppose that you regard the coin to be two-headed ($p = 1$) with probability .01 and fair ($p = 1/2$) with probability .99. On each of the 10 tosses the coin lands heads. What is your probability now that the coin is two-headed?

2. A machine produces defective items with probability p. When the machine is "in control," $p = .01$, and when it is not, $p = .10$. The hypothesis that the machine is in control has prior probability .80. Of the first nine items of the day's production, two are found to be defective. What is the posterior probability that the machine is in control?

3. A proportion p of graduates of Jackson High School meet the minimum requirements for admission to a certain college. Initially, we are willing to take as the distribution of p the frequency distribution of this percentage among all high schools in the United States, so the prior density of p is

$$g(p) = 72p\,(1 - p)^7, \qquad 0 < p < 1.$$

(This means that, for example, 20 per cent of *all* graduates meet the requirements.) Ten students are sampled randomly from the Jackson graduating class of 1000, and 6 are found to meet the requirements.

(a) Assuming quadratic loss, what is the Bayes estimate of p?
(b) What is the probability that an individual student from Jackson will meet the requirements?

4. Grasshoppers lay their eggs in sacs or pods, the number of eggs per pod depending on the species and individual. A new species of grasshopper has been discovered, and it is desired to estimate the mean number, μ, of eggs per pod for that species. Take as a prior distribution for μ the frequency of mean numbers of eggs among known species, which is approximately $\mathcal{N}(48, 121)$. Within the species the number of eggs laid, X, is known to be approximately $\mathcal{N}(\mu, 81)$; it cannot be exactly normal because of its integer nature. The number of eggs is observed for six members of the new species and found to be 58, 75, 63, 69, 60, and 53. Find the Bayes estimate of μ and give an interval for μ which contains 95 per cent of the (posterior) probability.

5. An urn contains five balls, R of which are red. Initially, the probability distribution of R is uniform:

$$g(r) = \frac{1}{6}, \qquad r = 0, 1, \ldots, 5.$$

Three balls are selected randomly *without replacement* and all are found to be red.

(a) What is the posterior distribution of R (the number of red in the box *originally*) given the information?
(b) What is the probability that the next ball selected is red?

6. I have two "coins." Coin A has probability 2/3 of heads and coin B has probability 1/3 of heads. I select coin A with probability θ and coin B with probability $1 - \theta$ and toss the coin selected n times obtaining the Bernoulli sample $(X_1, \ldots, X_n)$.

(a) Write the likelihood function of θ for this sample.
(b) If $g(\theta) = 1$, for $0 < \theta < 1$, $n = 2$ and $X_1 = X_2 = 1$ (both tosses result in heads), give the posterior density of θ and the Bayes estimate of θ corresponding to quadratic loss.
(c) For the prior and data in part (b), calculate the probability that the third toss of the coin selected results in heads.

7. Your company plans to market a product in a region where the demand, μ, of the product is unknown and wants to know how much to produce. Assume that the loss to the company is

$$\$(\mu - a)^2,$$

where a is the amount produced. Before the production stage your information about μ is negligible ($\tau^2 = \infty$), but you are able to get information about μ from a sample $(X_1, \ldots, X_n)$, where the X_i are independent and distributed $\mathcal{N}(\mu, 10^6)$. Unfortunately, observing each X costs \$1600.

(a) How many observations should you make?
(b) Suppose that you take $n = 25$ observations and find that $\bar{X} = 16{,}323$. How much of the product should your company produce? What is the probability that your estimate is within 500 of the true demand?

★ Sequential Testing

In making decisions by testing hypotheses based on samples of specified size, one is quickly aware of two rather disconcerting aspects of the process: (1) The test result can be inconclusive (even though a conclusion must be drawn) and so suggest the need for *more* data before taking any action. (2) The test may be so conclusive that a *smaller* sample would have sufficed. Such considerations lead to the notion of stepwise procedures.

After a first sample one might decide, depending on some statistic of that sample, to accept H_0 (and take the appropriate action), to reject H_0 (and act accordingly), *or to withhold judgment* until more data are obtained. After a second sample, one would again choose, on the basis of all the data collected so far, to accept H_0, to reject it, or to get more data, and so on. The process might terminate by design after a fixed number of stages, or it might go on indefinitely. The Dodge–Romig "double-sampling" plan for accepting lots of manufactured items is an early instance of the application of such a procedure, one with two stages.

This chapter will present only certain *sequential* schemes, those in which the three choices (accept, reject, or get more data) are considered after *each one* of a potentially infinite sequence of independent observations.

14.1 An Example

In a comic strip (Hi and Lois), two children debated whether or not a piece of bread "always" fell off the table with buttered side up. One claimed that it did so 90 per cent of the time; the other felt it was a matter of even odds. These hypotheses were

$$H_0: \ p = \frac{1}{2} \qquad \text{versus} \qquad H_1: \ p = \frac{9}{10},$$

where p is the probability of buttered side down (D).

Suppose in one trial it fell D. Now $P_0(D) = .5$, and $P_1(D) = .9$, and because D is less likely under H_0 than under H_1, one might decide that H_1 is true. However, "only 5/9 as likely" is not very conclusive against H_0, so suppose another trial is run. If the second trial also results in D, with

$$P_0(D, D) = .25 \qquad \text{and} \qquad P_1(D, D) = .81,$$

the ratio of likelihoods under H_0 and under H_1 is now 25/81, suggesting more strongly that H_1 is correct. But if not satisfied one might try again; and suppose the third time it falls U (buttered side up). Now the ratio is

$$\frac{P_0(D, D, U)}{P_1(D, D, U)} = \frac{.5}{.9} \times \frac{.5}{.9} \times \frac{.5}{.1} = \frac{.125}{.081},$$

and the data seem to support H_0.

Suppose this "sequential" consideration of results is continued. It is apparent that for each D the previous ratio will be multiplied by $.5/.9 = 5/9$ and for each U, by $.5/.1 = 5$. The ratio of probabilities under the two models, H_0 and H_1, will be large when there is a preponderance of U's, and small if D's predominate. If it is large enough, one would accept H_0, and if small enough, accept H_1.

"Large enough" and "small enough," of course, are arbitrary designations, and any particular limits that are chosen define a test procedure. Suppose it is decided that a ratio of 99 to 1 is large. That is, after a particular number of trials, if the probability of the observed result under H_1 is 99 times what it is under H_0, then H_0 is rejected. And if its probability under H_0 is 99 times what it is under H_1, then H_0 is accepted. The procedure which employs this rule at each stage, continuing the sampling until one or the other decision is reached, turns out to have error sizes of about $\alpha = \beta = .01$.

A ratio of 99 to 1 is quickly reached by powers of .5/.9 or of .5/.1. Thus $(5/9)^8 < 1/99$, and $5^3 > 99$; so eight D's in a row would say to reject $p = .5$ (as opposed to $p = .9$), and three U's in a row would say to accept $p = .5$ (as opposed to $p = .9$). More commonly, one would encounter sequences between these two extremes, sequences involving both D's and U's, and a decision would take somewhat longer.

To record the results graphically, as sampling proceeds, it is convenient to plot *logarithms* of probability ratios at each stage. The graph would also have lines at $\log A$ and $\log B$, where B is the upper limit, calling for rejection of H_0, and A is the lower limit, calling for acceptance of H_0. If the 99 : 1 ratio mentioned above is adopted, $\log B = -\log A = \log 99$. Figure 14-1 shows a record of four runs—that is, four sequences of trials. Two are trials in which $p = .5$, and in the other two, $p = .9$. In each of these four runs, the right decision was reached, although there is a small chance (.01) that the decision will be wrong, rejecting $p = .5$ when $p = .5$, or accepting $p = .5$ when $p = .9$. The sequence of U's and D's in a given graphical record can be read from the graph (although when several graphs are plotted on the same axes the record of an individual run may be obscured). Thus, in run 2, there were three U's in a row. In run 3, after some alternation, a string of nine D's brought about the decision to reject $p = .5$.

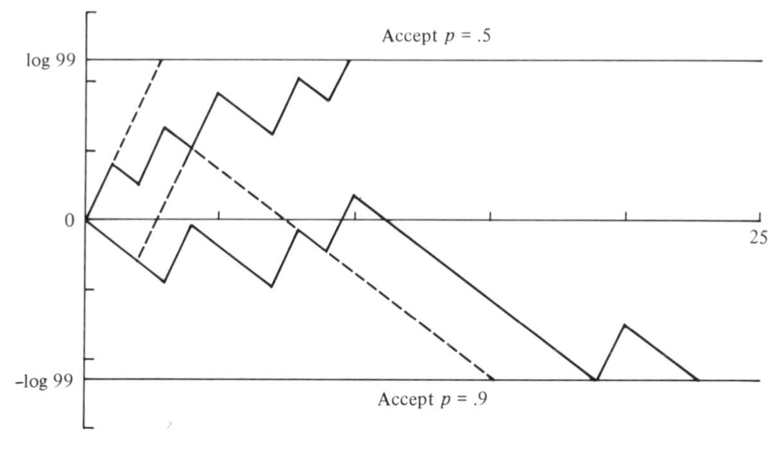

Figure 14-1 Testing $p = .5$ against $p = .9$.

14.2 Likelihood

As in the example, it appears natural and fruitful to base inferences on the relative probabilities of observed results under competing models. The probability of a result x whose probability in a particular discrete model is $f(x)$ is termed the *likelihood* of the model. In the case of a continuous model, an observed x has zero probability, so the likelihood of the model is defined to be the *density* at x; it could be taken to be the probability element $f(x)\,dx$, but in comparing likelihoods (and this is essentially all we do with them) the dx's would cancel out anyway.

The definition of likelihood, given the results $(x_1, \ldots, x_n)$ of a random sample, is as follows:

> The *likelihood* of the model $f(x)$ is the joint density or probability function of the sample:
>
> $$\prod_{i=1}^{n} f(x_i) = f(x_1) \cdots f(x_n).$$

When the model is one of a parametric class of models $\{f(x;\theta)\}$, the likelihood is attributed to the parameter θ. It is a function of θ (for given $x_1, \ldots, x_n$), called the *likelihood function* (see Section 10.8).

When there are just two competing models, $f_0(x)$ and $f_1(x)$, the ratio of their likelihoods is a statistic commonly used in choosing between them:

> *Likelihood ratio* statistic for the random sample $X_1, \ldots, X_n$:
>
> $$\Lambda = \frac{\prod f_0(X_i)}{\prod f_1(X_i)}.$$

It is fundamental to "likelihood inference" that a very large value of Λ is evidence in favor of H_0 [or $f_0(x)$] as opposed to H_1. If it is very small, this is evidence in favor of H_1. *If it is neither very large nor very small, the evidence is inconclusive.*

Example 14-a† I have two pair of dice; one pair is of the usual type, but the other is specially made. Of this special pair, one die has 5's on all faces, and the other has four 2's and two 6's. I toss one pair—you do not see which pair—and get a 7. Which pair did I toss?

† This same problem is considered in Example 13-a from a Bayesian viewpoint.

The probability of the 7 with the ordinary set is 1/6, and with the specially made set it is 2/3. The ratio of likelihoods is

$$\Lambda = \frac{2/3}{1/6} = 4;$$

and if you had to choose, you would be apt to say that it was the special pair that I tossed. If you were hard to convince, you might want more data. So suppose that I toss again and get another 7. Now the likelihood ratio is

$$\Lambda = \frac{(2/3)(2/3)}{(1/6)(1/6)} = 16;$$

that is, the observed result is 16 times as likely under the assumption that the dice are ordinary than it is under the hypothesis that they are the special ones. Knowing that I had the two sets of dice, most people would think I was using the specially made set. ◄

Example 14-b The likelihood of a random sample $(X_1, \ldots, X_n)$ of 0's and 1's from a Bernoulli population is

$$\prod (p^{X_i} q^{1-X_i}) = p^Y q^{n-Y}$$

where p is the probability of success and $Y \equiv \Sigma X_i$ is the number of successes among the n trials. The likelihood ratio for the models defined by a p_0 and a p_1 is

$$\Lambda = \frac{p_0^Y q_0^{n-Y}}{p_1^Y q_1^{n-Y}} = \left(\frac{p_0}{p_1} \cdot \frac{q_1}{q_0}\right)^Y \left(\frac{q_0}{q_1}\right)^n.$$

If $p_1 > p_0$, this is a decreasing function of Y. That is, a large number of successes means a small value of Λ and points to the larger of p_0 and p_1 as the probability of success.

In the case of the buttered bread mentioned in the discussion of the preceding section, $p_0 = .5$ and $p_1 = .9$, the likelihood ratio for Y successes in n trials is

$$\Lambda = \left(\frac{.5}{.9}\right)^Y \left(\frac{.5}{.1}\right)^{n-Y} = \frac{5^n}{9^Y}.$$

The larger the Y, the smaller the value of Λ and the stronger the evidence in favor of $p = .9$ (over $p = .5$). ◄

Example 14-c Suppose that X is normal with mean μ and known variance $\sigma^2 = 1$. The population density is

$$f(x) = \frac{1}{\sqrt{2\pi}} \exp\left[-\frac{1}{2}(x - \mu)^2\right],$$

and the likelihood function:

$$\prod_1^n f(X_i) = (2\pi)^{-n/2} \exp\left[-\frac{1}{2}\sum (X_i - \mu)^2 \right].$$

For two particular values μ_0 and μ_1, the likelihood ratio is

$$\Lambda = \frac{\exp[-\sum (X_i - \mu_0)^2/2]}{\exp[-\sum (X_i - \mu_1)^2/2]} = \exp\left\{ n(\mu_0 - \mu_1)\left[\bar{X} - \frac{1}{2}(\mu_0 + \mu_1) \right] \right\}.$$

If $\mu_0 < \mu_1$, a large $\bar{X}$ yields a small Λ and points to μ_1 as opposed to μ_0. A small $\bar{X}$ would suggest μ_0 as the more plausible of the two. ◄

14.3 The Sequential Likelihood Ratio Test

The sequential likelihood ratio test (SLRT) chooses between models $f_0(x)$ and $f_1(x)$ according to the indication of the likelihood ratio, recomputed after each observation is obtained in a sequence of independent observations. At each stage, the decision is made to continue sampling unless either f_0 or f_1 is strongly supported by the data obtained so far. In terms of the likelihood ratio,

$$\Lambda_n = \prod_{i=1}^n \frac{f_0(X_i)}{f_1(X_i)},$$

the rule is as follows:

Sequential likelihood ratio test of $f_0(x)$ versus $f_1(x)$:

After each X_n in the independent sequence $X_1, X_2, \ldots$ compute the likelihood ratio Λ_n. Then

(i) If $\Lambda_n \geq B$, accept H_0 [or $f_0(x)$].
(ii) If $\Lambda_n \leq A$, reject H_0 [and accept $f_1(x)$].
(iii) If $A < \Lambda_n < B$, take another observation.

The limits A and B, which define the test, define the error sizes of the test, α and β. But it is hard to compute α and β from given A and B.

Approximate formulas have been developed that work rather well if α and β are small:

Approximate error sizes for given limits:

$$\alpha = \frac{A(B-1)}{B-A}, \qquad \beta = \frac{1-A}{B-A}.$$

Limits to achieve specified error sizes:

$$A = \frac{\alpha}{1-\beta}, \qquad B = \frac{1-\alpha}{\beta}.$$

The approximation involved here is the assumption that the rejection and acceptance levels, when reached, are reached precisely. The approximation is better when a decision is not reached too early. (It can be shown that if A and B are defined in terms of specified α and β as above, then the *actual* error sizes α' and β' satisfy $\alpha' + \beta' \leq \alpha + \beta$.)

The sample size N required to reach a decision is *random*. Sometimes a decision is reached quickly, sometimes not. It depends on the particular sequence of observations.

The reader may wonder if he can be sure that a decision will ever be reached. Could it happen that the likelihood ratio *never* goes outside the specified A and B? The answer is, strictly speaking, that it could; however, it can be shown mathematically that if f is either f_0 or f_1, this is "almost sure" not to happen—the probability is zero. So N is a proper random variable with finite values; but what is more, it has a finite mean. The following formulas usually give good approximations to the expected sample size:

$$E(N|H_0) = \frac{\alpha \log A + (1-\alpha)\log B}{E[\log(f_0/f_1)|H_0]},$$

$$E(N|H_1) = \frac{(1-\beta)\log A + \beta \log B}{E[\log(f_0/f_1)|H_1]}.$$

Example 14-d Consider again the Bernoulli model, this time testing $p = .4$ against $p = .6$. Then

$$\frac{f_0(X)}{f_1(X)} = \frac{.4^X .6^{1-X}}{.6^X .4^{1-X}} = \left(\frac{.16}{.36}\right)^X (1.5)$$

with logarithm

$$\log \frac{f_0}{f_1} = \log 1.5 + X \log \frac{4}{9}.$$

If one takes $\alpha = \beta = .05$, then $A = 1/19$ and $B = 19$, and the expected sample size, for $H_0\,(p = p_0)$, is

$$E(N|H_0) = \frac{.05 \log (1/19) + .95 \log 19}{\log 1.5 + p_0 \log (4/9)} = \frac{.90 \times 2.944}{.4055 + .4 \times (-.8109)},$$

or about 33. (The computation for $p_1 = .6$ yields the same thing because of symmetry.) A test with fixed sample size would require about 65 observations to achieve the same α and β.

The likelihood ratio after n observations is

$$\Lambda_n = \left(\frac{4}{9}\right)^{\Sigma X_i} (1.5)^n,$$

so the inequality for continued sampling is

$$\log A = -2.94 < n \log (1.5) + \left(\sum_1^n X_i\right) \log \frac{4}{9} < 2.94 = \log B$$

or, equivalently,

$$.5n - 3.631 < \sum_1^n X_i < .5n + 3.631.$$

Figure 14-2 shows a record of four sequences of observations with $p = .4$, plotted according to the value of the log of the likelihood ratio. The upper and lower lines are at $\log B$ and $\log A$, respectively, as defined by error sizes $\alpha = \beta = .05$. As it happened, three of the four runs terminated in the correct decision before $n = 50$; one of them had not yet reached a decision after 50 observations. ◄

14.4 Composite Hypotheses

The testing of a specific parameter value θ_0 against another specific value θ_1 is rather rare in practice. Nature cannot usually be pinned down to two such precise values—values that one can specify accurately. However, whether the problem is sequential or calls for a fixed sample size, it is quite reasonable and practical to construct a test for θ_0 against θ_1 and then apply it to a one-sided problem, such as testing $\theta \leq \theta^*$ against $\theta > \theta^*$. In such application one would want to know how the test performs, not just at θ_0 and at θ_1 but at all admissible values of θ.

The power function $\pi(\theta)$ or the operating characteristic function $OC(\theta) = 1 - \pi(\theta)$ describes the performance of a test. Specifying an α corresponding to θ_0 and a β corresponding to θ_1 amounts to fixing two points on the power curve:

$$\pi(\theta_0) = \alpha, \qquad 1 - \pi(\theta_1) = \beta.$$

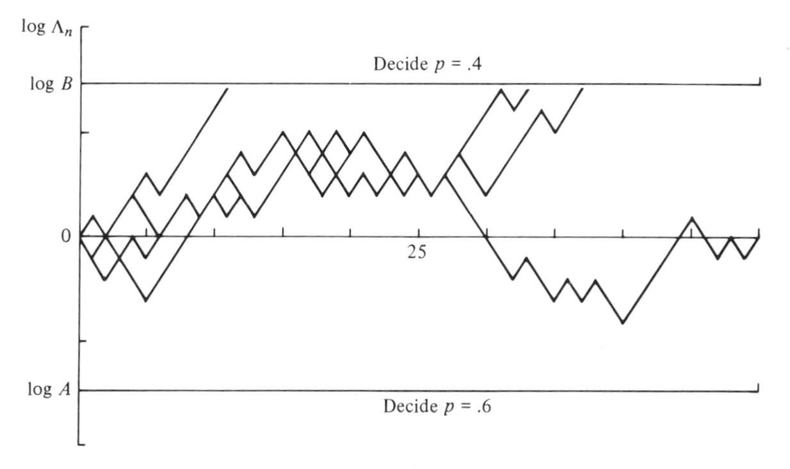

Figure 14-2 See Example 14-d.

The A and the B of a sequential test of θ_0 against θ_1 are fixed by the choice of α and β, and the rest of the power function is determined by the constants.

A. Wald, who first studied these matters (*Sequential Analysis*, New York: John Wiley & Sons, Inc., 1950) showed how to obtain the power function of an SLRT. It will not be given here in detail because a very acceptable version of $\pi(\theta)$ can be sketched from a plot of just five points. In the case of a Bernoulli experiment, these points are given as follows for the case $p_0 < p_1$.

p	$\pi(p)$
0	0
p_0	α
$\dfrac{\log(q_1/q_0)}{\log[p_0q_1/(p_1q_0)]}$	$\dfrac{\log B}{\log(B/A)}$
p_1	$1-\beta$
1	1

Example 14-e In testing $p_0 = .5$ against $p_1 = .9$, with $\alpha = \beta = .01$ (and therefore $A = 99$, $B = 1/99$), the five points are as follows:

p	$\pi(p)$
0	0
.5	.01
.732	.218
.9	.99
1	1

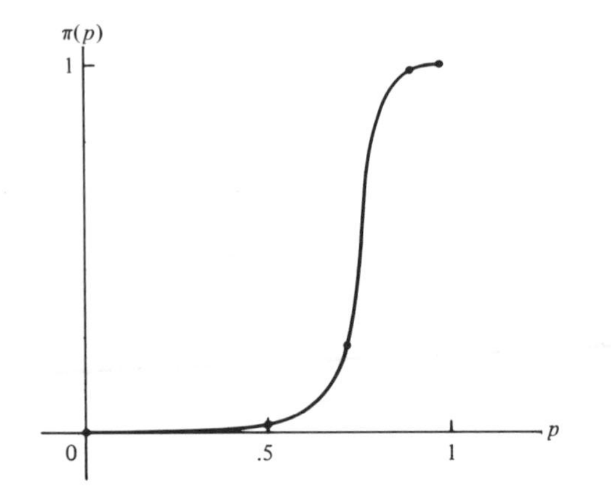

Figure 14-3 Power function for Example 14-e.

The power curve is sketched in Figure 14-3. This curve makes clear what the chances are of deciding to "reject H_0" for whatever p might be of concern. ◀

In the case of testing μ_0 against μ_1, when μ is the mean of a normal distribution with known variance, the five points for sketching the power curve are as follows (for $\mu_1 > \mu_0$):

μ	$\pi(\mu)$
$-\infty$	0
μ_0	α
$\dfrac{\mu_0 + \mu_1}{2}$	$\dfrac{\log B}{\log (B/A)}$
μ_1	$1 - \beta$
∞	1

The average sample size for parameter values not restricted to θ_0 and θ_1 is given by the formula

$$E(N|\theta) \doteq \frac{\pi(\theta) \log A + [1 - \pi(\theta)] \log B}{E[\log (f_0/f_1)|\theta]}.$$

It will tend to be larger for θ-values between the θ_0 and θ_1 used in computing the likelihood ratio. This is to be expected. For, in choosing between, say, $p = .4$ and $p = .6$, when the data are generated by $p = .5$, it should take longer to decide. On the other hand, for θ *outside* $\theta_0 < \theta < \theta_1$, the expected sample size is smaller. (For data generated by $p = .9$, one would tend to get lots of "heads" and quickly choose $p = .6$ rather than $p = .4$.)

Example 14-f Consider testing between $\mu = 0$ and $\mu = 1$ in a normal population with $\sigma^2 = 1$. The likelihood ratio statistic (see Example 14-c) is

$$\Lambda_n = \exp\left[-n\left(\bar{X} - \frac{1}{2}\right)\right].$$

And with $\alpha = \beta = .01$, so that $1/A = B = 99$, one can obtain the following values of the power function and of the expected sample size required for a decision:

μ	$\pi(\mu)$	$E(N\|\mu)$
$-\infty$	0	0†
0	.01	9
1/2	.50	21
1	.99	9
∞	1	0

[The calculation of $E(N|1/2)$ is rather tricky and not in order here; direct substitution in the formula given yields 0/0.] It is evident from the symmetry in the problem that the maximum expected sample size occurs at $\mu = \frac{1}{2}$.

The sample size required in the usual test ($\bar{X} > .5$), using a sample of fixed size with $\alpha = \beta = .01$, is about 21.65. The sequential test would usually require less data, and the power functions of the fixed sample-size test and the sequential test would agree at the five points listed. ◄

Problems

1. A computer scientist has proposed the following probabilistic device for checking whether a very large number n is prime. He picks a number $k < n$ and checks to see if a certain statement concerning the relation between k and n is true. If n is prime, the statement is false for all k. If n is not prime, he knows that the statement is true for at least half of all $k < n$. Assuming he can select randomly as many numbers $k < n$ as he needs, propose a sequential scheme using up to 10 steps for deciding whether n is prime. What are the 2 error sizes for your scheme?

2. Derive the SLRT in Example 14-a for H_0: $p = 2/3$ against H_1: $p = 1/3$ with $\alpha = .01$ and $\beta = .05$.

† The 0 here results from substitution in the formula for $E(N)$. It is clearly incorrect, since $N \geq 1$. The approximations leading to the formula are simply not good when tests tend to terminate early.

3. Use data from Table XIII in Appendix B to carry out a SLRT for $\mu = 0$ against $\mu = 1$ in a normal population with $\sigma^2 = 1$, using $\alpha = \beta = .02$.

4. A gasoline refinery is functioning properly if the mean number of impurities in a certain microscopic portion is $m = 5$. Sometimes this average gets too large and the filtration process must be improved; fix the alternative to be $m = 10$. Suppose the number of impurities X has probability function

$$f(x) = \frac{1}{x!}m^x e^{-m}, \qquad x = 0, 1, \ldots.$$

(Those who have read Chapter 7 will recognize this as the Poisson function.) Derive the SLRT of H_0: $m = 5$ against H_1: $m = 10$ with $\alpha = \beta = .01$.

5. Calculate $E(N|H_0)$ and $E(N|H_1)$ for the test of Problem 2.

6. Sketch the power function of the test in Problem 3 and compare $E(N|\mu = 0)$ and $E(N|\mu = 1)$ with the sample size required by a test with fixed sample size and the same error sizes (.02).

15

Tests for Categorical Populations

The term *categorical population* refers to one whose individual members are characterized as belonging to one of several categories or classes, not necessarily numerical or even necessarily ordered. A population may consist of Democrats, Republicans, and "other," for example, and each individual is in one of these classes.

The *model* for a categorical population consists of a list of the possible categories together with a probability (a population proportion) for each. *Data* are ordinarily presented in the form of a frequency distribution.

Model	
Category	Probability
A_1	p_1
A_2	p_2
$\vdots$	$\vdots$
A_k	p_k
	Sum: 1

Sample	
Category	Frequency
A_1	f_1
A_2	f_2
$\vdots$	$\vdots$
A_k	f_k
	Sum: n (sample size)

The statistical problem to be considered in this chapter is that of testing whether the data "fit" a specific model or a class of models. An important special case is that of testing whether two classification variables are related.

15.1 The Chi-Square Goodness-of-Fit Test

To test a particular set of probabilities $p_1, \ldots, p_k$, Karl Pearson proposed (1900) a test statistic based on a comparison of observed frequencies and expected frequencies:

$$\chi^2 = \sum_1^k \frac{(f_i - np_i)^2}{np_i}.$$

The product np_i is the expected number of occurrences of category A_i among n trials, if p_i is its probability. The statistic χ^2 will be small if and only if all the observed frequencies are close to the expected ones, and so a critical region of the form $\chi^2 >$ constant is used as a basis for rejecting the model defined by the p's.

As always, if the statistic χ^2 is to be used in testing, the nature of its distribution—at least under the null model—must be understood. Although this distribution is generally quite complicated and dependent on the model being tested, these difficulties, remarkably, disappear as the sample size becomes large. The distribution of χ^2 under H_0 for large n is approximately one of the family of chi-square distributions, depending on the number of categories but otherwise not on the model being tested.

The chi-square family of distributions, introduced in connection with estimating the variance of a normal population (Section 10.7), is indexed by a parameter called the *number of degrees of freedom*. In the present application, this parameter is $k - 1$ when k is the number of categories. Selected percentiles of the chi-square distribution, for small numbers of degrees of freedom, are given in Table II. (A formula for finding percentiles for larger numbers of degrees of freedom is given at the foot of the table, an approximate formula based on the fact that the chi-square distribution tends to the normal as the parameter becomes infinite.)

> Large-sample goodness-of-fit test:
>
> Reject the model with probabilities $p_1, p_2, \ldots, p_k$ if
>
> $$\chi^2 \equiv \sum_1^k \frac{(f_i - np_i)^2}{np_i} > \text{constant}.$$
>
> The null distribution of χ^2 is approximately chi-square $(k - 1)$.

Example 15-a A wheel of fortune with eight stopping positions (as used in selecting musical styles in the random opera described in Problem 3, Chapter 2) was spun 100 times to test its fairness, with these results:

Position	*a*	*b*	*c*	*d*	*e*	*f*	*g*	*h*
Frequency	10	11	16	13	17	9	12	12

The expected frequency of each category or cell is $100 \times 1/8 = 12.5$, if the wheel is "fair" or ideal. Then

$$\chi^2 = \frac{(10 - 12.5)^2}{12.5} + \frac{(11 - 12.5)^2}{12.5} + \cdots + \frac{(12 - 12.5)^2}{12.5} = 4.32.$$

This value is not at all extreme, in terms of what can happen when the fair model governs the spins; it is close to the 30th percentile of chi-square(7). (See Figure 15-1.) The null hypothesis is not rejected at the usual significance levels. ◄

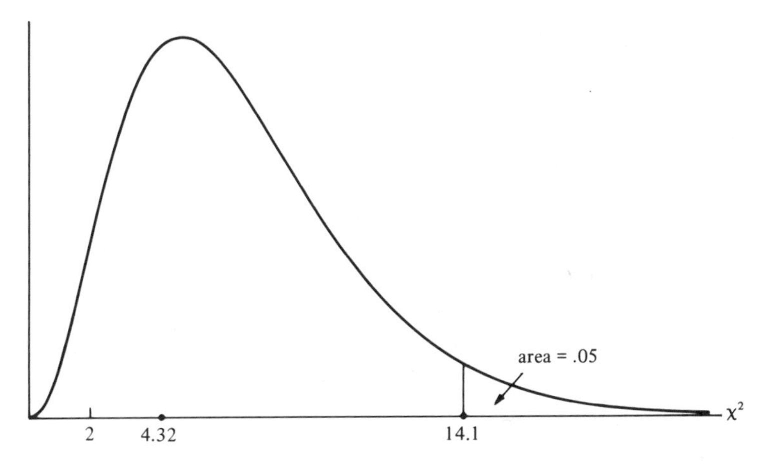

Figure 15-1 Chi-square density—7 degrees of freedom.

Although the chi-square test is essentially a test for a particular model for a *discrete* experiment, it can be modified to test the fit of a continuous model using data that are summarized in a frequency distribution. This is done by treating the experiment as discrete, with outcomes defined by the class intervals. Each class interval has a probability p_i under the model to be tested, and again the observed f_i and the expected np_i are compared using the Pearson χ^2 statistic. In this continuous case, however, there are tests that seem more appropriate and more effective in testing goodness of fit. (For example, see Section 11.7.)

It deserves specific mention that the χ^2-statistic presumes a *finite* number of categories (or class intervals, in the continuous case). An approximate test for a model with an infinite number of categories (such as the Poisson) can be accomplished by lumping those in the tail (or tails) into a single category. The process of combining categories is recommended whenever there are many categories with small expected frequencies. For the chi-square approximation to work reasonably well, one should have at least 5 to 10 times as many observations as categories, and very few categories with expected frequencies as low as 1. (The problem with a small expected frequency is that it appears in the denominator and may unduly inflate χ^2. An expected frequency of .05, for example, multiplies the squared difference in the numerator by 20.)

15.2 Models with Unknown Parameters

The chi-square test has the important feature of being adaptable to testing an incompletely specified model—a class of distributions depending on a parameter θ:

$$H_0: \quad P(A_1) = p_1(\theta), \ldots, P(A_k) = p_k(\theta).$$

Of course, expected frequencies could only be calculated from a knowledge of θ; but if θ is estimated from the sample in a way such as the maximum likelihood method of Chapter 10, the χ^2 statistic is again asymptotically chi-square under H_0, but with a number of degrees of freedom reduced by 1 (or by r, if r parameters must be estimated).

Chi-square statistic for testing

$$H_0: \quad P(A_i) = p_i(\theta) \qquad (i = 1, \ldots, k):$$

$$\chi^2 \equiv \sum_1^k \frac{[f_i - np_i(\hat{\theta})]^2}{np_i(\hat{\theta})},$$

where $\hat{\theta}$ is a maximum likelihood estimate of θ. Null distribution of χ^2 is asymptotically $\chi^2(k-2)$.

Example 15-b A genetic theory leads to this model:

	A	a	
B	θ^2	$\theta(1-\theta)$	θ
b	$\theta(1-\theta)$	$(1-\theta)^2$	$1-\theta$
	θ	$1-\theta$	1

Data are summarized in cell frequencies:

	A	a	
B	f_1	f_2	$f_1 + f_2$
b	f_3	f_4	$f_3 + f_4$
	$f_1 + f_3$	$f_2 + f_4$	n

It seems reasonable that both $(f_1 + f_2)/n$ and $(f_1 + f_3)/n$ should be good estimates of θ. Indeed, their average turns out to be an even better estimate (it is the maximum likelihood estimator of Section 10.8):

$$\hat{\theta} = \frac{2f_1 + f_2 + f_3}{2n}.$$

Suppose that 100 observations are made, with these results:

	A	a
B	45	28
b	22	5

The estimate of θ is $140/200 = .70$, so with 100 observations the expected cell frequencies are estimated to be

	A	a
B	49	21
b	21	9

and

$$\chi^2 = \frac{(49 - 45)^2}{49} + \frac{(21 - 28)^2}{21} + \frac{(21 - 22)^2}{21} + \frac{(9 - 5)^2}{9} = 5.4.$$

The 95th percentile of $\chi^2(2)$ is 5.99, so one would not reject the model being tested at $\alpha = .05$. ◄

15.3 Bernoulli Populations

An important special case of a categorical variable is that in which there are just two categories—a Bernoulli variable. The model for such a variable is defined by a parameter p, the population proportion of individuals of one category. (The probability of the other is automatically $1 - p \equiv q$.) Moreover, this parameter can be thought of as a population mean, simply by assigning 0 and 1 to the categories so that $P(1) = p$. This idea was encountered and exploited earlier (Chapter 10) in studying the estimation of the population proportion p by the proportion of 1's (or successes) in a large sample. This sample proportion is the sample mean.

Clearly, a large sample test for $p = p_0$ can be defined in terms of the statistic

$$Z = \frac{\hat{p} - p_0}{\sqrt{\dfrac{p_0(1 - p_0)}{n}}},$$

where $\hat{p} = f/n$ is the proportion of 1's in a random sample of size n. The distribution of Z is approximately standard normal if $p = p_0$, for large n. The square of Z, fortuitously, is precisely the same as the χ^2 statistic used in testing the Bernoulli model defined by p_0:

$$Z^2 = \frac{(f/n - p_0)^2}{p_0(1 - p_0)/n} = \frac{(f - np_0)^2}{np_0} + \frac{[n - f - n(1 - p_0)]^2}{n(1 - p_0)} = \chi^2.$$

Thus the one-sided region $\chi^2 > k^2$ is equivalent to the *two*-sided critical region $|Z| > k$, and the distribution of $Z^2 = \chi^2$ is asymptotically chi-square with 1 degree of freedom. (The chi-square distribution with 1 degree of freedom happens to be the distribution of the square of a standard normal variate.) On the other hand, using the Z statistic allows for a *one*-sided test. which would be appropriate for testing in a one-sided problem of the form H_0: $p \le p^*$ against H_A: $p > p^*$.

Example 15-c Problem 3 in Chapter 5 described the comparing of two makes of automobile by 100 drivers. The hypothesis that a driver is as likely to prefer one as the other is $p = 1/2$, and the results reported showed that 54 favor make M. The statistic for testing $p = 1/2$ is

$$Z = \frac{.54 - .50}{\sqrt{\dfrac{.50 \times .50}{100}}} = .8,$$

or alternatively,

$$\chi^2 = \frac{(54 - 50)^2}{50} + \frac{(46 - 50)^2}{50} = (.8)^2.$$

The 97.5 percentile of Z is 1.96 and the 95th percentile of $\chi^2(1)$ is $3.84 = (1.96)^2$. The value $p = .50$ would not be rejected at $\alpha = .05$, because $|.8| < 1.96$—or because $.64 < 3.84$. ◄

Large-sample comparisons of two population proportions p_1 and p_2 can be carried out using a Z statistic based on the difference in *sample* proportions, $\hat{p}_1 - \hat{p}_2$. This difference, a difference of means, is asymptotically normal with mean $p_1 - p_2$ and variance

$$\text{var}\,\hat{p}_1 + \text{var}\,\hat{p}_2 = \frac{p_1(1 - p_1)}{n_1} + \frac{p_2(1 - p_2)}{n_2}.$$

Under the null hypothesis $p_1 = p_2$, the numerators become equal and can be factored out. The common value is not known, of course, but is approximated by the corresponding proportion of successes in the combined sample.

Statistic for testing $p_1 = p_2 = p$:

$$Z \equiv \frac{\hat{p}_1 - \hat{p}_2}{\sqrt{\hat{p}(1 - \hat{p})\left(\dfrac{1}{n_1} + \dfrac{1}{n_2}\right)}},$$

where $\hat{p}$ is the proportion of successes in the *combined* sample. For large n_1 and n_2, Z is approximately standard normal under H_0.

Again, it turns out that Z^2 is precisely the same as a goodness-of-fit statistic, the one that will be presented in the next section. The approach there permits generalization to the comparison of several categorical populations, and the present case of two Bernoulli populations can be handled by the more general method as a special case (if the alternative is two-sided).

Example 15-d Poll A reports 53 per cent in favor of a certain charter amendment, while poll B gives 49 per cent as the proportion in favor. If both polls use 1800 individuals, is it possible that they are sampling the same population?

The test statistic for testing $p_1 = p_2$ is computed as follows:

$$Z = \frac{.53 - .49}{\sqrt{.51 \times .49\left(\dfrac{1}{1800} + \dfrac{1}{1800}\right)}} = 2.4,$$

where the .51 under the radical sign is computed as the proportion in favor in the combined sample:

$$\hat{p} = \frac{.53 \times 1800 + .49 \times 1800}{3600}.$$

The value 2.4 is a bit far out in the tail of the standard normal distribution for $p_1 = p_2$ to be credible (the *P*-value is .008). ◄

15.4 Independence

The classification of population members into categories defines a random variable in a general (nonnumerical) sense. Thus eye color and sex are two variables associated with members of a population. Each has a certain number of nonnumerical categories, and a particular person falls into one category (of each variable).

As in the case of numerical variables, it is natural to ask about possible relationships between categorical variables, or the lack of relationships. Two categorical variables are said to be *independent* if the events defined by categories of one variable are always independent of events defined by categories of the other:

> Variable A with categories $A_1, \ldots, A_c$ and variable B with categories $B_1, \ldots, B_r$ are *independent* if and only if
>
> $$P(A_i \cap B_j) = P(A_i)P(B_j)$$
>
> for all i, j.

This definition repeats what was given in Chapter 4 as a definition of independent experiments. It says that the two-way probability table for two independent variables can be constructed by multiplication of marginal probabilities.

Example 15-e The *suit* of a card drawn at random from a deck of playing cards is independent of whether or not it is an *honor card* (A, K, Q, J, or 10). The probabilities are as follows:

	Spades	Hearts	Clubs	Diamonds	
Honor	5/52	5/52	5/52	5/52	5/13
Nonhonor	8/52	8/52	8/52	8/52	8/13
	1/4	1/4	1/4	1/4	1

Clearly, multiplying marginal totals gives back the table entries.

If one thinks of the population of 52 cards from which the card is drawn, this population divides among the various cells in proportion to the probabilities:

	Spades	Hearts	Clubs	Diamonds	
Honor	5	5	5	5	20
Nonhonor	8	8	8	8	32
	13	13	13	13	52

In the body of this (and the earlier) table, the rows are in the same proportion (1:1:1:4) as the row of column totals, and the columns are all in the same proportion (5:8:13) as the column of row totals. Observe, too, that each entry in the table is the product of corresponding marginal totals divided by the grand total. For instance,

$$5 = \frac{13 \times 20}{52} = \frac{13}{52} \times 20 = 13 \times \frac{20}{52}.$$

That is, 13/52 of the 20 honor cards are spades, and 20/52 of the 13 spades are honor cards. (*To be continued.*) ◄

Each category of one variable can be thought of as defining a subpopulation, which in turn is classified according to categories of the other variable. In a particular subpopulation, the probabilities of those categories are obtained as conditional probabilities, given that subpopulation. If these are the same for all subpopulations, the subpopulations are termed *homogeneous*; and this is indeed the case when the two variables involved are independent.

Example 15-e (continued) The population of 52 cards in the deck can be thought of as broken into four subpopulations according to suit: spades, hearts, diamonds, and clubs, with 13 in each. The spades, say, are classified as honor or nonhonor, with probabilities (in the subpopulation of spades)

$$\frac{5}{13} = P(\text{honor}|\text{spade}) = \frac{P(\text{spade honor})}{P(\text{spade})} = \frac{5/52}{13/52}.$$

The independence of honor and suit means that the probability of an honor is the same in each suit—and the same as in the whole deck:

$$P(\text{honor}|\text{spade}) = P(\text{honor}|\text{heart}) = P(\text{honor}|\text{diamond}) = P(\text{honor}|\text{club}) = P(\text{honor}).$$

These subpopulations are homogeneous with regard to honor or nonhonor. ◄

The lack of a relationship between two categorical variables can be thought of either as independence of those variables, *or* as homogeneity of the subpopulations defined by one of them.

Which way to think depends on how the data are gathered. To test independence, one takes a sample from the basic population and classifies each member of the sample according to each of the two variables, in a *contingency table*:

	A_1	A_2	$\cdots$	A_c	
B_1	n_{11}	n_{12}	$\cdots$	n_{1c}	n_{1+}
$\vdots$	$\vdots$	$\vdots$		$\vdots$	$\vdots$
B_r	n_{r1}	n_{r2}	$\cdots$	n_{rc}	n_{r+}
	n_{+1}	n_{+2}	$\cdots$	n_{+c}	n

The notational convention for the marginal totals is that a + indicates a summation over the index in that position:

$$n_{i+} = \sum_j n_{ij}, \qquad n_{+j} = \sum_i n_{ij}, \qquad n_{++} = \sum_j \sum_i n_{ij} \equiv n.$$

The marginal totals n_{i+} sum to n, as do the column totals n_{+j}.

To test homogeneity, one takes a sample of size n_1 from A_1, n_2 from A_2, ..., and n_c from A_c classifying each member of each sample according to B, and obtaining again a contingency table:

	A_1	A_2	$\cdots$	A_c	
B_1	n_{11}	n_{12}	$\cdots$	n_{1c}	
$\vdots$	$\vdots$	$\vdots$		$\vdots$	
B_r	n_{r1}	n_{r2}	$\cdots$	n_{rc}	
	n_1	n_2	$\cdots$	n_c	n

Here the column totals are the c sample sizes, fixed in the sampling scheme. The row totals, however, are again random. But, as before, the row totals and the column totals both add to n.

Example 15-f For the purpose of examining the possible relationship between smoking and a student's college of enrollment, data can be collected in at least two ways. In one way, each student is classified according to each of the two variables and recorded as a tally in a cell of a contingency table. Frequencies summarizing the results for 200 students might turn out as follows:

	Arts	Technology	Agriculture	Education
Smoker	23	16	9	12
Nonsmoker	27	34	41	38

Alternatively, one might take samples of given sizes from each of the four colleges included in the study, classifying each individual according to whether or not he smokes. Again, the results might turn out just as in the above table, the four columns summarizing the four samples (of 50 each in this example).

Just looking at a table of observed frequencies such as the above does not reveal which sampling method was used. However, in this particular instance, the equality of the totals in the lower margin suggests that perhaps four samples of 50 were taken; equality of the four column totals would otherwise be an unusual happenstance. And colleges do maintain lists from which samples of 50 could be drawn. On the other hand, it would appear not to be feasible for a sample of 60 to be drawn from the population of students who smoke, and one of 140 from the population of nonsmokers. If, however, such populations could be identified, that approach would be proper. The statistical questions, in these three approaches, are as follows: (1) Is the variable "smoking" independent of the variable "college of enrollment"? (2) Are the four populations (colleges) homogeneous with respect to the variable "smoking"? (3) Are the populations of smokers and the population of nonsmokers distributed the same with respect to college of enrollment? These are essentially equivalent ways of asking, "Is there some kind of relationship between smoking and the college of enrollment?" ◄

15.5 Testing for a Relationship

In a two-way table for variables A and B, let n_{ij} denote the frequency of the A_iB_j-cell and p_{ij} its probability. If the sample contingency table turns out to have the property that its entries are formed by multiplying marginal totals:

$$n_{ij} = \frac{n_{i+}n_{+j}}{n},$$

as is the case in a finite population in which the two variables involved *are* independent, then one could easily accept the null hypothesis of independence (or equivalently, homogeneity of subpopulations):

$$H_0: \quad p_{ij} = p_{i+}p_{+j}.$$

A χ^2 statistic is formed by comparing the observed n_{ij} with the "expected" values, $n_{i+}n_{+j}/n$.

Chi-square statistic for independence (large n):

$$\chi^2 \equiv \sum_{j=1}^{c} \sum_{i=1}^{r} \frac{[n_{ij} - (n_{i+}n_{+j}/n)]^2}{n_{i+}n_{+j}/n}.$$

Reject independence if $\chi^2 > 100(1 - \alpha)$ percentile of chi-square $[(r - 1)(c - 1)]$.

The "expected" values are actually *estimates* of the cell expectations under H_0, that is, estimates of the form

$$n\hat{p}_{ij} = n\hat{p}_{i+}\hat{p}_{+j} \doteq n\left(\frac{n_{i+}}{n}\right)\left(\frac{n_{+j}}{n}\right),$$

the marginal relative frequencies being natural estimates of marginal cell probabilities; so what one might first think to be $rc - 1$ degrees of freedom must be reduced by the number of estimated parameters. Thus, the distribution of χ^2, when the cell expectations are computed from marginal relative frequencies as above, is asymptotically *chi-square* with $rc - 1 - (r - 1) - (c - 1) = (r - 1)(c - 1)$ degrees of freedom, under the null hypothesis of homogeneity.

Example 15-g The data on smoking and college of enrollment for 200 university students given in Example 15-f are repeated here:

	Liberal arts	Technology	Agriculture	Education	
Smoker	23	16	9	12	60
Nonsmoker	27	34	41	38	140
	50	50	50	50	200

A table with these marginal totals but with proportional rows and proportional columns is obtained by multiplying marginal totals and dividing by 200:

$$\frac{50 \times 60}{200} = 15, \qquad \frac{50 \times 140}{200} = 35,$$

with this result:

15	15	15	15	60
35	35	35	35	140
50	50	50	50	200

The entries in the cells of this table are estimates of the cell expected frequencies under H_0, the hypothesis of independence. Then χ^2 is constructed as usual with terms corresponding to the eight cell frequencies:

$$\chi^2 = \frac{(23 - 15)^2}{15} + \frac{(16 - 15)^2}{15} + \cdots + \frac{(38 - 35)^2}{35} \doteq 10.5.$$

Because this exceeds 7.81, the 95th percentile of $\chi^2(3)$, the null hypothesis would be rejected at $\alpha = .05$. ◀

Although the examples given above involve variables that are strictly categorical, with unordered categories, the concepts and tests can be applied to variables with ordered categories, or to numerical variables (with values automatically ordered), or even to continuous variables, by approximating them in terms of discrete class intervals. For instance, the variable "family income" might be given in terms of the categories less than \$10,000, between \$10,000 and \$20,000, and more than \$20,000. And the variable "instructor rating" might have the categories: excellent, good, fair, poor, abominable.

Problems

1. In a controlled experiment, the geneticist Mendel crossed peas in a way that, according to his theory, should produce round and yellow, round and green, wrinkled and yellow, and wrinkled and green peas in the ratio 9:3:3:1. His results gave frequencies 315, 108, 101, and 32. Test his model using his data, at $\alpha = .05$. (It is felt by some that Mendel's laboratory assistants may have doctored the data. Account for this feeling.)

2. A sample of 1000 families with eight children resulted in the following distribution of number of boys:

Number of boys	Frequency
0	10
1	37
2	115
3	215
4	243
5	223
6	111
7	33
8	13

(a) Test the hypothesis that the distribution of the number of boys in families of eight children is binomial. (If the distribution is *not* binomial, this could mean that not all families have the same p!)

(b) There are 8000 children in all, in the group of families being studied. Test the hypothesis that $p = 1/2$, using this sample of children. (Is this a meaningful exercise?)

3. A large industrial plant compiled the following summary of industrial accidents by months:

Number of accidents in a month	Frequency of months
0	22
1	6
2	2
3	2

One way of formulating the idea that accidents occur "randomly" in time is embodied in the *Poisson* distribution (Chapter 7). In that distribution the probabilities would be as follows:

$$P(k \text{ accidents in a month}) = \frac{m^k}{k!}e^{-m},$$

where m is the *average* number per month (i.e., the population mean). The cumulative probabilities for this discrete distribution are given in Table XII of Appendix B. Use the sample data to calculate $\vec{X}$, use this as an estimate of m, and enter Table XII with that m to read out all probability estimates. Taking "3 or more" as one category, use the chi-square statistic to test the hypothesis that the distribution is in the Poisson family.

4. A box of 400 thumbtacks is spilled onto the floor. If 228 of them land with point up, would you accept the hypothesis that $p = 1/2$, where p is the probability that a single tack lands point up?

5. To test the theory that people tend to postpone their death dates until after their birth dates, birth–death data were collected for 348 notables and 1202 athletes. Consider the hypothesis of equal likelihood for months $b, b + 1, b + 2, \ldots, b + 11$ as death months, when b is the birth month. Under this hypothesis, the probability of $b - 1$ (or $b + 11$) is $1/12 = .0833$. Test $P(b - 1) = 1/12$ as the null hypothesis in the following cases:

(a) Using the 348 notables, given that 16 died in month $b - 1$.
(b) Using the 1202 athletes, given that 102 died in month $b - 1$.

6. Determine the sample size needed to distinguish between $p = .50$ (H_0) and $p = .52$ (H_1) with 5 per cent accuracy ($\alpha = \beta = .05$).

7. A poll showed that of the 996 individuals surveyed, 27 per cent favored Representative F, 31 per cent favored Governor A, and 42 per cent were divided among several others as favored to replace Senator H if he should resign to run for President. Test the hypothesis that Representative F and Governor A were really "neck and neck," or equally favored in the population of voters.

(HINT: Ignore individuals not favoring either F or A.)

8. Police records in Hennepin County showed that of the 114 drivers killed in auto accidents in 1973, 67 were legally intoxicated, whereas among the 83 killed in 1974, only 40 were legally intoxicated. Is the difference statistically significant? (Authorities used the data as evidence that their new program of enforcement in 1974 was effective, stating that "it is unlikely the marked change was caused by a statistical fluke.")

9. It has been suggested that a person should go to college if he wants to get a good job. A random sample of 100 people is taken and the results are summarized in this frequency table:

	Blue collar	Professional	Executive	Other
College degree	20	10	6	4
No college degree	50	0	4	6

Test the hypothesis that one's occupation is independent of whether one has a degree. Supposing that independence is rejected, do you think that this should be taken as meaning that if one goes to college, one is more likely to get a better job? Explain.

10. The Gallup Poll sampled 1520 people in the United States and asked each if he or she favored a constitutional amendment that would prohibit abortions except when the mother's life was in danger. The following table presents the results by sex:

	Favor	Oppose	No opinion
Men	320	380	60
Women	360	360	40

(The data are real except that the numbers are rounded to make computation easier.)

In reporting the results, George Gallup says, "Analysis of the findings . . . reveals that: Men tend to be more pro-abortion than women." Analyze these data and give a correct conclusion. (Does Gallup's statement fairly summarize the finding?)

11. In a student evaluation of instruction, the responses were classed according to the rating of the professor and according to the rating of the teaching assistant section leader:

		Professor		
		Excellent	Good	Fair or poor
Teaching assistant	Excellent	4	6	4
	Good	1	11.5	3
	Fair or poor	0	5.5	0

(The .5 came about because one student divided his response between fair and good.) On the basis of this, do you think there is a relationship between a student's rating of his professor and his rating of the teaching assistant?

12. Data were collected on smoking and education as part of the profile obtained for each family registered in a certain health-care facility. A random sample of 192 fathers yielded data as shown in the following table:

		Education			
		1	2	3	4
Smoking	S	2	32	22	7
	N	8	36	27	28
	Q	0	13	8	9

The code is as follows: 1, less than high school graduate; 2, high school graduate; 3, some college; 4, college graduate; S, currently smoking; N, never smoked; Q, smoked once but quit. Test the hypothesis of no relationship between the level of education and smoking habits.

16

Correlation and Regression

An earlier chapter considered the question of describing and detecting a relationship between a pair of categorical variables. We now take up similar questions for pairs of numerical variables. The structure of numerical value spaces makes the questions both more diverse and more complex. But the essence of those questions is this: What kind of relationship is suggested by the data?

As in the case of categorical variables, the existence of a relationship does not necessarily mean that one variable causes the other, although it may suggest the possibility that such is the case.

Some examples of pairs of possibly related numerical variables are these:

X	Y
Dose of a drug	Change in blood pressure
Temperature	Yield of a reactor
Advertising	Sales
Amount of carbon	Hardness of an alloy
Height of father	Height of son
High school rank	College performance
Frequency of cricket's chirps	Temperature

In most of these examples one suspects some kind of cause mechanism and would be terribly surprised if he found that the variables were not related.

A mathematical relationship between numerical variables is a *function* $y = f(x)$. A statistical relationship may be one of a tendency, which, even though not precise, may be strong enough to suggest causation or to permit prediction of a sort.

As in the case of categorical variables, there are two ways of obtaining bivariate data. One is to fix one variable at two or more levels and make measurements of the other variable at each of these levels. The other is to make pairs of measurements without controlling either. Sometimes it is the first way that is more natural, and sometimes it is the second. Among the examples cited earlier, the first way is natural in the first four, and the second is natural in the last three.

16.1 Prediction

A random variable cannot be perfectly predicted (except by chance); and if one predicts Y to have the value k, the mean-squared prediction error is

$$\text{m.s.p.e.} = E[(Y - k)^2] = \text{var } Y + [E(Y) - k]^2.$$

This is clearly smallest if the predicted value is taken to be $k = E(Y)$, and then the *minimum* m.s.p.e. is σ_Y^2.

Suppose, in the random pair (X, Y), the value of X, say, has been observed: $X = x_0$. In this case the best predictor of Y in the mean-square sense is the *conditional* mean, given $X = x_0$.

Minimum mean-square-error prediction of Y, given $X = x_0$:

$$\text{Predict } Y \text{ to be } E(Y|x_0).$$

$$\text{Minimum m.s.p.e.} = \text{var } (Y|x_0).$$

The conditional mean $E(Y|x)$, as a function of the given value x, is called the *regression function of Y on X*. It is useful to think of this as describing the tendency in the joint variation of Y with X, when X is thought of as the primary variable and Y the variable to be predicted.

The true regression function may be complicated, but it is often convenient and adequate to use a *linear* predictor, a linear function of x. The m.s.p.e. in predicting Y by the linear function $a + bx$ is

$$\begin{aligned}\text{m.s.p.e.} &= E[Y - (a + bX)]^2 \\ &= (a - \mu_Y + b\mu_X)^2 + \sigma_X^2\left(b - \frac{\sigma_{XY}}{\sigma_X^2}\right)^2 + \sigma_Y^2(1 - \rho^2),\end{aligned}$$

where ρ is the population *correlation coefficient*:

$$\rho \equiv \frac{\sigma_{XY}}{\sigma_X \sigma_Y}.$$

(The identity given above for m.s.p.e. can be checked by expanding the squared error and comparing coefficients of a^2, b^2, a, b, and ab.) From the expression on the right for m.s.p.e. it is apparent that this can be minimized by choosing a and b to make the first two terms zero. Thus the best *linear* predictor of Y in terms of X is given by

$$\mu_Y + \frac{\sigma_{XY}}{\sigma_X^2}(X - \mu_X).$$

This may also be written with the slope coefficient expressed in terms of ρ:

$$\mu_Y + \rho\frac{\sigma_Y}{\sigma_X}(X - \mu_X).$$

If the regression function *is* linear, then the best linear predictor is the best predictor (least-square sense). Also, the regression function *is* linear in an important bivariate model, that called *normal.* The generalization of the univariate normal model (Chapter 8) will not be studied here. Suffice it to say that the density is bell-shaped in the three-dimensional sense, although with possibly elliptical contours; that the regression functions of Y on X and of X on Y are linear; and that the conditional variances (of Y given X and of X given Y) are constant.

The mean-square prediction error involved in using the best linear predictor is $\sigma_Y^2(1 - \rho^2)$, this being the only term remaining when a and b are given their minimizing values. This shows, in a sense, the degree to which a linear prediction given X is an improvement over the prediction μ_Y that is best when X is not known:

> The r.m.s.p.e. in predicting Y can be reduced by a factor $\sqrt{1 - \rho^2}$ if X is known and one uses as a predictor the linear function
>
> $$\mu_Y + \frac{\sigma_{XY}}{\sigma_X^2}(X - \mu_X).$$

Observe that $1 - \rho^2$ must be nonnegative, which means that $\rho^2 \le 1$. Moreover, if $\rho = \pm 1$, a *perfect* prediction is possible with a linear function. The intermediate case, $\rho = 0$, is one in which it does not pay to use the value of X, even if it is known. The parameter ρ is thus a measure of *linear* correlation. (It may be, for instance, that Y is really a quadratic function of X—perfectly predictable from X, whereas ρ might be 0. That is, a linear function of X might afford no better prediction than μ_Y.)

Example 16-a Suppose it has been found, from much experience, that the regression function of first-year college G.P.A. (Y) on high school rank in per cent (X) is given by

$$y = 1.5 + .02x,$$

with a linear correlation coefficient of .60. Then by using this function to predict the G.P.A. of a student with rank x, one can reduce the r.m.s. prediction error from σ_Y (which it would be using the predictor μ_Y) to $\sigma_Y\sqrt{1 - .6^2} = .8\sigma_Y$. ◄

The variation in Y can be thought of as consisting of two components, expressed by writing the variance as a sum of two terms:

$$\sigma_Y^2 = (1 - \rho^2)\sigma_Y^2 + \rho^2\sigma_Y^2.$$

The first term measures the variation about the best-fitting straight line. The second term is the portion of σ_Y^2 attributable to a linear regression; it is large if there is a high degree of linear correlation.

16.2 The Method of Least Squares

It is natural to gather bivariate data by fixing one variable at various levels, when that variable can be controlled. The other variable is then thought of as having a distribution, for each level of the controlled variable, about a mean value that is, by definition, the regression function:

$$g(x) \equiv E(Y_x) = E(Y|X = x).$$

(The value of Y at $X = x$ will be denoted by Y_x when thinking of X as a deterministic variable that can be controlled, and by $\{Y|X = x\}$ when thinking of X as random but observed to have the value x.)

One might think of the regression function, in the controlled variable context, as the true relation between x and y, with randomness entering when the ideal response $g(x)$ is corrupted by a random error:

$$Y_x = g(x) + \varepsilon.$$

This random error ε, in a simple model, has mean zero and a variance independent of x:

Regression model:

$$Y_x = \{Y|X = x\} = g(x) + \varepsilon,$$

where $E(\varepsilon) = 0$ and σ_ε^2 is the same for all x.

Data of the form $(x_1, y_1), \ldots, (x_n, y_n)$ are assumed available for inference. The x's may be random, or they may be fixed, by design, as values of a controlled X. (Because of this, we do not attempt to keep the distinction clear by capitalization.)

One aim of regression analysis is to find the simplest regression function that fits a given set of bivariate data reasonably well. Here we treat only the simplest model, that in which the regression function is linear:

Linear regression model:

$$Y_x = \alpha + \beta x + \varepsilon,$$

where $E(\varepsilon) = 0$, var $Y_x \equiv \sigma^2$, and all observations on Y_x are independent.

This model is often useful when X is a controlled variable; and in the case of bivariate sampling, it *is* the model implied by the assumption of a bivariate normal distribution for (X, Y). The problems in linear regression are to estimate the parameters α and β as well as the error variance σ^2, perhaps to test hypotheses about the parameters, and to predict Y given $X = x_0$, using the estimated linear regression function.

A method of estimating the regression line is suggested by the fact that

$$\text{var } \varepsilon = E[(Y_x - \alpha - \beta x)^2]$$

is the *smallest* second moment of Y_x. If a and b are well chosen, then each squared "residual" $[y_i - (a + bx_i)]^2$ will tend to be small. The method of *least squares*, proposed by K. F. Gauss, is to choose a and b to minimize the average squared residual:

$$\frac{1}{n}\sum (y_i - a - bx_i)^2.$$

(This weighs the terms equally, which seems reasonable, since σ_ε^2 is the same at each x.) Let $\hat{\alpha}$ and $\hat{\beta}$ be the minimizing values of a and b, respectively. The corresponding "best" linear function is $\hat{\alpha} + \hat{\beta}x \equiv \tilde{y}$. The minimum average squared residual, about the fitted value $\tilde{y}_i$, is

$$\hat{\sigma}^2 \equiv \frac{1}{n}\sum (y_i - \tilde{y}_i)^2.$$

This is a natural estimator of σ^2.

The minimizing values $\hat{\alpha}$ and $\hat{\beta}$ can be found by methods of algebra or of calculus. The following algebraic identity, analogous to one given in the preceding section in seeking a best linear predictor, can be established† in much the same way:

$$\frac{1}{n}\sum [y_i - (a + bx_i)]^2 \equiv (a - \bar{y} + b\bar{x})^2 + S_x^2\left(b - \frac{S_{xy}}{S_x^2}\right)^2 + S_y^2(1 - r^2),$$

where S_{xy} is the covariance of the x's and y's and r the corresponding sample correlation coefficient:

$$S_{xy} = \frac{1}{n}\sum (x_i - \bar{x})(y_i - \bar{y}), \qquad r \equiv \frac{S_{xy}}{S_x S_y}.$$

Examination of the identity reveals that the mean-squared residual is minimized by choosing a and b so that each of the two terms involving a and b vanishes. The minimum mean-squared residual is then what is left, namely, the third term of the identity.

Least-squares linear regression function: $\hat{\alpha} + \hat{\beta}x$, where

$$\hat{\beta} \equiv \frac{S_{xy}}{S_x^2} = r\frac{S_y}{S_x},$$

$$\hat{\alpha} \equiv \bar{y} - \hat{\beta}\bar{x}.$$

The mean-squared "error" about this best line is

$$\hat{\sigma}^2 \equiv S_y^2(1 - r^2).$$

Example 16-b Consider the following hypothetical data points, chosen to be simple in order to illustrate the method without obscuring it with arithmetic: (0, 0), (1, 1),

† Indeed, it *has* been established, mathematically; the earlier demonstration *includes* the present case, as one in which probability or "mass" $1/n$ is assigned to each of the n data points.

(1, 0), (2, 1). Putting them in tabular form allows one to add columns for needed products and squares:

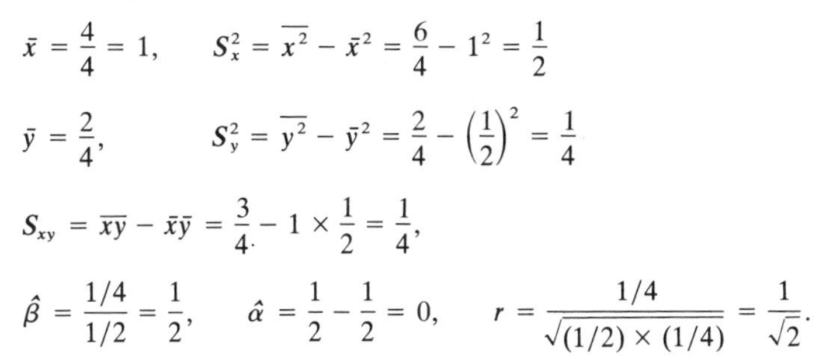

	x	y	x^2	y^2	xy	$(y - x/2)^2$
	0	0	0	0	0	0
	1	1	1	1	1	$(1/2)^2$
	1	0	1	0	0	$(-1/2)^2$
	2	1	4	1	2	0
Sum	4	2	6	2	3	1/2

The computations are then as follows:

$$\bar{x} = \frac{4}{4} = 1, \qquad S_x^2 = \overline{x^2} - \bar{x}^2 = \frac{6}{4} - 1^2 = \frac{1}{2}$$

$$\bar{y} = \frac{2}{4}, \qquad S_y^2 = \overline{y^2} - \bar{y}^2 = \frac{2}{4} - \left(\frac{1}{2}\right)^2 = \frac{1}{4}$$

$$S_{xy} = \overline{xy} - \bar{x}\bar{y} = \frac{3}{4} - 1 \times \frac{1}{2} = \frac{1}{4},$$

$$\hat{\beta} = \frac{1/4}{1/2} = \frac{1}{2}, \qquad \hat{\alpha} = \frac{1}{2} - \frac{1}{2} = 0, \qquad r = \frac{1/4}{\sqrt{(1/2) \times (1/4)}} = \frac{1}{\sqrt{2}}.$$

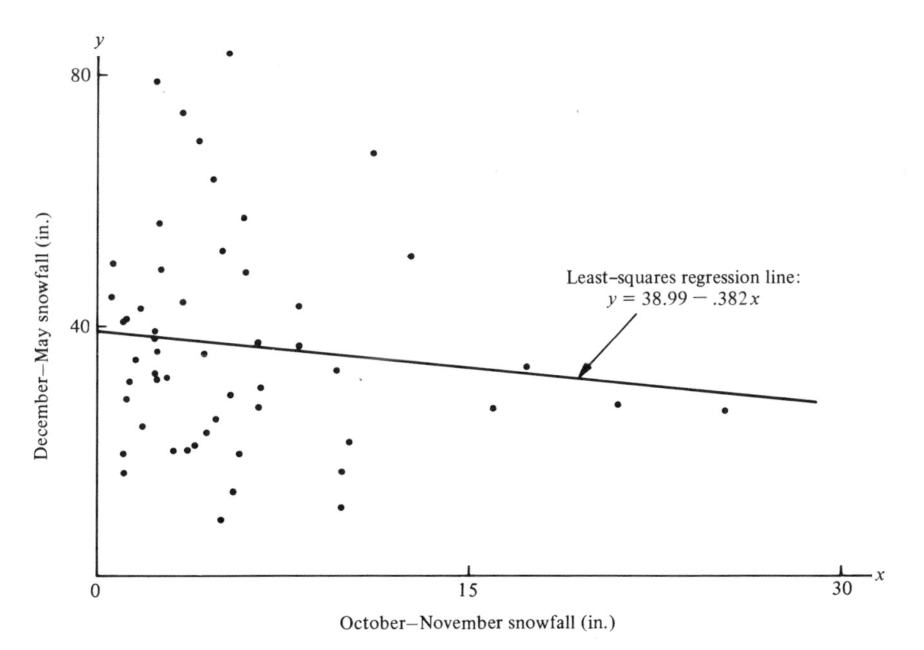

Figure 16-1

The least-squares regression line is $y = x/2$, and the average squared residual about it—the error variance estimate—is

$$\hat{\sigma}^2 = \frac{1}{4}\sum\left(y - \frac{x}{2}\right)^2 = \frac{1}{8} = S_y^2(1 - r^2). \blacktriangleleft$$

Example 16-c A TV weather announcer asserted in early December that because the snowfall through November had been light, we could expect a lot of snow during the rest of the season. The weather almanac for the area gives the following snowfall figures, year by year since 1920:

Oct.–Nov.	Dec.–May.	Oct.–Nov.	Dec.–May	Oct.–Nov.	Dec.–May
1.1	19.5	2.4	39.2	6.8	32.3
16.6	26.7	.6	44.5	10.3	10.9
1.2	41.0	26.3	26.2	5.7	13.4
1.4	31.0	3.8	20.1	10.6	21.2
3.2	20.0	2.9	31.5	2.4	37.8
6.9	30.1	10.3	16.6	2.5	78.8
5.0	25.1	2.5	31.4	5.6	28.9
5.3	51.8	4.5	35.4	0	28.9
3.6	43.8	4.1	20.9	4.3	69.4
4.6	22.9	21.8	27.3	1.6	34.5
5.2	9.0	2.5	35.8	3.6	73.8
0.7	45.9	2.7	48.9	1.1	16.4
2.4	32.3	5.6	83.3	4.9	63.2
6.0	19.6	11.6	67.4	6.2	57.2
8.5	43.1	10.1	32.8	6.3	48.4
2.6	56.2	1.9	23.8	13.2	51.0
1.8	42.8	6.8	27.1	1.1	40.6
1.2	28.3	8.5	36.7	18.0	33.2

The least-squares coefficients are computed to be as follows:

$$\hat{\beta} = -0.382, \qquad \hat{\alpha} = 38.99.$$

The data are plotted in Figure 16-1, together with the least-squares line. The coefficient of correlation is

$$r = -.119.$$

The factor $\sqrt{1 - r^2}$, an estimate of $\sqrt{1 - \rho^2}$, has the value .992, so not much reduction in prediction error is afforded by the use of the least-squares prediction of December–May snowfall, given the October–November snowfall.

16.3 Inference for Linear Regression

The least-squares coefficients $\hat{\alpha}$ and $\hat{\beta}$ can be thought of as estimates of the true coefficients in the regression function, $\alpha + \beta x$. To see how good they are as estimators, one has to study their distribution, and in order that this be possible, the specification of the regression model must include a distribution for the error component ε. This is usually assumed to be *normal.*

Normal linear regression model:

$$Y_x = \alpha + \beta x + \varepsilon,$$

where ε is $\mathcal{N}(0, \sigma^2)$. Observations on Y_x are independent.

Because the least-squares estimators $\hat{\beta}$ and $\hat{\alpha}$ are *linear* functions of the y's, given the x's, it is not hard to show that in the normal model their distributions are also normal, with means and variances as follows.

Distribution of the least-squares estimators for the normal model:

$$\text{(i) } \hat{\alpha} \text{ is } \mathcal{N}\left(\alpha, \frac{\sigma^2}{n}\left(1 + \frac{\bar{x}^2}{S_x^2}\right)\right).$$

$$\text{(ii) } \hat{\beta} \text{ is } \mathcal{N}\left(\beta, \frac{\sigma^2}{nS_x^2}\right).$$

Since both variances involve the unknown σ^2, an estimate of σ^2 is needed for defining standard errors. The minimum mean-squared error provides such an estimate:

$$\hat{\sigma}^2 = \frac{1}{n}\sum (y_i - \hat{\alpha} - \hat{\beta}x_i)^2 = S_y^2(1 - r^2).$$

This estimator has $n - 2$ "degrees of freedom," in that the residuals upon which it is based satisfy two linear relations:

$$\sum (y_i - \hat{\alpha} - \hat{\beta}x_i) = 0,$$

$$\sum x_i(y_i - \hat{\alpha} - \hat{\beta}x_i) = 0.$$

(These *normal equations* of the least-squares process are the result of setting derivatives with respect to a and b equal to 0, when calculus is used to minimize the sum of squared residuals.) It is said that "2 degrees of freedom are used up in estimating α and β."

Standard errors for $\hat{\alpha}$ and $\hat{\beta}$ as estimates of α and β are then obtained by using $\hat{\sigma}$ in place of σ in the formulas for their standard deviations.

Standard error of $\hat{\alpha}$:

$$\hat{\sigma}\sqrt{\frac{1}{n} + \frac{\bar{x}^2}{nS_x^2}}.$$

Standard error of $\hat{\beta}$:

$$\frac{\hat{\sigma}}{\sqrt{nS_x^2}}.$$

It can be shown that if σ is replaced by $\hat{\sigma}$ in the formulas for standardized $\hat{\alpha}$ and $\hat{\beta}$, and n by $n - 2$, the resulting variables have the t-distribution with $n - 2$ degrees of freedom. This means that confidence intervals can be constructed as follows:

Confidence interval for α:

$$\hat{\alpha} \pm k\frac{\hat{\sigma}}{\sqrt{n - 2}}\sqrt{1 + \frac{\bar{x}^2}{S_x^2}},$$

and for β:

$$\hat{\beta} \pm k\frac{\hat{\sigma}}{\sqrt{(n - 2)S_x^2}},$$

where k is a percentile of $t(n - 2)$, defined by confidence coefficient η:

$$k = t_{(1+\eta)/2}(n - 2).$$

(The confidence coefficient does *not* apply to an assertion that the intervals simultaneously cover α and β, respectively, but to *one or the other.*)

Directions for testing hypotheses about α or about β will not be given here explicitly. However, it is clear that the confidence intervals given above can be used as the basis for testing particular parameter values.

The particular null hypothesis $\beta = 0$, corresponding to independence of X and Y in the bivariate normal model, can be thought of as $\rho = 0$. The sample correlation r is a natural test statistic for hypotheses about ρ, and it can be shown that *if* $\rho = 0$, the statistic

$$T = \sqrt{n - 2}\frac{r}{\sqrt{1 - r^2}}$$

has a $t(n - 2)$ distribution. A t-test using this statistic is actually equivalent to the t-test based on $\hat{\beta}$.

Example 16-d The snowfall data of Example 16-c led to the least-squares linear regression function $38.99 - .382x$. The approximate 95 per cent confidence limits for β are

$$\hat{\beta} \pm \frac{2\hat{\sigma}}{\sqrt{(n - 2)S_x^2}} = -.382 \pm .835,$$

so at a 5 per cent significance level one could not reject the value $\hat{\beta} = 0$ (corresponding to the model in which X is of no value in predicting Y). In terms of testing for $\rho = 0$ (equivalent to $\beta = 0$) one has

$$T = \sqrt{(n - 2)}\frac{r}{\sqrt{1 - r^2}} = .864,$$

not close to usual rejection limits in the $t(52)$-distribution. ◄

To predict the value of Y at a particular value $X = x_0$, the regression function $E(Y|x_0)$ is the best predictor in the mean-square sense. If the regression function is not known but is assumed to be linear, then it can be approximated by the value of Y on the least-squares line (fitted to given data) at $X = x_0$. If this approximate value, $\hat{\alpha} + \hat{\beta}x_0$, is used as a predictor, the mean prediction error is zero, and the mean-squared prediction error is

$$\text{m.s.p.e.} = E[(Y_{x_0} - \hat{\alpha} - \hat{\beta}x_0)^2] = \text{var}\,(Y_{x_0} - \hat{\alpha} - \hat{\beta}x_0).$$

To calculate this variance would require knowing cov $(\hat{\alpha}, \hat{\beta})$. This is not especially difficult to derive, but we give only the final result:

$$\text{m.s.p.e.} = \sigma^2\left[1 + \frac{1}{n} + \frac{(x_0 - \bar{x})^2}{nS_x^2}\right].$$

The error variance $\sigma^2 = \sigma_{Y|x}^2$ can be approximated here by $\hat{\sigma}^2 = S_y^2(1 - r^2)$, permitting an assessment of the m.s.p.e. in terms of the data and the distance from x_0 to the center of the data. Observe that the m.s.p.e. will be large if one tries to predict Y at a point x_0 far from the center of the data, because of the term $(x_0 - \bar{x})^2$.

Problems

1. From hundreds of thousands of scores on a certain exam it is known that the score (X) in Part A and the score (Y) in Part B are correlated with $\rho = .8$. Suppose that scores in both parts are scaled so that $\sigma_X = \sigma_Y = 10$, and $\mu_X = \mu_Y = 70$.

(a) What is the best linear prediction of Y given $X = 80$?
(b) Determine the r.m.s. prediction error involved in this prediction.

2. Four observations on (X, Y) are obtained as follows: $(1, 2)$, $(2, 0)$, $(3, 0)$, and $(4, 2)$.

(a) Determine the least-squares regression line of Y on X.
(b) What value do you predict for Y when $X = 5$?
(c) Determine the correlation coefficient r.
(d) It happens that the four points given lie on the parabola $y = (x - 5/2)^2 - 1/4$, a "regression function" that fits perfectly. How does one account for such a small coefficient of correlation?

3. Given these midterm and final exam scores for 12 students in a certain class:

Midterm	Final
71	83
49	62
80	76
73	77
93	89
85	74
58	48
82	78
64	76
32	51
87	73
80	89

(a) Make a scatter diagram [plot of the data points (x, y)] and try to guess r and the best-fitting line.
(b) Compute r.
(c) Determine the least-squares regression line.

4. The following data give the effects of 15 daily doses of thyroxine, in various amounts, on the kidney transamidinase activities of thyroidectomized rats. The quantity X is the logarithm of the daily dose per 100 grams of body weight, and Y is a measure whose units are micromoles G.A. formed per gram of kidney per hour (where G.A. is guanidinoacetic acid).

X	-2	-1.7	-1.4	-1.1	$-.8$	$-.5$
Y	18.6	21.8	27.0	35.7	41.1	46.9

Obtain the least-squares regression line.

5. In studying the reaction rate of a certain synthetase of a bovine lens, it is found that the reciprocal of this rate appears to be linearly related to the reciprocal of the substrate concentration. Obtain the least-squares estimate of this linear relationship, given the following data:

Reciprocal of substrate concentration (nanomoles of product formed/minute)	Reciprocal of reaction rate (micromoles/liter)
24	.429, .444
20	.293, .293
16	.251, .268
12	.207, .216
8	.239, .218
6	.180, .199
2	.156, .167

6. Obtain a 90 per cent confidence interval for the slope β of the true regression line in Problem 4. (From the computations of the problem, $S_x = .5123$, $S_y = 10.22$, and $\hat{\beta} = 19.82$.)

7. Ten couples comparing notes gathered the following data on the number of years of college attended:

Husband	4	1	7	6	5	4	7	5	4	6
Wife	4	0	2	4	4	0	6	2	2	4

Determine the coefficient of correlation. (Would a t-statistic be appropriate for testing $\rho = 0$?)

8. (a) From the data of Problem 4, predict the value of y corresponding to a log daily dose of -1.5, and give the rms error of prediction. (The least-squares line was $y = 56.6 + 19.82x$.)

(b) If in Problem 4 one observed a response of $Y = 24$, what would you estimate to be the log daily dose that produced this response?

9. An educator has data on 500 students consisting of a grade-point average (G.P.A.) after two years of college and a preentrance aptitude score for each student. The G.P.A.'s are scattered about a mean $\bar{Y} = 2.72$ with a standard deviation $S_Y = .36$, and the aptitude scores have mean $\bar{X} = 70$ and standard deviation $S_X = 10$.

(a) Without using any relationship between X and Y, the predicted G.P.A. for any new student would be $Y = 2.72$. What is the rms error involved in this prediction?

(b) The educator computes $r = .60$ and a least-squares regression function $y = .2 + .036x$. Using these, what G.P.A. would he predict for a new student who scores 82 on the aptitude test, and what is the r.m.s. error involved in this prediction?

Analysis of Variance

This chapter introduces a technique for analyzing numerical data from several populations. These populations are defined by the application of different treatments to experimental units. The treatments may be thought of as levels or categories of a variable factor. The development is related to several earlier problems. It extends the notion of comparing means, considered for two populations in Chapter 12, to several populations. As in problems of regression, a controlled variable defines the several populations; but here the controlled variable is either categorical or (if numerical) treated as categorical. Moreover, the analysis of variance, really an analysis of sums of squares to test equality of means, is a technique that is applicable also to regression problems; indeed, the tests suggested in Chapter 16 are special cases of such an analysis.

17.1 A Single Factor

Such things as the kind of fertilizer, the operator of a production machine, the type of pedagogical method, the brand of typewriter, the type of advertising medium, and so on, are thought of as *factors* that may (or may not) contribute to a *response*—growth, output, achievement, sales, or whatever measure is of concern in a particular situation. These factors are usually categorical, as are the examples just cited, but they may be naturally ordered, as are, e.g., levels of education. The different categories of a factor will be referred to as *levels*, even when they are not naturally ordered.

In this section we consider problems in which there is a single factor that will be taken into account and controlled. This factor will be applied at various levels. A level of the controlled factor defines a *treatment*, applied to experimental units, and the problem is to determine whether or not a change in level brings about a change in response. If it does, there is said to be a *treatment effect* or *factor effect*.

Other factors that may be present are assumed to contribute to response in an unknown fashion (although some of them may be identifiable). Taken together, these factors result in a component of response that is thought of and modeled as *random*, or as *random error*. The presence of random errors makes the problem a statistical one.

As will soon become evident, repeated observations are necessary to sort out the treatment effect from the error, so it will be assumed that each level of the factor or treatment is applied to several experimental units. Let X_{ij} denote the jth observed response to factor level i, $j = 1, \ldots, n_i$. This response is thought of as consisting of a component associated with the treatment and an additive random error:

$$X_{ij} = \mu_i + \varepsilon_{ij}.$$

The errors ε_{ij} are assumed to be *independent, normal* random variables with mean 0 and variance σ^2 independent of i and j. The mean μ_i can be further decomposed into a common component μ plus a deviation τ_i, chosen so that $\sum \tau_i = 0$. The deviation $\tau_i = \mu_i - \mu$ is the *effect* at level i.

Single-factor model:

$$X_{ij} = \mu + \tau_i + \varepsilon_{ij}, \qquad \begin{cases} i = 1, \ldots, k, \\ j = 1, \ldots, n_i, \end{cases}$$

where $\sum \tau_i = 0$ and the ε_{ij} are independent $\mathcal{N}(0, \sigma^2)$.

With this model, the null hypothesis of no treatment effect is that $\mu_i \equiv \mu$, equivalent in turn to $\tau_i \equiv 0$:

$$H_0\colon \quad \mu_1 = \mu_2 = \cdots = \mu_k.$$

To see how to test this null hypothesis, it will be helpful to look at some data of known structure. Ordinarily, the parameters (μ, τ_i) of the model are not and never will be known. However, to illustrate the model here we shall assign values to the parameters and then construct artificial data by adding in random errors from a known population.

It will be helpful to have a concrete situation in mind, even though the data will be artificial. Suppose that an operator is an essential part of a production process, and the question has arisen as to whether or not it makes a difference in yield (X) which operator is used. Suppose there are three operators; these are the levels of the operator factor, or treatment, so in the above notation $k = 3$. Suppose further that, without an error component, the average yield for the three operators is 10, with operator 1 deviating from that by +3, operator 2 by −1, and operator 3 by −2:

$$\mu = 10; \qquad \mu_1 - \mu \equiv \tau_1 = 3, \qquad \tau_2 = -1, \qquad \tau_3 = -2.$$

So, except for random errors, the yield would always be $\mu_1 = \mu + \tau_1 = 13$ for operator 1 (or level 1, or population 1), always $\mu + \tau_2 = 9$ for operator 2, and always 8 for operator 3.

Suppose next that four production runs are made by each operator, that is, four observations at each level ($n_1 = n_2 = n_3 = 4$), and that the random errors involved are distributed as $\mathcal{N}(0, 1)$. A sample of 12 such ε's is simulated by taking 12 successive numbers from Table XIV (normal random deviates):

$$i = 1: \quad -.22, 1.04, .03, -1.04,$$

$$i = 2: \quad .40, -2.03, -.15, 1.43,$$

$$i = 3: \quad 1.63, 2.09, -.18, -.32.$$

The observed data will then be composed of the means μ_i plus the random errors: $X_{ij} = \mu_i + \varepsilon_{ij} = 10 + \tau_i + \varepsilon_{ij}$. These are listed in the following table along with sample means:

Level (i)	τ_i	Sample (X_{ij})	Mean ($\bar{X}_i$)
1	3	12.78, 14.44, 13.03, 11.96	13.0525
2	−1	9.40, 6.97, 8.85, 10.43	8.9125
3	−2	9.63, 10.09, 7.82, 7.68	8.805

But if there were not a treatment effect, that is, if $\tau_i \equiv 0$, then the data would vary randomly about the common mean $\mu = 10$, as in the following table:

Level (i)	τ_i	Sample (X_{ij})	Mean ($\bar{X}_i$)
1	0	9.78, 11.44, 10.03, 8.96	10.0525
2	0	10.40, 7.97, 9.88, 11.43	9.9125
3	0	11.63, 12.09, 9.82, 9.68	10.805

The reader has been in on the construction of the data and knows which set of data is which. But at this point he should study the two sets of *data* for a clue to which set involves a treatment effect. It is evident that, although the sample means are not all 10 in the latter set, they are closer to each other than are the means of the sample from the populations with treatment differences included.

If the population means are the same (no treatment effect), all equal to μ, then the "grand mean" or mean of *all* the observations,

$$\bar{X} \equiv \frac{1}{n}\sum n_i\bar{X}_i = \frac{1}{n}\sum\sum X_{ij}$$

(where $n = \sum n_i$), is an estimate of μ. The individual sample means $\bar{X}_i$, even then, will generally equal neither $\bar{X}$ nor μ, and the amount of their variation about $\bar{X}$ can be used in estimating the error variance σ^2.

When there is no treatment effect ($\mu_i = \mu$), the expected value of each $\bar{X}_i$ is μ, so that $\bar{X}_i - \bar{X}$ is approximately a deviation of $\bar{X}_i$ about its mean. But the variance of the ith sample mean is var $\bar{X}_i = \sigma^2/n_i$, so it is the weighted squared deviations $n_i(\bar{X}_i - \bar{X})^2$ whose average gives an estimate of σ^2, valid under H_0:

$$\hat{\sigma}_0^2 = \frac{1}{k}\sum_{i=1}^{k} n_i(\bar{X}_i - \bar{X})^2.$$

A slight modification of this, used in testing for the presence of a treatment effect, is called the *treatment mean square*:

$$\text{MSTr} \equiv \frac{1}{k-1}\sum_{i=1}^{k} n_i(\bar{X}_i - \bar{X})^2 = \frac{1}{k-1}\sum_{i=1}^{k}\sum_{j=1}^{n_i}(\bar{X}_i - \bar{X})^2 = \frac{\text{SSTr}}{k-1},$$

where SSTr is called the *treatment sum of squares.*

If the population means are *not* the same, that is, if there *is* a treatment effect, the treatment mean square MSTr will not be a valid estimate of σ^2. This is because the differences in population means (the τ_i's) will introduce more variation among the sample means $\bar{X}_i$ than can be attributed to randomness in the sample observations. This inflation is seen in the data sets constructed above; the estimates are

$$\text{MSTr} = \begin{cases} 23.46, & \text{with treatment effect,} \\ 0.92, & \text{no treatment effect.} \end{cases}$$

As anticipated, the first (with treatment) is much larger than the second. Thus intuition suggests that when MSTr is large, one should reject H_0 and conclude that there is a treatment effect. But how can one decide what is large, when (as is the case in practice) no data set without τ's is available for comparison?

Because there are several observations at each factor level, other estimates of the error variance σ^2 *are* available, estimates whose usefulness in estimation *does not depend on the absence of a treatment effect.* At each level, of course, there is a sample of several observations that would provide an estimate of error variance, namely, the sample variance; but since the error variance has been assumed to be the same at all treatment levels, a better estimate is obtained by *pooling* these sample variances, much as was done in the two-sample case with equal population variances. Pooling is accomplished by weighting the individual sample variances with the sample sizes and then dividing by something (to be explained later, but in the spirit of the $n - 1$ sometimes used in defining sample variance):

$$\text{MSE} \equiv \frac{1}{n-k} \sum n_i S_i^2 = \frac{1}{n-k} \sum_i \sum_j (X_{ij} - \bar{X}_i)^2.$$

The second expression for this *error mean square* shows that it may be thought of as an average of the squared deviations of all the sample observations, each about the mean of the sample it belongs to. It is also written as $\text{SSE}/(n-k)$, where

$$\text{SSE} \equiv \sum\sum (X_{ij} - \bar{X}_i)^2,$$

called the *error sum of squares*. The estimate MSE is referred to as the "within-samples" estimate of the error variance, in contrast to the "between-samples" (more properly, "among-samples") estimate MSTr. The within-samples estimate is the same for both of the above data sets, because the τ_i's present in one set are eliminated in the subtractions $X_{ij} - \bar{X}_i$:

$$\text{MSE} = 1.564.$$

(This is a sample estimate of the common population variance, which we happen to know is $\sigma^2 = 1$, because the errors were taken from Table XIV.)

The ratio of the within-samples and the between-samples variance estimates is the test statistic usually used to decide whether or not there is a factor effect:

$$F \equiv \frac{\text{MSTr}}{\text{MSE}} = \begin{cases} \dfrac{23.46}{1.564} = 15.0, & \text{data with treatment effect,} \\[2ex] \dfrac{.92}{1.564} = .589, & \text{data with no treatment effect.} \end{cases}$$

Fixing a critical value of F for a specified significance level requires knowing the distribution of F under H_0. This distribution depends only on the numbers of "degrees of freedom" (df) in its numerator and denominator. Table IV gives 95th and 99th percentiles of $F(r, s)$, the "F-distribution with r degrees of freedom in the numerator and s degrees of freedom in the denominator." A test with $\alpha = .05$ would call for rejecting H_0 if F exceeds the 95th percentile of $F(k - 1, n - k)$, where $k - 1$ is the numerator df and $n - k$ is the denominator df.

Test for H_0: $\mu_1 = \cdots = \mu_k$ (no factor effect). Define

$$F \equiv \frac{\sum n_i(\bar{X}_i - \bar{X})^2/(k - 1)}{\sum\sum (X_{ij} - \bar{X})^2/(n - k)},$$

and reject H_0 if $F > F_{1-\alpha}(k - 1, n - k)$, where α is the desired significance level (Table IV).

In the illustrative example above, the degrees of freedom for F are $k - 1 = 2$ and $n - k = 12 - 3 = 9$, and the 95th percentile of the corresponding F-distribution is $F_{.95}(2, 9) = 4.26$. So in the case of the data without a treatment component, with $F = .589 < 4.26$, the hypothesis of no treatment effect is accepted at $\alpha = .05$. The data with treatment component included gave $F = 15.0 > 4.26$; because this value of F is in the critical region of the test, H_0 would be rejected.

In defining the variance estimates used in the test ratio F, we divide the sum of squares by the corresponding numbers of degrees of freedom, in order that the traditional and available tables may be used.

The degrees of freedom in the present case are as follows†:

† These correspond to the number of algebraically independent quantities whose squares are the summands in the various sums of squares.

Treatment: df(Tr) = 2 = $k - 1$ (number of samples minus 1),

Error: df(E) = 9 = $n - k$
(number of observations minus number of samples).

The $n - k$ can be thought of as $\Sigma (n_i - 1)$, the sum of the degrees of freedom in the sample variances that are pooled in the estimate MSE.

The factor or treatment sum of squares SSTr and the error sum of squares SSE are related to the sum of squared deviations about $\bar{X}$, called the total sum of squares or SST, by the following identity:

$$\Sigma\Sigma (\bar{X}_{ij} - \bar{X})^2 \equiv \Sigma\Sigma (\bar{X}_i - \bar{X})^2 + \Sigma\Sigma (X_{ij} - \bar{X}_i)^2$$

or

$$\text{SST} \equiv \text{SSTr} + \text{SSE}.$$

This decomposition of total sum of squares into components attributable to error and to treatment, and the comparison of these components in terms of their degrees of freedom, is what is called an *analysis of variance*. It is significant that the degrees of freedom add, correspondingly:

$$\text{df(T)} = \text{df(E)} + \text{df(Tr)},$$

where df(T) = $n - 1$, the number of degrees of freedom in SST [the sum of squares of the n quantities $X_i - \bar{X}$, which are constrained by the single relation $\Sigma (X_i - \bar{X}) = 0$].

The calculation of the variances is often displayed in an *ANOVA* (ANalysis Of VAriance) table. For the treatment data given earlier, it is as follows:

Source	Sum of squares	Degrees of freedom	Mean square	F-ratio
Treatment	46.92	2	23.46	15.0
Error	14.08	9	1.56	—
Total	61.00	11		

Example 17-a A tensile test measures the quality of a spot weld of an aluminum-clad material. To determine whether there is a machine effect when welding a material of specified gauge, samples from the three machines used in the welding process are tested, with these results:

	Machine A	Machine B	Machine C
	3.2	4.9	3.0
	4.1	4.5	2.9
	3.5	4.5	3.7
	3.0	4.0	3.5
	3.1	4.2	4.2
Means	3.38	4.42	3.46

The overall or grand mean is $\bar{X} = 3.753$, and the sum of squares (SS) about this mean is decomposed as shown in the following ANOVA table:

Source	Sum of squares	Degrees of freedom	Mean square
Machine	3.349	2	1.675
Error	2.388	12	0.199
Total	5.737	14	

The test ratio is then

$$F = \frac{3.349/2}{2.388/12} = \frac{1.675}{.199} = 8.42.$$

This exceeds $3.89 = F_{.95}(2, 12)$, from Table IVa, so it would be concluded (with $\alpha = .05$) that there is a machine effect. (Looking at the sample means suggests that machine B is better than the other ones. The analysis given does not permit assigning a significance level to such a conclusion.) ◄

As mentioned in this last example, a conclusion about factor effect is drawn properly only about the levels of treatment (the machines, operators, brands, etc.) that are actually used in the experiment. That is, the treatment-effect components τ_i are constants associated with the specific treatment levels used; the corresponding model is referred to as a *fixed-effects* model. Sometimes a model is used, called a random-effects model, in which the τ_i's are considered as random variables, corresponding to machines (or operators or whatever) drawn randomly from a population of machines. Curiously, the test for no treatment effect is the same with this model as with the fixed-effects model discussed here, although the interpretation of the conclusion is different—as is the study of power.

The power or sensitivity of an ANOVA test is certainly of interest, but to calculate it requires extensive tables or charts (as in, e.g., W. J. Dixon and F. J. Massey, Jr., *Introduction to Statistical Analysis*, 3rd ed., New York: McGraw-Hill, 1969).

The testing of the null hypothesis of no treatment effect may be just the beginning of an analysis. For instance, it is natural to ask about the magnitudes of the various differences and which one or more machines are the better ones. For approaches to such questions, see texts on ANOVA under the topic of "contrasts."

17.2 Two Factors

Suppose there are two identifiable and controlled factors, A and B, and that it is desired to test whether, say, there is a factor B effect. One approach is to *fix* the level of factor A, but then any conclusions about B would only be valid when A enters at the level used in the experiment. The analysis to be taken up here permits determining whether there is a B-effect while allowing A to vary, or conversely.

Suppose the factor A is applied at r levels and B at c levels. There are then rc distinct treatments, or combinations of a level of A with a level of B. One might make several measurements of response for each of these rc treatments, but (perhaps surprisingly) under appropriate assumptions it is possible to learn something about the individual factors, even if only one observation per treatment or cell is obtained. This is the case that will be considered here.

More than one observation per cell would, in fact, permit a more sensitive test, and would allow for a more complicated model; however, there may be a dearth of experimental units, and being able to separate factor effects with a minimal amount of data is important. Let X_{ij} denote the single response observed when factor A enters at level i and factor B enters at level j; it is usually entered in the ij-cell of a two-way table with r rows and c columns corresponding to the factor levels. It will be assumed that the mean response μ_{ij} is composed of an overall average μ, a deviation τ_i caused by A, and a deviation β_j caused by B, all combining additively. Random errors, assumed independent and normal with mean 0 and the same variance σ^2, also combine additively with μ_{ij} to yield the observed X_{ij}.

Model for two factors (no interaction):

$$X_{ij} = \mu + \tau_i + \beta_j + \varepsilon_{ij}, \qquad \begin{cases} i = 1, \ldots, r, \\ j = 1, \ldots, c, \end{cases}$$

where $\Sigma\, \tau_i = \Sigma\, \beta_j = 0$, and the ε_{ij} are independent $\mathcal{N}(0, \sigma^2)$.

(The condition of "no interaction" in this simple model means that there is no provision for the contribution of factor A to be influenced by or dependent on the level at which B is operating. An analysis based on a more complicated model which includes an interaction component may be more realistic, but it requires more than one observation per cell.)

As usual, the presence of random errors makes it harder to sort out the effects, but to see how to do this it will be helpful to consider first what responses would be like *without* errors. Suppose, then, that $r = c = 3$, and that the means μ_{ij} are given by $\mu = 10$ and by

$$\tau_1 = -2, \quad \tau_2 = 6, \quad \tau_3 = -4,$$

$$\beta_1 = 2, \quad \beta_2 = -1, \quad \beta_3 = -1.$$

With no errors, the responses would be equal to the means μ_{ij}:

			β_j		Row
		2	−1	−1	means
τ_i	−2	10	7	7	8
	6	18	15	15	16
	−4	8	5	5	6
Column means		12	9	9	10

The entry μ_{21} is constructed as follows:

$$\begin{aligned} \mu_{21} &= \tau_2 + \beta_1 + \mu \\ &= (16 - 10) + (12 - 10) + 10 = 16 + 12 - 10 = 18. \end{aligned}$$

Similarly, *each* entry in the body of the table is computed as the sum of the marginal entries ($\tau_i + \mu$ and $\beta_j + \mu$) minus the entry in the corner (μ). Given only the body of the table, one can calculate the τ's and β's and decide if there is an A-effect ($\tau_i \neq 0$) or a B-effect ($\beta_j \neq 0$)—and how much it is. But, in practice the entries μ_{ij} will be obscured by errors ε_{ij}.

Suppose that we take nine normal random deviates from Table XIV, that is, nine independent errors from $\mathcal{N}(0, 1)$:

−.61	.80	−2.48
−1.34	.07	.34
−.54	1.07	−.73

Combining these with the μ_{ij} yields observations of the sort one might actually make, given next along with row and column means. These means are denoted $\bar{X}_{i\cdot}$ and $\bar{X}_{\cdot j}$, respectively, the dot in the subscript indicating which subscript has been "averaged out."

		Levels of B 1	2	3	Row means ($\bar{X}_{i\cdot}$)
Levels of A	1	9.39	7.80	4.52	7.237
	2	16.66	15.07	15.34	15.690
	3	7.46	6.07	4.27	5.933
Column means ($\bar{X}_{\cdot j}$)		11.170	9.647	8.043	9.620 ($\bar{X}_{\cdot\cdot}$)

The existence of variations in the row means might suggest a factor A effect, and in the column means, a factor B effect; but even with zero factor components τ_i and β_j, these row and column means would have *some* variation about $\bar{X}$, each being a mean of three observations: var $\bar{X}_{i\cdot} = \sigma^2/3$, and var $\bar{X}_{\cdot j} = \sigma^2/3$.

The sum of squares of deviations of the row means about the overall mean,

$$\text{SSA} \equiv \sum_i \sum_j (\bar{X}_{i\cdot} - \bar{X})^2 = \sum_i 3(\bar{X}_{i\cdot} - \bar{X})^2,$$

will tend to be large if τ_i is *not* identically 0; and similarly, the sum

$$\text{SSB} = \sum_i \sum_j (\bar{X}_{\cdot j} - \bar{X})^2 = \sum_j 3(\bar{X}_{\cdot j} - \bar{X})^2$$

will tend to be large if $\beta_j \not\equiv 0$. For the data above, SSA = 168.35 and SSB = 14.67. Without knowing about either τ_i or β_j, there is no way (so far) of judging whether these sums of squares are inordinately large.

An estimate of the error variance σ^2 that is valid even in the presence of factor A and factor B effects is constructed as follows. Since $\bar{X}_{i\cdot} - \bar{X}$ is an estimate of τ_i and $\bar{X}_{\cdot j} - \bar{X}$ is an estimate of β_j, and since $\bar{X}$ estimates μ, one can estimate the value of ε_{ij}:

$$\hat{\varepsilon}_{ij} = X_{ij} - (\hat{\mu} + \hat{\tau}_i + \hat{\beta}_j) = X_{ij} - [\bar{X} + (\bar{X}_{i\cdot} - \bar{X}) + (\bar{X}_{\cdot j} - \bar{X})]$$
$$= X_{ij} - \bar{X}_{i\cdot} - \bar{X}_{\cdot j} + \bar{X}.$$

(With no factor contributions this combination of observation and marginal means, as seen earlier, would be small.) So σ^2 should be given approximately by an average of the squares of these estimates. The *sum* of squares is

$$\text{SSE} \equiv \sum\sum (X_{ij} - \bar{X}_{i\cdot} - \bar{X}_{\cdot j} + \bar{X})^2.$$

Test ratios are now defined as ratios of variance estimates—mean squares with degrees of freedom as divisors:

$$F_A = \frac{\text{MSA}}{\text{MSE}} = \frac{\text{SSA/df(A)}}{\text{SSE/df(E)}}, \qquad F_B = \frac{\text{MSB}}{\text{MSE}} = \frac{\text{SSB/df(B)}}{\text{SSE/df(E)}}$$

If F_A is large, one takes that as evidence against the hypothesis of no factor A effect: $\tau_i \equiv 0$. If F_B is large, this is considered evidence against $\beta_j \equiv 0$. The ANOVA table is laid out as follows, using the above data for illustration:

Source	Sum of squares	Degrees of freedom	Mean square	F-ratio
Factor A	168.350	$r - 1 = 2$	84.175	79.58
Factor B	14.667	$c - 1 = 2$	7.333	6.93
Error	4.231	$(r - 1)(c - 1) = 4$	1.058	
Total	187.248	$rc - 1 = 8$		

The hypothesis that $\tau_i \equiv 0$ is rejected if $F_A > F_{1-\alpha}(2, 4)$, since under this null hypothesis, F_A has again an F-distribution, with $r - 1$ degrees of freedom in the numerator and $(r - 1)(c - 1)$ in the denominator.

Test for H_0: no factor A effect (assuming no interaction). Define

$$F_A \equiv \frac{\sum_i c(\bar{X}_i - \bar{X})^2/(r-1)}{\sum_i \sum_j (X_{ij} - \bar{X}_{i\cdot} - \bar{X}_{\cdot j} + \bar{X})^2/[(r-1)(c-1)]},$$

and reject H_0 if $F_A > F_{1-\alpha}[r-1, (r-1)(c-1)]$, where α is the desired significance level.

In the example considered, the 5 per cent critical boundary for each F-ratio is $F_{.95}(2, 4) = 6.94$. The F-ratio (79.58) for factor A exceeds this, and also exceeds the 1 per cent limit of 18.0 by a considerable margin. The evidence is strong that there is a factor A effect, that is, that not all τ_i's are zero.

The ratio for factor B is 6.93, barely inside ($6.93 < 6.94$) the acceptance region for a 5 per cent significance level. *We* know, from how we constructed the data, that the factor components β_j were not all zero. They were $(2, -1, -1)$, of magnitudes not far from the σ of the errors, and they have been almost hidden by the errors. The 5 per cent test reaches the wrong conclusion (accept $\beta_j \equiv 0$), but a test at the 6 per cent significance level would correctly reject $\beta_j \equiv 0$.

Example 17-b A wear-testing machine, with four weighted brushes under which samples of fabrics are fixed, is used to determine resistance to abrasion, measured as loss of weight after a given number of cycles. Data are obtained as follows:

		Brush position				
		1	2	3	4	Mean
Fabric	1	1.93	2.38	2.20	2.25	2.190
	2	2.55	2.72	2.75	2.70	2.680
	3	2.40	2.68	2.31	2.28	2.418
	4	2.33	2.40	2.28	2.25	2.315
Mean		2.303	2.545	2.385	2.370	2.401

To illustrate the computation of a term in SSE, let $i = j = 1$;

$$(X_{11} - \bar{X}_{1\cdot} - \bar{X}_{\cdot 1} + \bar{X})^2 = (1.93 - 2.19 - 2.303 + 2.401)^2 = (-.162)^2.$$

The quantity SS used in testing for fabric effect begins with the term

$$(\bar{X}_{1\cdot} - \bar{X})^2 = (2.19 - 2.401)^2 = .00445.$$

The results are given in the ANOVA table:

Source	Sum of squares	Degrees of freedom	Mean square	F-ratio
Fabric	.520	3	.173	13.319
Brush	.127	3	.042	3.242
Error	.117	9	.013	
Total	.764	15	—	—

Since $F_{.95}(3, 9) = 3.86 < 13.319$, the conclusion would be drawn (at $\alpha = .05$) that there *is* a fabric difference. On the other hand, $3.86 > 3.242$, so the differences among column means are not significant (at $\alpha = .05$).

In view of the latter conclusion, one might wonder if it is worth separating out the B-effect. If the data were treated as one-factor data with four replications at each level—incorporating any B-effect into the "error," the ANOVA table would be as follows:

Source	Sum of squares	Degrees of freedom	Mean square
Fabric	.520	3	.173
Error	.244	12	.0203
Total	.764	15	

The F-ratio is $.173/.0203 = 8.53 > 3.49 = F_{.95}(3, 12)$. The hypothesis of no fabric effect is again rejected at $\alpha = .05$, but the case is not as strong. Failure to take brush position into account when possible makes it less likely that a fabric effect will be detected. ◄

A word of caution: The test statistics F_A and F_B are not independent as random variables, so probabilities associated with simultaneous inference about the two factors are not readily calculable.

17.3 On Analysis and Design

In presenting a technique for analyzing certain simple one-factor and two-factor experiments, it was assumed that one would be working with data that were collected according to certain rules. These define what is termed the *design* of the experiment. The design of the abrasion-resistance experiment of Example 17-b calls for independent, single observations for each treatment, defined as a pairing of a brush position with a fabric type. If that is indeed how the data were obtained, then there is hope that the technique of analysis presented, suggested by the normal model for the given design, will yield useful results.

In Chapter 12, a test for comparing population means was given for paired data. As discussed there, pairing is a device—part of the *design* of the experiment—for obtaining greater power with a given amount of data in certain kinds of situations. Before analyzing samples $(X_1, \ldots, X_n)$ and $(Y_1, \ldots, Y_n)$, it is essential to know the regimen under which they were collected—the experimental design. If they were obtained by pairing, then the single sample test on the n differences is appropriate; if the samples are independent, the two-sample tests given in Chapter 12 may be used.

Ideally, an experiment should be designed before the collection of data, and designed so that some specified model will represent the collection of data. Then the data may be analyzed by a technique appropriate to that model. In practice, a statistician is often presented with data and told "Here, analyze this." He will then need to inquire about how the data were collected, for without such knowledge, any inference or evaluation of results would be hard to defend.

We have presented here only the simplest of designs, and the corresponding analyses by the most popular technique. Real problems often require more complicated designs, sometimes using a mixture of regression and classification variables. The art of designing experiments is discussed in, for example, *Fundamental Concepts in the Design of Experiments*, by Charles R. Hicks (New York: Holt, Rinehart and Winston, 2nd ed., 1973), and *Design of Experiments, a Realistic Approach*, by Virgil L. Anderson and R. A. McLean (New York: Marcel Dekker, Inc., 1974).

Problems

1. Three observations are made on each of three populations, defined by the treatments A, B, and C, respectively.

(a) If the sample results are as follows, test the null hypothesis of no treatment effect at the 5 per cent significance level:

Treatment	Sample
A	2, 6, 4
B	12, 6, 6
C	4, 6, 8

(b) Suppose the results had been as follows:

Treatment	Sample
A	4, 4, 4
B	8, 8, 8
C	6, 6, 6

Now what is the pooled variance (or error variance estimate)? Would you accept the null hypothesis at any level? [How is it, with the same variation among sample means in parts (a) and (b), that you would reach different conclusions?]

2. Redo Problem 4 of Chapter 12 as an ANOVA problem—one factor with two levels, and verify this relation (which is true in general) between the statistics used: $T^2(k - 1) = F(1, k - 1)$.

3. To compare the effectiveness of three different types of phosphorescent coating of airplane instrument dials, eight dials each are coated with the three types. Then the dials are illuminated by an ultraviolet light, and the following are the number of minutes each glowed after the light source was shut off:

Type 1	Type 2	Type 3
52.9	58.4	71.3
62.1	55.0	66.6
57.4	59.8	63.4
50.0	62.5	64.7
59.3	64.7	75.8
61.2	59.9	65.6
60.8	54.7	72.9
53.1	58.4	67.3

Test the null hypothesis that there is no difference in the effectiveness of the three coatings at the 1 per cent significance level.

4. To study the performance of three different detergents at three different water temperatures, the following "whiteness" readings were obtained with specially designed equipment:

	Detergent		
	A	B	C
Cold water	45	43	55
Warm water	37	40	56
Hot water	42	44	46

Perform a two-way analysis of variance using the level of significance $\alpha = 0.05$.

5. Treat the data in Problem 4 as three samples corresponding to "levels" of the brand of detergent, disregarding the factor of water temperature. Carry out the F-test for the hypothesis of no brand differences. (Is the result surprising? Comment.)

18 Some Industrial Applications

Various statistical methods are used in business and industry. This chapter takes up two that are important and are demonstrated to be of great value in the area of production "quality control." They involve ramifications of the basic ideas of testing hypotheses.

The quality of lots of manufactured items is a matter of considerable importance both to the producer and the consumer. Quality can be determined by a 100 per cent inspection, but this is often too expensive and not necessarily perfect. Moreover, it may be destructive in some instances. It is common practice to inspect only the items in a sample from the lot and to judge the whole lot on the basis of the sample quality. Quality may be measured by some numerical characteristic, or it may be defined by whether individual items are "good" or "defective." The latter case, where the method is referred to as "sampling of attributes," will be discussed in Section 18.1.

One way to maintain acceptable quality is to monitor a production process by sampling the output periodically to see the critical variables are maintained according to specifications. The control chart, to be discussed in Section 18.2, is a statistical tool for accomplishing this.

18.1 Sampling Inspection by Attributes

Articles may be classed as either *defective* or *good*, and the quality of a lot is measured by the proportion p of defectives in the lot. Small p means high quality, and one might arbitrarily set a standard of the form $p \leq p^*$ for a lot to be acceptable.

A sample is assumed to be drawn from the lot at random, without replacement. A lot is to be *accepted* or *rejected* on the basis of the fraction or proportion defective in that sample. A rule for doing this is essentially a test of the hypothesis that the lot is acceptable, although it is not necessary, in defining a rule, to specify a precise range of population proportions (such as $p \leq p^*$) as constituting H_0. The obvious type of decision rule to use is one of the form: "Accept the lot if the number of defectives in the sample is less than or equal to c, where c is a specific constant that defines the rule.

Notation:

N = lot size

n = sample size

Y = number of defectives in a sample

c = acceptance number (accept the lot if $Y \leq c$)

p = lot fraction defective.

Now, if there is a null hypothesis here, it is surely not simple; that is, it does not specify a particular value of p and so does not define unique probabilities. So a "significance level" is not defined, and the performance of the test or rule must be characterized by its power function or, as is customary in industrial applications, by a related function, the *operating characteristic*:

Operating characteristic for the rule with acceptance number c:

$$OC(p) \equiv P(\text{lot is accepted}|p) = P(Y \leq c|p).$$

The notation $P(E|p)$ means the probability of the event E in the model defined by the parameter value p. [The vertical bar has also been used to denote a conditional probability, and the usage here is not really inconsistent with this when viewed by a Bayesian, who thinks of p as a random variable (as in Chapter 13).]

The operating characteristic function is clearly the complement of the power function

$$OC(p) = 1 - \pi(p),$$

since one is the acceptance and the other the rejection probability. The OC-function (acceptance probability) should be large for good lots (small p) and small for bad lots (large p).

The exact probability that $Y \leq c$, which depends on the lot fraction defective p, is hypergeometric:

$$OC(p) = \sum_{k=0}^{c} \frac{\binom{M}{k}\binom{N-M}{n-k}}{\binom{N}{n}},$$

where $M = Np$ is the *number* of defectives in the lot. If the sample size is small compared to the lot size: $n \ll N$, then the hypergeometric probabilities may be approximated by binomial probabilities:

$$OC(p) \doteq \sum_{k=0}^{c} \binom{n}{k} p^k (1-p)^{n-k}.$$

This last expression does not involve the population size N.

If lot qualities are generally good, the acceptance number c will usually be small, and the significant portion of the OC-curve is that in which p is small. For small p, the approximate binomial probabilities may in turn be approximated by Poisson probabilities:

$$OC(p) \doteq \sum_{k=0}^{c} e^{-np} \frac{(np)^k}{k!}.$$

These cumulative probabilities are given in Table XII.

Example 18-a Consider a lot of size eight, and the plan defined by an acceptance number $c = 1$ in a sample of size four. The operating characteristic function is then

$$OC(p) = \sum_{k=0}^{1} P(Y = k|p) = \frac{\binom{M}{0}\binom{8-M}{4}}{\binom{8}{4}} + \frac{\binom{M}{1}\binom{8-M}{3}}{\binom{8}{4}},$$

where $p = M/8$. This is defined for $M = 0, 1, \ldots, 8$, and its values are given in the following table:

M	0	1	2	3	4	5	6	7	8
OC	1	1	$\frac{55}{70}$	$\frac{35}{70}$	$\frac{17}{70}$	$\frac{5}{70}$	0	0	0

◄

Example 18-b Consider the more realistic situation in which $N = 1000$, $n = 50$, and $c = 2$. In this case the binomial approximation is fairly good, and for small p the Poisson approximation applies:

$$OC(p) = \sum_{k=0}^{2} \frac{\binom{M}{k}\binom{1000 - M}{50 - k}}{\binom{1000}{50}} \doteq \sum_{k=0}^{2} \binom{50}{k} p^k (1 - p)^{50-k}$$

$$\doteq e^{-50p}(1 + 50p + 1250p^2).$$

This is plotted in Figure 18-1 (as though p varied continuously on the interval from 0 to 1, whereas actually $p = M/1000$). The same operating characteristic would be obtained, for practical purposes, if the lot size had been 10,000, with the sample size of $n = 50$. ◄

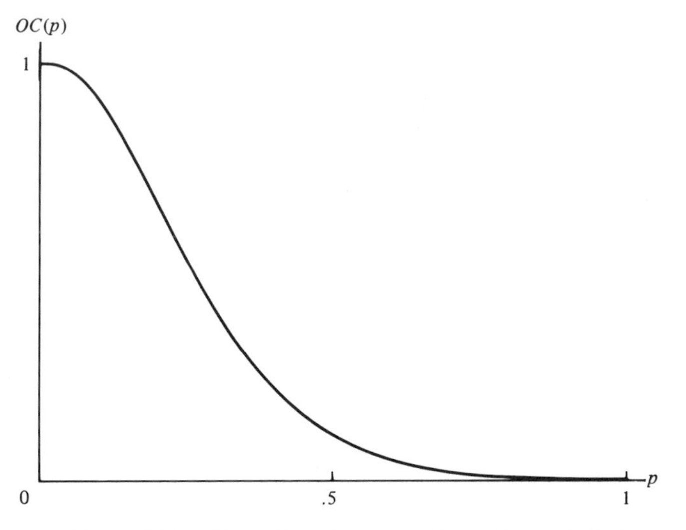

Figure 18-1 Operating characteristic for Example 18-b.

As suggested earlier, a problem with two choices of action such as we have here (accept the lot, reject the lot) can be considered as a problem of hypothesis testing. Since acceptable lots would be those with a p no greater than some given amount p^*, the "hypothesis" would be of the form

$$H_0: \quad p \leq p^* \qquad \text{(lots are of acceptable quality)},$$

$$H_A: \quad p > p^* \qquad \text{(lots should be rejected)}.$$

The testing problem is one-sided, and the tests considered above ($Y > c$) are correspondingly one-sided. But neither H_0 nor H_A pins down p so that probabilities can be computed—they are *composite* hypotheses. Thus there is neither a unique α nor a unique β in the usual sense, and the operating characteristic, giving probabilities as functions of p, is needed to tell the story.

Sometimes, however, it is convenient to define an α and a β of a sort, in terms of really good lots and definitely bad lots. That is, suppose p_1 is chosen so that a lot with $p \leq p_1$ ought to be accepted, i.e., rejected with very low probability. The probability p_1 is termed the acceptable quality level (AQL). The maximum probability of rejection for such acceptable lots is of particular interest to the producer:

Producer's risk:

$$\alpha \equiv \max_{p \leq p_1} [1 - OC(p)] = 1 - OC(p_1).$$

This is a bound on type I error sizes for $p \leq p_1$; it is achieved by $1 - OC(p)$ at $p = p_1$ because the OC-function (as can be shown mathematically) is a decreasing function of p. Similarly, the consumer, who does not want to accept bad lots, may select a p_2 such that lots with $p > p_2$ should seldom be accepted. This value p_2 is called the rejectable quality level (RQL). The maximum probability of accepting such lots is

Consumer's risk:

$$\beta \equiv \max_{p \geq p_2} OC(p) = OC(p_2).$$

The risks α and β defined in this manner are then determined by given values of n and c, the sample size and acceptance number. However, in designing a sampling plan, one may want to specify these two constants along with quality levels p_1 and p_2; doing so imposes two conditions on $OC(p)$, which are simultaneous equations for the design constants n and c.

The equations $OC(p_1) = 1 - \alpha$ and $OC(p_2) = \beta$, for given α and β, can only be solved for n and c by numerical means. Figure 18-2 was constructed using such numerical solutions and permits the approximate determination of the values of n and c that will yield $OC(p_1) = .95$ and $OC(p_2) = .05$.

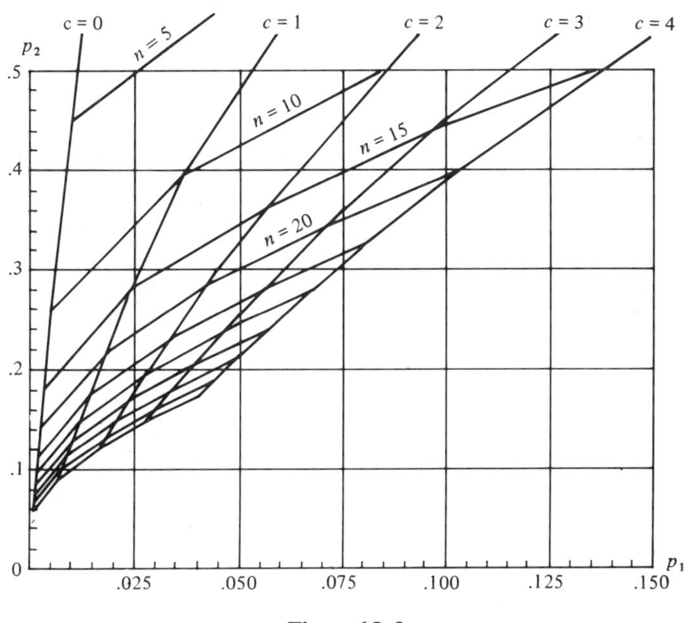

Figure 18-2

The curved grid lines correspond to $c = 0, 1, 2, 3, 4$ and to $n = 5, 10, \ldots, 50$. To use the chart, enter it at the chosen values of p_1 and p_2 and read the values of n and c close to the point (p_1, p_2). For example, suppose that it is desired to have a probability .95 of acceptance when $p = .05$ (p_1) and a probability .05 of acceptance when $p = .25$ (p_2). The point (.05, .25) lies on the $c = 3$ curve between $n = 25$ and $n = 30$. The actual α and β for the plan with $n = 28$ and $c = 3$ are as follows:

$$\alpha = 1 - OC(.05) = 1 - \sum_0^3 \binom{28}{k} (.05)^k (.95)^{28-k} \doteq .054.$$

$$\beta = OC(.25) = \sum_0^3 \binom{28}{k} (.25)^k (.75)^{28-k} \doteq .062.$$

18.2 Control Charts

A *control chart*, in its rudimentary form, is simply a graphical tool for conducting periodic significance tests, right at the scene of the production process, as an aid to maintaining acceptable standards of quality. However quality is defined, some specified variation in quality from piece to piece is usually accepted as a compromise between the high cost (and perhaps unattainability) of perfection and the consequences of marketing a product of uncontrolled quality. When the probability distribution assumed to describe piece by piece variation remains constant as the production proceeds, the process is said to be *in control*. A process that goes "out of control" because of *assignable* causes may be investigated and corrected. In an industrial environment there is a continuing effort to bring a process into acceptable control and then monitor its behavior.

The basis of a control chart procedure is a sequence of samples from the production line, taken at regular intervals. The sample size is usually small, on the order of 2 to 10, partly because of expediency, and partly because of the presumption that the process is not changing much during the production of that small sample. A statistic measuring some aspect of quality is plotted on what is called a *control chart* as a function of sample number (or equivalently, time). The chart will have *control limits* shown as horizontal lines. If the statistic falls between them, the process is assumed to be in control; if the statistic falls outside the control lines, action is taken to restore control.

Deciding the state of control after each sample is obtained is essentially testing the null hypothesis that the process is in control. The region outside the control limits is the critical region of the test, and its probability under the hypothesis of control is the significance level α of the test. For a given sample size, the fixing of the control limits determines α. As in testing generally, however, fixing them too loosely to achieve a very small α has the effect of increasing the tendency of the test to miss a real change in control, or increasing the type II error sizes.

The control chart involves a sequence of tests; if the control lines are set so that $\alpha = .05$ for a single test, then one would expect about 5 per cent of the points to fall outside the control lines when the process is actually in control. It is not easy to relate the α for a single test using given control lines to the performance of the control chart method over a sequence of tests. In any event, the optimal setting of control lines would vary from one situation to another, depending on the costs of readjusting production needlessly and of allowing too many unusable items to be produced out of control.

Standard practices have developed over the years, practices that can be used as a starting point, with modifications then made as experience in a particular situation suggests. In the United States, the standard control lines are set at the mean ± 3 standard deviations, referring to moments of the test statistic being used. In Great Britain, standard practice is to set warning limits at ± 1.96 standard deviations, and rejection limits at ± 3.09 standard deviations from the mean (corresponding to $\alpha = .05$ and $\alpha = .002$, respectively, for each point if the test statistic is normal).

Quality may be measured by a variable or variables that are categorical or numerical. Here we discuss only the case in which it is given by a single numerical variable—a dimension or weight, for example. The charts usually kept in the case of a single continuous variable are the $\bar{X}$- and R-charts.

In controlling quality to predetermined standards it is assumed that past experience provides parameter values for the distribution describing the production process when it is proceeding acceptably—in control. Since $E(\bar{X}) = \mu$ and $\sigma_{\bar{X}} = \sigma/\sqrt{n}$, the center line for the X-bar chart is taken to be μ and the control limits $\mu \pm 3\sigma/\sqrt{n}$.

The distribution of the range R is known when the population is *normal.* In this case, constants a_n and b_n are defined by

$$a_n \equiv E\left(\frac{R}{\sigma}\right), \qquad b_n \equiv \sqrt{\operatorname{var}\frac{R}{\sigma}},$$

and given in Table V for various sample sizes. The expected value $a_n\sigma$ is often taken as the center line of an R-chart, even if the population is not exactly normal. And since $\sigma_R = b_n\sigma$, the control limits are placed at $a_n\sigma \pm 3b_n\sigma$.

In summary, for controlling a process to specified mean and variance, the charts are defined below in terms of standard notation.† The center line is designated by CL and the upper and lower control limits by UCL and LCL. (In the LCL for the range, $a_n - 3b_n$ is negative for small n; but since $R \geq 0$ there is no point in drawing the line below $R = 0$.)

† American Society of Quality Control.

Controlling to specified μ and σ^2.

$\bar{X}$-chart:

$$\text{UCL} = \mu + A\sigma \qquad (A = 3/\sqrt{n}),$$

$$\text{CL} = \mu,$$

$$\text{LCL} = \mu - A\sigma.$$

R-chart:

$$\text{UCL} = D_2\sigma \qquad (D_2 = a_n + 3b_n),$$

$$\text{CL} = d_2\sigma \qquad (d_2 = a_n),$$

$$\text{LCL} = D_1\sigma \qquad [D_1 = \max(0, a_n - 3b_n)].$$

Example 18-c For purposes of illustration, a sequence of samples of size 5 was generated on a computer, the means and ranges being plotted on $\bar{X}$- and R-charts. The control limits were computed according to the formulas given above.

Figure 18-3 shows 100 samples from a process that is in control. As it happened, none of the plotted points fell outside the control lines. In Figure 18-4, the first 25

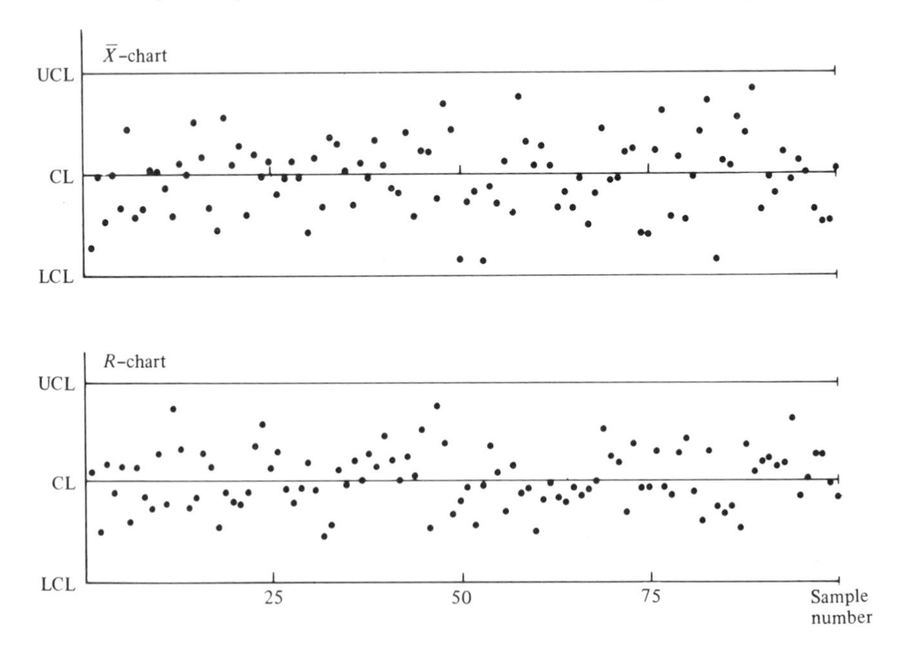

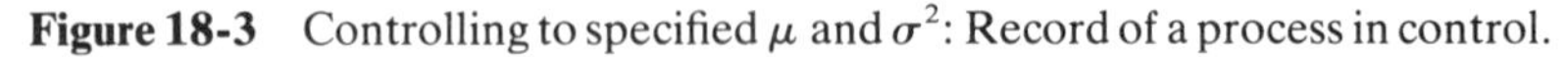

Figure 18-3 Controlling to specified μ and σ^2: Record of a process in control.

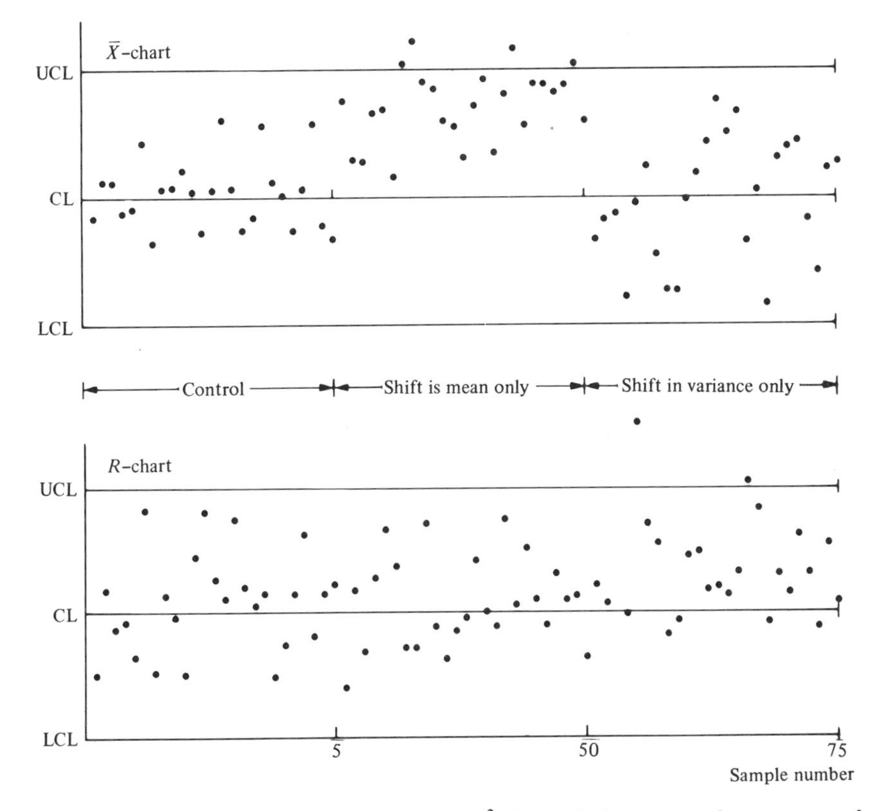

Figure 18-4 Controlling to specified μ and σ^2: Record of a process that goes out of control.

samples were recorded while the process was in control. In the next section, samples 26–50, the mean was shifted; this does become evident in the $\bar{X}$-chart where $\bar{X}$ goes above the UCL. The R-chart does not noticeably change when only the mean is changed. In the final section (51–75) the mean was restored but the variance was increased. Both the R- and $\bar{X}$-charts give evidence of this; the range actually exceeds the UCL, and the pattern of $\bar{X}$-values clearly has a greater variability since $\sigma_{\bar{X}}$ is proportional to σ, although the process level is not changed.

Figure 18-5 shows a record of 75 samples with 1–25 in control, but with a linear trend in the mean from 26 to 75 (only the $\bar{X}$-chart is shown because the ranges are not affected by a shift in μ). Figure 18-6 shows a record of 75 samples with 1–25 in control but with a linearly increasing σ. The increase in variability shows up in both $\bar{X}$ and R.

If the means were changed but, at the same time, the variance decreased, the $\bar{X}$-record might not go outside the limits. However, this combination of changes would be noticed in the R-chart, and the R's may even go below the LCL; it might also be noticed in the $\bar{X}$-chart, since the variability in $\bar{X}$ is proportional to σ^2.

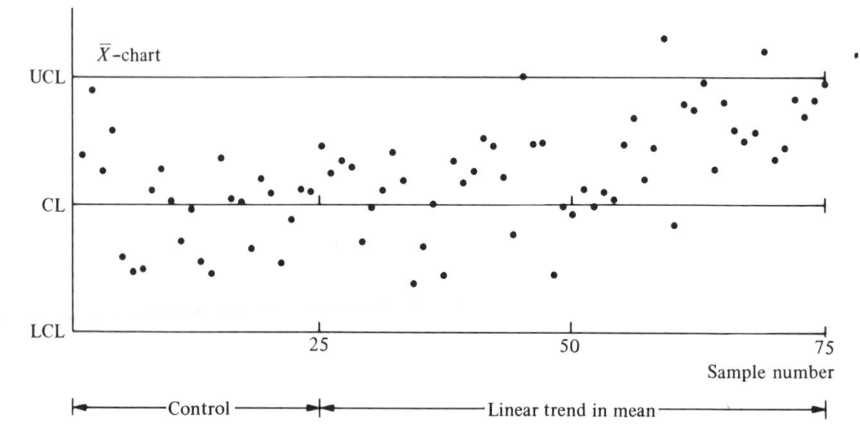

Figure 18-5

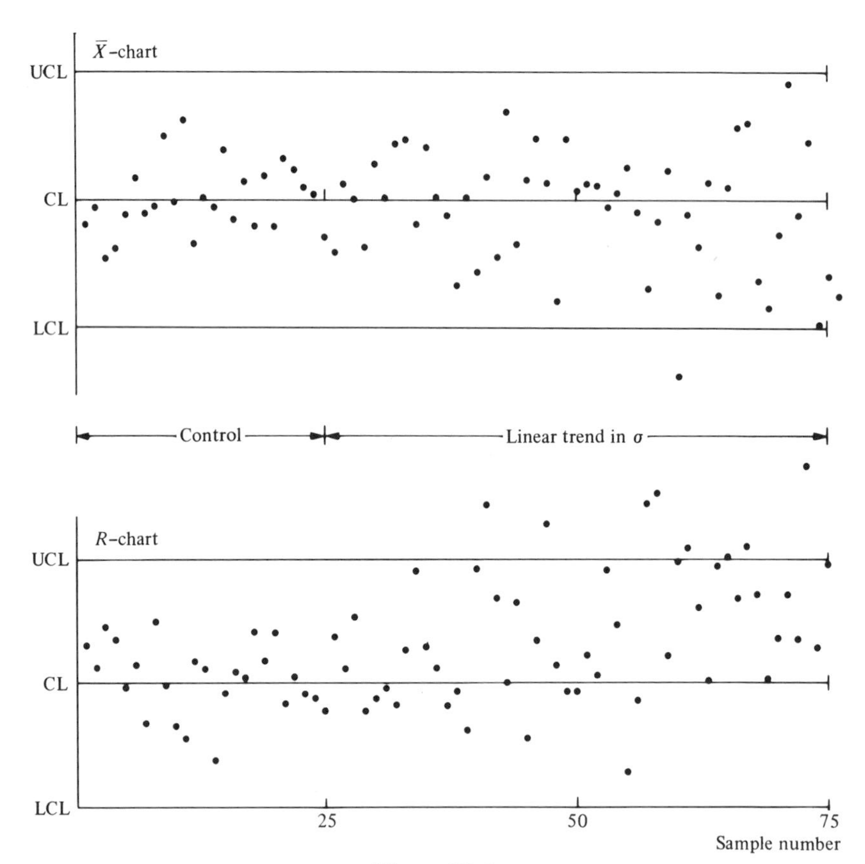

Figure 18-6

Anyone using these control charts would soon see that they contain more information than just whether or not the statistic goes outside the control limits. Even the eye can pick up variations in time that will point to process changes. Too many points above a center line, for instance, or a sequence of several points steadily rising on the chart, would alert one to the possibility that something is going on. Some of these ideas can be formalized into tests based on "runs," discussed in more comprehensive treatments of the subject. ◄

Control charts are also used to check control of a process for which standards have not yet been set. "In control" would simply mean, in such contexts, that the population is not changing. Again small samples are taken from the process periodically. Several values of $\bar{X}$ and R are entered on the charts for a number of samples before the center lines and control limits are determined. A decision as to whether or not there is control will be made after the control lines are in place.

The center line for the $\bar{X}$-chart should be at $\mu = E(\bar{X})$, but since μ is unknown, the mean $\bar{\bar{X}}$ of the sample means is used to estimate μ. The control limits should be at $\mu \pm 3\sigma/\sqrt{n}$, but since σ is not known, $\bar{R}/a_n$ is used in its place, where $\bar{R}$ is the mean of the ranges of the samples on hand.

For the R-chart, the center line should be at $a_n\sigma$; this is estimated by $\bar{R}$. The control limits would then extend $3\sigma_R$ on either side; but an estimate of σ_R must be used:

$$\hat{\sigma}_R = b_n\hat{\sigma} = \frac{b_n}{a_n}\bar{R}.$$

Again in terms of standard notation, the center lines and control limits are defined as follows:

Checking control with current data:

$\bar{X}$-chart:

$$\text{UCL} = \bar{\bar{X}} + A_2\bar{R} \qquad (A_2 = 3/a_n\sqrt{n}),$$
$$\text{CL} = \bar{\bar{X}},$$
$$\text{LCL} = \bar{\bar{X}} - A_2\bar{R}.$$

R-chart:

$$\text{UCL} = D_4\bar{R} \qquad (D_4 = 1 + 3b_n/a_n),$$
$$\text{CL} = \bar{R},$$
$$\text{LCL} = D_3\bar{R} \qquad [D_3 = \max(0, 1 - 3b_n/a_n)].$$

Control-chart constants for both kinds of charts presented here are given in Table 18.1. If a process is found to be in control using these charts, and if the levels of mean and variability are acceptable, the control lines that have been established can be used for checking the process as it proceeds further.

Table 18.1 Control-chart constants

	Sample size 3	4	5	6	7	8	9	10
$d_2(=a_n)$	1.693	2.059	2.326	2.534	2.704	2.847	2.970	3.078
b_n	.888	.880	.864	.848	.833	.820	.808	.797
A_1	1.732	1.500	1.342	1.225	1.134	1.061	1.000	.949
A_2	1.023	.729	.577	.483	.419	.373	.337	.308
D_1	0	0	0	0	.205	.387	.546	.687
D_2	4.358	4.698	4.918	5.078	5.203	5.307	5.394	5.469
D_3	0	0	0	0	.076	.136	.184	.233
D_4	2.574	2.282	2.114	2.004	1.924	1.864	1.816	1.777

Example 18-d Twenty-five samples of 5 are to be checked for control, and it is found that the overall mean is $\bar{\bar{X}} = 32.36$, and the mean range is $\bar{R} = 4.136$. The control lines are then as follows:

$$\bar{X}:\quad 32.36 \pm .577 \times 4.136 \qquad \text{or} \qquad 29.97 \quad \text{and} \quad 34.75,$$

$$R:\quad 0 \quad \text{and} \quad 8.74 = D_4\bar{R}.$$

The 25 samples are plotted in the $\bar{X}$-control chart in Figure 18-7 and are evidently in control. The same control lines are then used for observing the process further, and the means of 15 additional samples of 5 are plotted on the same chart. These dip below the lower control line, and it is also apparent from the overall pattern that something has happened to the process. ◄

Problems

1. A company will buy a shipment of 500 items if its inspector finds at most two defectives in a sample of 75. If the shipment contains 10 per cent defective items, what is the probability that it is accepted?

2. For a lot size $N = 8$ and samples size $n = 3$, consider the two sampling plans defined by acceptance numbers $c = 0$ and $c = 1$.

(a) Determine the *OC* function for each plan.
(b) Determine, for each plan, the consumer's and producer's risk, using $p_1 = 1/8$ and $p_2 = 1/2$.

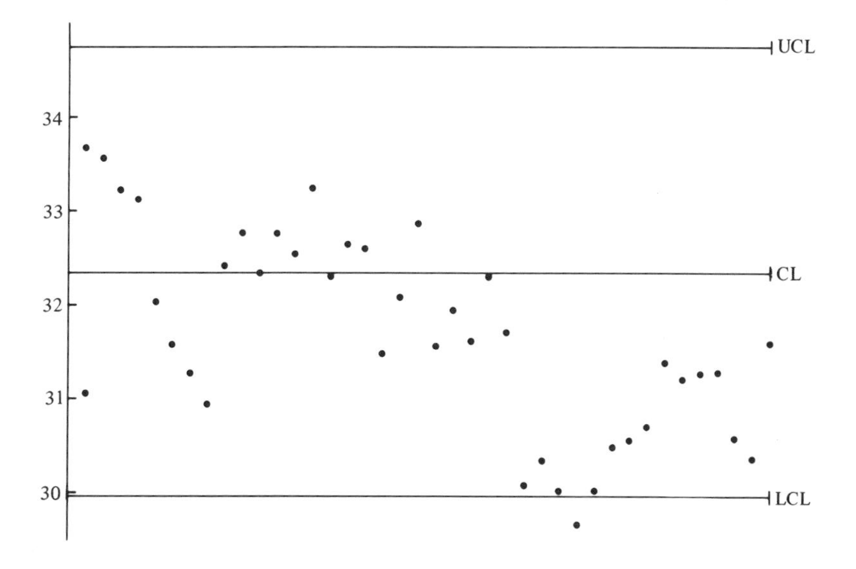

Figure 18-7 Control chart for Example 18-d.

3. Compare the operating characteristic of the first plan of Problem 2 ($N = 8$, $n = 3$, $c = 0$) with that for the situation $N = 100$, $n = 3$, $c = 0$.

4. Determine the *OC* function for the plan with acceptance number $c = 1$, when $N = 2000$, $n = 25$.

5. A process is known to have been in control with mean 15 and standard deviation 3.5. Samples of size 8 are obtained to check for continued control.

(a) Determine the center line and control limits for the $\bar{X}$- and R-charts.
(b) Suppose that the mean shifts to 16 with no change in standard deviation. Compute the probability that this shift will be detected on the first sample after the change (assuming a normal population). Determine also the probability that the shift is detected in either the first or second sample after the change.

6. Construct $\bar{X}$- and R-charts for samples of size 4 from a process that is supposed to be normal with mean 0 and variance 1. Take successive observations starting from a random point in Table XIV and use there (in groups of 4) as samples for checking control. At some point, add .5 to each observation and see how effective the charts are in picking up this shift.

7. Consider the following six subgroups of four observations, each taken in sequence (down in each column, then across). The data were obtained by measuring the power consumption of a unit and are given in kilowatt hours per 24 hours. No standards have been determined for the process.

3.79	3.53	3.59	3.83	3.75	3.73
3.53	3.71	3.35	3.84	3.90	3.82
3.86	3.89	3.69	3.59	3.67	3.75
3.57	3.85	3.77	3.49	3.63	3.65

(a) Estimate the mean power consumption.
(b) Determine the control limits for the $\bar{X}$-control chart.
(c) Determine the control limits for the R-control chart.
(d) Remark about the statistical control from the data above.

APPENDIX

A

Generating Functions

The idea of a generating function has been found most fruitful in the study of probability distributions. A generating function is a transform of density and probability functions that simplifies analyses for certain sums of random variables. Transforms of various kinds are suited to different needs and situations.

A.1 The Moment-Generating Function

Suppose that a random variable X has finite moments of all orders: $E(X^0)$, $E(X^1)$, $E(X^2), \ldots$. This sequence of numbers can be used as coefficients in a power series, with the kth moment as the coefficient of $t^k/k!$. Formally,

$$EX^0 + (EX)t + (EX^2)\frac{t^2}{2!} + (EX^3)\frac{t^3}{3!} + \cdots$$

$$= E\left[1 + Xt + \frac{(Xt)^2}{2!} + \frac{(Xt)^3}{3!} + \cdots\right] = E(e^{tX}).$$

This mean value is dependent on t—is a *function* of t:

Moment generating function of X:

$$\psi_X(t) \equiv E(e^{tX}).$$

Now, from calculus it is known that if a function has a Maclaurin series expansion—an expansion "about $t = 0$" in nonnegative powers of t:

$$\psi(t) = a_0 + a_1 t + a_2 \frac{t^2}{2} + \cdots,$$

then the coefficients are uniquely given by the values of its derivatives at $t = 0$: $a_k = \psi^{(k)}(0)$. But this would mean that the moments, used earlier as coefficients, are given by these derivatives:

$$E(X^k) = \psi^{(k)}(0), \qquad k = 0, 1, \ldots.$$

This fact is also evident upon formal differentiation under the $E(\ )$:

$$\psi'(t) = \frac{d}{dt} E(e^{tX}) = E(Xe^{tX}), \qquad \psi'(0) = E(X).$$

Similarly, continuing the differentiation, $\psi''(t) = E(X^2 e^{tX})$, and so on. In summary:

The moments of a probability distribution can be calculated as coefficients in the Maclaurin expansion of the m.g.f., or equivalently as values of its derivatives at $t = 0$.

Example A-a The model for the spinning pointer—a probability distribution with constant density on, say, $0 < x < 1$, has the m.g.f.:

$$E(e^{tX}) = \int_{-\infty}^{\infty} e^{tX} f(x)\, dx = \int_0^1 e^{tx}\, dx = \frac{e^t - 1}{t}.$$

Now, since

$$e^t = 1 + t + \frac{t^2}{2!} + \frac{t^3}{3!} + \cdots,$$

it follows that

$$\psi(t) = \frac{e^t - 1}{t} = 1 + \frac{t}{2!} + \frac{t^2}{3!} + \cdots.$$

The mean is then the coefficient of t: $EX = 1/2$, and the mean square is the coefficient of $t^2/2!$, $EX^2 = 1/3$. More generally, the kth moment about 0 is

$$EX^k = \frac{1}{k+1}. \quad ◀$$

A.2 Factorial Moment-Generating Functions

Another useful generating function of a random variable X is defined as the mean value of t^X:

> *Factorial moment generating function* of X:
>
> $$\eta_X(t) = E(t^X).$$

Its derivatives are as follows:

$$\eta'(t) = E[Xt^{X-1}],$$
$$\eta''(t) = E[X(X-1)t^{X-2}],$$
$$\eta'''(t) = E[X(X-1)(X-2)t^{X-3}]$$

and so on, with values at $t = 1$ given by

$$\eta'(1) = E[X],$$
$$\eta''(1) = E[X(X-1)],$$
$$\eta'''(1) = E[X(X-1)(X-2)].$$

The general formula, yielding the kth *factorial moment*, is

$$\eta^{(k)}(1) = E[X(X-1)(X-2)\cdots(X-k+1)].$$

Clearly, a knowledge of the factorial moments is equivalent to a knowledge of the ordinary moments. Thus, for example,

$$E[X(X-1)] = EX^2 - EX,$$
$$E[X(X-1)(X-2)] = EX^3 - 3EX^2 + 2EX,$$

so that

$$EX^2 = E[X(X-1)] + EX,$$
$$EX^3 = E[X(X-1)(X-2)] - 3E[X(X-1)] - E(X).$$

Notice that the m.g.f. and f.m.g.f. are simply related:

Relation between m.g.f. and f.m.g.f.:

$$\eta(t) = \psi(\log t).$$

The choice between them is largely a matter of whether t^x or e^{tx} combines most readily with $f(x)$ in evaluating the necessary integral or sum.

Example A-b Let X have Bernoulli distribution:

Value	Probability
1	p
0	$q = 1 - p$

Then

$$\eta(t) = E(t^X) = \sum t^x P(X = x) = t \cdot p + t^0 \cdot q = pt + q.$$

The derivatives are

$$\eta'(t) = p, \qquad \eta^{(k)}(t) = 0 \qquad (k \geq 2)$$

so that all factorial moments after the first are 0. Then

$$\sigma_X^2 = EX^2 - (EX)^2 = E[X(X-1)] + EX - (EX)^2 = 0 + p - p^2 = pq. \blacktriangleleft$$

Example A-c Let X have a Poisson distribution with mean value m:

$$P(X = k) = \frac{m^k}{k!}e^{-m}, \qquad k = 0, 1, \ldots.$$

The f.m.g.f. is a sum over the values of X:

$$\eta(t) = \sum_0^\infty t^k \frac{m^k}{k!} e^{-m} = e^{-m} \sum_0^\infty \frac{(mt)^k}{k!}$$
$$= e^{-m} e^{mt} = e^{m(t-1)}$$
$$= \sum_0^\infty \frac{[m(t-1)]^k}{k!}.$$

The coefficients of powers of $t - 1$, in this Taylor's expansion about $t = 1$, are the derivatives at $t = 1$ divided by $k!$. Thus,

$$\eta^{(k)}(1) = m^k.$$

In particular,

$$EX = \eta'(1) = m, \qquad E(X^2 - X) = \eta''(1) = m^2,$$

and so on. ◄

A.3 Sums of Independent Variables

If X and Y are independent, the m.g.f. of the sum $X + Y$ factors as follows:

$$E(e^{t(X+Y)}) = E(e^{tX}e^{tY}) = E(e^{tX})E({}^{tY}).$$

Similarly, the f.m.g.f. also factors:

$$E(t^{X+Y}) = E(t^X)E(t^Y).$$

Generating functions of the sum of independent X, Y:

$$\psi_{X+Y}(t) = \psi_X(t)\psi_Y(t),$$

$$\eta_{X+Y}(t) = \eta_X(t)\eta_Y(t).$$

Moreover, these properties—that the generating function of a sum is the product of the generating function of the summands—extend by induction to any finite number of independent summands.

Example A-d Let $X_1, X_2, \ldots, X_n$ be independent Bernoulli trials with common parameter p (the probability of success at each trial). Then if $S_n = X_1 + \cdots + X_n$, the f.m.g.f. of S_n is given by

$$\eta_{S_n}(t) = \eta_{X_1}(t) \cdots \eta_{X_n}(t) = (pt + q)^n.$$

But S_n, the number of successes in the n trials, has a binomial distribution. That is, the f.m.g.f. of the binomial distribution is given by $(pt + q)^n$. And then

$$\eta'_{S_n}(t) = np(pt + q)^{n-1}, \qquad \eta''_{S_n}(t) = np^2(pt + q)^{n-2},$$

from which, upon substitution of $t = 1$, one obtains

$$ES_n = \eta'_{S_n}(1) = np, \qquad E[S_n(S_n - 1)] = \eta''_{S_n}(t) = np^2,$$

and

$$\text{var } S_n = np^2 + ES_n - (ES_n)^2 = npq. \quad ◀$$

A.4 Probability-Generating Functions

Let X be a random variable with *integer* values, namely, $0, 1, 2, 3, \ldots$. Let $p_i = P(X = i)$. Then

$$\eta(t) = E(t^X) = \sum_0^\infty t^k p_k.$$

But this is (because X has *only* the values $0, 1, 2, \ldots$) a *power series in t*. It is the (unique) Maclaurin expansion of $\eta(t)$ *about* $t = 0$. So, whereas expanding $\eta(t)$ about $t = 1$ yields the factorial moments as coefficients, expanding it about $t = 0$ yields the probabilities.

Example A-e Let $X_1, X_2, \ldots, X_n$ be independent Bernoulli trials, with $S_n = \Sigma X_i$ giving the number of successes in the n trials. Then, as seen in Example A-d,

$$\eta_{S_n}(t) = (pt + q)^n = \sum_0^n \binom{n}{k}(pt)^k(q)^k$$

$$= \sum_0^n \left[\binom{n}{k} p^k q^{n-k}\right] t^k.$$

This binomial expansion of $(pt + q)^n$ is a power series in t. It is the Maclaurin series of $\eta_{S_n}(t)$; and the coefficients in this series are probabilities for S_n:

$$P(S_n = k) = \binom{n}{k} p^k q^{n-k}.$$

This is an alternative derivation of the binomial formula obtained earlier. ◀

Example A-f If $X_1, X_2, \ldots, X_n$ are independent Poisson variables with $EX_i = m_i$, then the f.m.g.f. of the sum $S_n = \Sigma X_i$ is obtained as a product:

$$\eta_{S_n}(t) = \prod \eta_{X_i}(t) = \prod e^{m_i(t-1)} = e^{(t-1)\Sigma m_i}.$$

This is clearly the f.m.g.f. of a Poisson variable with mean Σm_i; and because the f.m.g.f. uniquely defines probabilities of individual values (the probabilities are Maclaurin coefficients), it follows that S_n is Poisson, with mean Σm_i. ◀

Example A-g The f.m.g.f. for the outcome X of the toss of a fair die is

$$\eta(t) = E(t^X) = \sum_1^6 \frac{t^k}{6} = \frac{t(1-t^6)}{6(1-t)}.$$

The sum of the points on two dice (assuming independence) has then the f.m.g.f.

$$\eta_{X_1+X_2}(t) = [\eta(t)]^2 = \frac{t^2(1-t^6)^2}{36(1-t)^2}.$$

Now,

$$\frac{1}{(1-t)^2} = \frac{d}{dt}\left(\frac{1}{1-t}\right) = \frac{d}{dt}\sum_0^\infty t^k = \sum_1^\infty kt^{k-1},$$

and, by the binomial theorem,

$$(1-t^6)^2 = \sum_0^2 \binom{2}{j}(-t^6)^j.$$

Hence

$$\eta_{X_i+X_2}(t) = \frac{t^2}{36}\sum_0^2 \binom{2}{j}(-t^6)^j \sum_1^\infty kt^{k-1}$$

$$= \frac{1}{36}\sum_{j=0}^2 \sum_{k=1}^\infty \binom{2}{j}(-1)^j kt^{1+6j+k}.$$

The probability that the total number of points is, say, 10 is the coefficient of t^{10}. But $1 + 6j + k = 10$ when $k = 3$ and $j = 1$, and also when $k = 9$ and $j = 0$. Therefore,

$$P(X_1 + X_2 = 10) = \frac{1}{36}\left[3\binom{2}{1}(-1)^1 + \binom{2}{0}(-1)^0 9\right] = \frac{3}{36}.$$

This could have been computed by enumeration—there are 3 among the 36 equally likely pairs (i, j) with sum of 10, namely, (4, 6) (5, 5), and (6, 4). However, the technique illustrated is useful in more complicated problems (e.g., tossing four dice), where direct enumeration becomes much more difficult. ◀

Problems

Familiarity with these series expansions is useful for some of the problems:

(a) $1 + x + x^2 + \cdots = \dfrac{1}{1-x}$ (for $|x| < 1$).

(b) $e^x = 1 + x + \dfrac{x^2}{2!} + \dfrac{x^3}{3!} + \cdots.$

1. The exponential density $\lambda e^{-\lambda x}$, for $x > 0$, was encountered in Chapter 7 as the waiting-time density in a Poisson process. Obtain, by integration, the moment-generating function of the distribution. From its series expansion, read out the first two moments EX and EX^2.

2. The integral giving the moment-generating function of the standard normal distribution involves this exponent in the integrand:

$$-\frac{1}{2}x^2 + tx = -\frac{1}{2}(x - t)^2 + \frac{1}{2}t^2.$$

Use the right-hand form (obtained by "completing the square") to carry out the integration, observing that t is a *constant* in the integration process.

(NOTE: The area under a normal curve centered at $x = t$ is the same as for a normal curve centered at $x = 0$.) Expand the m.g.f. in a series and obtain the first four moments: EX^k, for $k = 1, 2, 3, 4$.

3. Write out the factorial moment-generating function for the number of points showing in the toss of a fair die. Use it to obtain EX and $E[X(X - 1)]$.

4. If X denotes the number of failures prior to the first success in a sequence of independent Bernoulli trials, the probability function for X is

$$f(x) = pq^k, \qquad k = 0, 1, 2, \ldots,$$

where p is the probability of success at each trial. Write out the series for the factorial m.g.f. and use one of the expansions above to express it in "closed" form. Obtain EX by evaluating $\eta'(1)$.

5. Let $X_1, X_2, \ldots, X_n$ denote independent, standard normal variables. Using the result in Problem 2, obtain the m.g.f. of the sum ΣX_i. Expand it in a power series and read out the mean and variance of the sum.

6. Let X and Y be independent binomial variables with the same p. Obtain the f.m.g.f. of the sum and recognize it as the m.g.f. of another binomial distribution. (Explain in terms of Bernoulli trials.)

7. Let $X_1, \ldots, X_n$ denote independent variables, each with the waiting time distribution of Problem 1. The sum ΣX_i is the time to the nth occurrence of an event in the corresponding Poisson process. Obtain the m.g.f. for the sum and obtain the mean of the sum by differentiation.

8. Use the technique of Example A-g to obtain the probability that the total number of points showing in the toss of three dice is 12.

APPENDIX B

Tables

I. The Standard Normal Distribution
II. Percentiles of the Chi-Square Distribution
III. Percentiles of the t-distribution
IV. Percentiles of the F-distribution
V. Distribution of the Standardized Range
VI. Acceptance Limits for the Kolmogorov–Smirnov Test of Goodness of Fit
VII. Cumulative Probabilities for the Wilcoxon Signed-Rank Statistic
VIII. Cumulative Probabilities for the Rank-Sum Statistic
IX. Binomial Coefficients
X. Binomial Probabilities
XI. Natural Logarithms
XII. Cumulative Poisson Probabilities
XIII. Random Digits
XIV. Gaussian Deviates

Table I Values of the Standard Normal Distribution Function

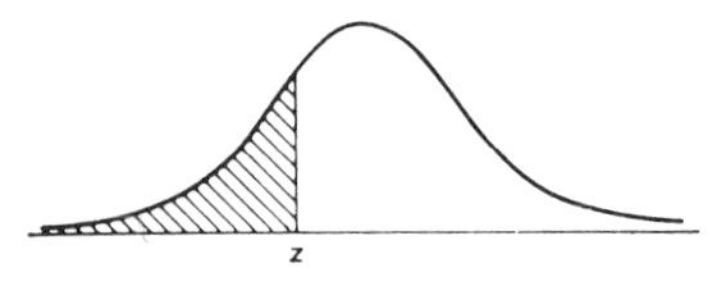

z	0	1	2	3	4	5	6	7	8	9
−3.	.0013	.0010	.0007	.0005	.0003	.0002	.0002	.0001	.0001	.0000
−2.9	.0019	.0018	.0017	.0017	.0016	.0016	.0015	.0015	.0014	.0014
−2.8	.0026	.0025	.0024	.0023	.0023	.0022	.0021	.0021	.0020	.0019
−2.7	.0035	.0034	.0033	.0032	.0031	.0030	.0029	.0028	.0027	.0026
−2.6	.0047	.0045	.0044	.0043	.0041	.0040	.0039	.0038	.0037	.0036
−2.5	.0062	.0060	.0059	.0057	.0055	.0054	.0052	.0051	.0049	.0048
−2.4	.0082	.0080	.0078	.0075	.0073	.0071	.0069	.0068	.0066	.0064
−2.3	.0107	.0104	.0102	.0099	.0096	.0094	.0091	.0089	.0087	.0084
−2.2	.0139	.0136	.0132	.0129	.0126	.0122	.0119	.0116	.0113	.0110
−2.1	.0179	.0174	.0170	.0166	.0162	.0158	.0154	.0150	.0146	.0143
−2.0	.0228	.0222	.0217	.0212	.0207	.0202	.0197	.0192	.0188	.0183
−1.9	.0287	.0281	.0274	.0268	.0262	.0256	.0250	.0244	.0238	.0233
−1.8	.0359	.0352	.0344	.0336	.0329	.0322	.0314	.0307	.0300	.0294
−1.7	.0446	.0436	.0427	.0418	.0409	.0401	.0392	.0384	.0375	.0367
−1.6	.0548	.0537	.0526	.0516	.0505	.0495	.0485	.0475	.0465	.0455
−1.5	.0668	.0655	.0643	.0630	.0618	.0606	.0594	.0582	.0570	.0559
−1.4	.0808	.0793	.0778	.0764	.0749	.0735	.0722	.0708	.0694	.0681
−1.3	.0968	.0951	.0934	.0918	.0901	.0885	.0869	.0853	.0838	.0823
−1.2	.1151	.1131	.1112	.1093	.1075	.1056	.1038	.1020	.1003	.0985
−1.1	.1357	.1335	.1314	.1292	.1271	.1251	.1230	.1210	.1190	.1170
−1.0	.1587	.1562	.1539	.1515	.1492	.1469	.1446	.1423	.1401	.1379
− .9	.1841	.1814	.1788	.1762	.1736	.1711	.1685	.1660	.1635	.1611
− .8	.2119	.2090	.2061	.2033	.2005	.1977	.1949	.1922	.1894	.1867
− .7	.2420	.2389	.2358	.2327	.2297	.2266	.2236	.2206	.2177	.2148
− .6	.2743	.2709	.2676	.2643	.2611	.2578	.2546	.2514	.2483	.2451
− .5	.3085	.3050	.3015	.2981	.2946	.2912	.2877	.2843	.2810	.2776
− .4	.3446	.3409	.3372	.3336	.3300	.3264	.3228	.3192	.3156	.3121
− .3	.3821	.3783	.3745	.3707	.3669	.3632	.3594	.3557	.3520	.3483
− .2	.4207	.4168	.4129	.4090	.4052	.4013	.3974	.3936	.3897	.3859
− .1	.4602	.4562	.4522	.4483	.4443	.4404	.4364	.4325	.4286	.4247
− 0	.5000	.4960	.4920	.4880	.4840	.4801	.4761	.4721	.4681	.4641

Table I *(continued)*

z	0	1	2	3	4	5	6	7	8	9
.0	.5000	.5040	.5080	.5120	.5160	.5199	.5239	.5279	.5319	.5359
.1	.5398	.5438	.5478	.5517	.5557	.5596	.5636	.5675	.5714	.5753
.2	.5793	.5832	.5871	.5910	.5948	.5987	.6026	.6064	.6103	.6141
.3	.6179	.6217	.6255	.6293	.6331	.6368	.6406	.6443	.6480	.6517
.4	.6554	.6591	.6628	.6664	.6700	.6736	.6772	.6808	.6844	.6879
.5	.6915	.6950	.6985	.7019	.7054	.7088	.7123	.7157	.7190	.7224
.6	.7257	.7291	.7324	.7357	.7389	.7422	.7454	.7486	.7517	.7549
.7	.7580	.7611	.7642	.7673	.7703	.7734	.7764	.7794	.7823	.7852
.8	.7881	.7910	.7939	.7967	.7995	.8023	.8051	.8078	.8106	.8133
.9	.8159	.8186	.8212	.8238	.8264	.8289	.8315	.8340	.8365	.8389
1.0	.8413	.8438	.8461	.8485	.8508	.8531	.8554	.8577	.8599	.8621
1.1	.8643	.8665	.8686	.8708	.8729	.8749	.8770	.8790	.8810	.8830
1.2	.8849	.8869	.8888	.8907	.8925	.8944	.8962	.8980	.8997	.9015
1.3	.9032	.9049	.9066	.9082	.9099	.9115	.9131	.9147	.9162	.9177
1.4	.9192	.9207	.9222	.9236	.9251	.9265	.9278	.9292	.9306	.9319
1.5	.9332	.9345	.9357	.9370	.9382	.9394	.9406	.9418	.9430	.9441
1.6	.9452	.9463	.9474	.9484	.9495	.9505	.9515	.9525	.9535	.9545
1.7	.9554	.9564	.9573	.9582	.9591	.9599	.9608	.9616	.9625	.9633
1.8	.9641	.9648	.9656	.9664	.9671	.9678	.9686	.9693	.9700	.9706
1.9	.9713	.9719	.9726	.9732	.9738	.9744	.9750	.9756	.9762	.9767
2.0	.9772	.9778	.9783	.9788	.9793	.9798	.9803	.9808	.9812	.9817
2.1	.9821	.9826	.9830	.9834	.9838	.9842	.9846	.9850	.9854	.9857
2.2	.9861	.9864	.9868	.9871	.9874	.9878	.9881	.9884	.9887	.9890
2.3	.9893	.9896	.9898	.9901	.9904	.9906	.9909	.9911	.9913	.9916
2.4	.9918	.9920	.9922	.9925	.9927	.9929	.9931	.9932	.9934	.9936
2.5	.9938	.9940	.9941	.9943	.9945	.9946	.9948	.9949	.9951	.9952
2.6	.9953	.9955	.9956	.9957	.9959	.9960	.9961	.9962	.9963	.9964
2.7	.9965	.9966	.9967	.9968	.9969	.9970	.9971	.9972	.9973	.9974
2.8	.9974	.9975	.9976	.9977	.9977	.9978	.9979	.9979	.9980	.9981
2.9	.9981	.9982	.9982	.9983	.9984	.9984	.9985	.9985	.9986	.9986
3.	.9987	.9990	.9993	.9995	.9997	.9998	.9998	.9999	.9999	1.0000

1. If a normal variable X is not standard, its values must be standardized: $Z = (X - \mu)/\sigma$, i.e., $P(X \leq x) = \Phi[(x - \mu)/\sigma]$.
2. For $z \geq 4$, $\Phi(z) = 1$ to four decimal places; for $z \leq -4$, $\Phi(z) = 0$ to four decimal places.
3. Entries opposite 3 are for 3.0, 3.1, 3.2, etc.

Table Ia Percentiles of the Standard Normal Distribution

$P(Z \leq z)$	z
.001	−3.0902
.005	−2.5758
.01	−2.3263
.02	−2.0537
.03	−1.8808
.04	−1.7507
.05	−1.6449
.10	−1.2816
.15	−1.0364
.20	−.8416
.30	−.5244
.40	−.2533
.50	0
.60	.2533
.70	.5244
.80	.8416
.85	1.0364
.90	1.2816
.95	1.6449
.96	1.7507
.97	1.8808
.98	2.0537
.99	2.3263
.995	2.5758
.999	3.0902

Table Ib Two-Tailed Probabilities for the Standard Normal Distribution

$P(\|Z\| > K)$	K
.001	3.2905
.002	3.0902
.005	2.80703
.01	2.5758
.02	2.3263
.03	2.1701
.04	2.0537
.05	1.9600
.06	1.8808
.08	1.7507
.10	1.6449
.15	1.4395
.20	1.2816
.30	1.0364

Table Ic Multipliers of the Standard Error for Confidence Interval Construction

Confidence coefficient	Multiplier
.80	1.282
.90	1.645
.95	1.960
.99	2.576

Table II Percentiles of the Chi-Square Distribution

Degrees of freedom	p									
	.01	.025	.05	.10	.70	.80	.90	.95	.975	.99
1	.000	.001	.004	.016	1.07	1.64	2.71	3.84	5.02	6.63
2	.020	.051	.103	.211	2.41	3.22	4.61	5.99	7.38	9.21
3	.115	.216	.352	.584	3.66	4.64	6.25	7.81	9.35	11.3
4	.297	.484	.711	1.06	4.88	5.99	7.78	9.49	11.1	13.3
5	.554	.831	1.15	1.61	6.06	7.29	9.24	11.1	12.8	15.1
6	.872	1.24	1.64	2.20	7.23	8.56	10.6	12.6	14.4	16.8
7	1.24	1.69	2.17	2.83	8.38	9.80	12.0	14.1	16.0	18.5
8	1.65	2.18	2.73	3.49	9.52	11.0	13.4	15.5	17.5	20.1
9	2.09	2.70	3.33	4.17	10.7	12.2	14.7	16.9	19.0	21.7
10	2.56	3.25	3.94	4.87	11.8	13.4	16.0	18.3	20.5	23.2
11	3.05	3.82	4.57	5.58	12.9	14.6	17.3	19.7	21.9	24.7
12	3.57	4.40	5.23	6.30	14.0	15.8	18.5	21.0	23.3	26.2
13	4.11	5.01	5.89	7.04	15.1	17.0	19.8	22.4	24.7	27.7
14	4.66	5.63	6.57	7.79	16.2	18.2	21.1	23.7	26.1	29.1
15	5.23	6.26	7.26	8.55	17.3	19.3	22.3	25.0	27.5	30.6
16	5.81	6.91	7.96	9.31	18.4	20.5	23.5	26.3	28.8	32.0
17	6.41	7.56	8.67	10.1	19.5	21.6	24.8	27.6	30.2	33.4
18	7.01	8.23	9.39	10.9	20.6	22.8	26.0	28.9	31.5	34.8
19	7.63	8.91	10.1	11.7	21.7	23.9	27.2	30.1	32.9	36.2
20	8.26	9.59	10.9	12.4	22.8	25.0	28.4	31.4	34.2	37.6
21	8.90	10.3	11.6	13.2	23.9	26.2	29.6	32.7	35.5	38.9
22	9.54	11.0	12.3	14.0	24.9	27.3	30.8	33.9	36.8	40.3
23	10.2	11.7	13.1	14.8	26.0	28.4	32.0	35.2	38.1	41.6
24	10.9	12.4	13.8	15.7	27.1	29.6	33.2	36.4	39.4	43.0
25	11.5	13.1	14.6	16.5	28.2	30.7	34.4	37.7	40.6	44.3
26	12.2	13.8	15.4	17.3	29.2	31.8	35.6	38.9	41.9	45.6
27	12.9	14.6	16.2	18.1	30.3	32.9	36.7	40.1	43.2	47.0
28	13.6	15.3	16.9	18.9	31.4	34.0	37.9	41.3	44.5	48.3
29	14.3	16.0	17.7	19.8	32.5	35.1	39.1	42.6	45.7	49.6
30	15.0	16.8	18.5	20.6	33.5	36.2	40.3	43.8	47.0	50.9
40	22.1	24.4	26.5	29.0	44.2	47.3	51.8	55.8	59.3	63.7
50	29.7	32.3	34.8	37.7	54.7	58.2	63.2	67.5	71.4	76.2
60	37.5	40.5	43.2	46.5	65.2	69.0	74.4	79.1	83.3	88.4

Note: For degrees of freedom $k > 30$, use $\chi_p^2 = \frac{1}{2}(z_p + \sqrt{2k - 1})^2$, where z_p is the corresponding percentile of the standard normal distribution.

Table III Percentiles of the t-Distribution

Degrees of Freedom	p					
	.8	.9	.95	.975	.99	.995
1	1.38	3.08	6.31	12.7	31.8	63.7
2	1.06	1.89	2.92	4.30	6.96	9.92
3	.978	1.64	2.35	3.18	4.54	5.84
4	.941	1.53	2.13	2.78	3.75	4.60
5	.920	1.48	2.01	2.57	3.36	4.03
6	.906	1.44	1.94	2.45	3.14	3.71
7	.896	1.42	1.90	2.36	3.00	3.50
8	.889	1.40	1.86	2.31	2.90	3.36
9	.883	1.38	1.83	2.26	2.82	3.25
10	.879	1.37	1.81	2.23	2.76	3.17
11	.876	1.36	1.80	2.20	2.72	3.11
12	.873	1.36	1.78	2.18	2.68	3.06
13	.870	1.35	1.77	2.16	2.65	3.01
14	.868	1.34	1.76	2.14	2.62	2.98
15	.866	1.34	1.75	2.13	2.60	2.95
16	.865	1.34	1.75	2.12	2.58	2.92
17	.863	1.33	1.74	2.11	2.57	2.90
18	.862	1.33	1.73	2.10	2.55	2.88
19	.861	1.33	1.73	2.09	2.54	2.86
20	.860	1.32	1.72	2.09	2.53	2.84
21	.859	1.32	1.72	2.08	2.52	2.83
22	.858	1.32	1.72	2.07	2.51	2.82
23	.858	1.32	1.71	2.07	2.50	2.81
24	.857	1.32	1.71	2.06	2.49	2.80
25	.856	1.32	1.71	2.06	2.48	2.79
26	.856	1.32	1.71	2.06	2.48	2.78
27	.855	1.31	1.70	2.05	2.47	2.77
28	.855	1.31	1.70	2.05	2.47	2.76
29	.854	1.31	1.70	2.04	2.46	2.76
30	.854	1.31	1.70	2.04	2.46	2.75
∞	.842	1.28	1.64	1.96	2.33	2.58

1. The area to the right of the table entry is $1 - p$.
2. The distribution is symmetric. For example, for 10 degrees of freedom,

$$P(-1.37 < t < 1.37) = .9 - .1.$$

Table IVa Ninety-fifth Percentiles of the F-Distribution

Denominator Degrees of Freedom	Numerator Degrees of Freedom 1	2	3	4	5	6	8	10	12	15	20	24	30
1	161	200	216	225	230	234	239	242	244	246	248	249	250
2	18.5	19.0	19.2	19.2	19.3	19.3	19.4	19.4	19.4	19.4	19.4	19.5	19.5
3	10.1	9.55	9.28	9.12	9.01	8.94	8.85	8.79	8.74	8.70	8.66	8.64	8.62
4	7.71	6.94	6.59	6.39	6.26	6.16	6.04	5.96	5.91	5.86	5.80	5.77	5.75
5	6.61	5.79	5.41	5.19	5.05	4.95	4.82	4.74	4.68	4.62	4.56	4.53	4.50
6	5.99	5.14	4.76	4.53	4.39	4.28	4.15	4.06	4.00	3.94	3.87	3.84	3.81
7	5.59	4.74	4.35	4.12	3.97	3.87	3.73	3.64	3.57	3.51	3.44	3.41	3.38
8	5.32	4.46	4.07	3.84	3.69	3.58	3.44	3.35	3.28	3.22	3.15	3.12	3.08
9	5.12	4.26	3.86	3.63	3.48	3.37	3.23	3.14	3.07	3.01	2.94	2.90	2.86
10	4.96	4.10	3.71	3.48	3.33	3.22	3.07	2.98	2.91	2.85	2.77	2.74	2.70
11	4.84	3.98	3.59	3.36	3.20	3.09	2.95	2.85	2.79	2.72	2.65	2.61	2.57
12	4.75	3.89	3.49	3.26	3.11	3.00	2.85	2.75	2.69	2.62	2.54	2.51	2.47
13	4.67	3.81	3.41	3.18	3.03	2.92	2.77	2.67	2.60	2.53	2.46	2.42	2.38
14	4.60	3.74	3.34	3.11	2.96	2.85	2.70	2.60	2.53	2.46	2.39	2.35	2.31
15	4.54	3.68	3.29	3.06	2.90	2.79	2.64	2.54	2.48	2.40	2.33	2.29	2.25
16	4.49	3.63	3.24	3.01	2.85	2.74	2.59	2.49	2.42	2.35	2.28	2.24	2.19
17	4.45	3.59	3.20	2.96	2.81	2.70	2.55	2.45	2.38	2.31	2.23	2.19	2.15
18	4.41	3.55	3.16	2.93	2.77	2.66	2.51	2.41	2.34	2.27	2.19	2.15	2.11
19	4.38	3.52	3.13	2.90	2.74	2.63	2.48	2.38	2.31	2.23	2.16	2.11	2.07
20	4.35	3.49	3.10	2.87	2.71	2.60	2.45	2.35	2.28	2.20	2.12	2.08	2.04
21	4.32	3.47	3.07	2.84	2.68	2.57	2.42	2.32	2.25	2.18	2.10	2.05	2.01
22	4.30	3.44	3.05	2.82	2.66	2.55	2.40	2.30	2.23	2.15	2.07	2.03	1.98
23	4.28	3.42	3.03	2.80	2.64	2.53	2.37	2.27	2.20	2.13	2.05	2.01	1.96
24	4.26	3.40	3.01	2.78	2.62	2.51	2.36	2.25	2.18	2.11	2.03	1.98	1.94
25	4.24	3.39	2.99	2.76	2.60	2.49	2.34	2.24	2.16	2.09	2.01	1.96	1.92
30	4.17	3.32	2.92	2.69	2.53	2.42	2.27	2.16	2.09	2.01	1.93	1.89	1.84
40	4.08	3.23	2.84	2.61	2.45	2.34	2.18	2.08	2.00	1.92	1.84	1.79	1.74
60	4.00	3.15	2.76	2.53	2.37	2.25	2.10	1.99	1.92	1.84	1.75	1.70	1.65

Note: The fifth percentiles are obtainable as follows:

$$F_{.05}(r, s) = \frac{1}{F_{.95}(s, r)}$$

Table IVb Ninety-ninth Percentiles of the F-Distribution

Denominator Degrees of Freedom	Numerator Degrees of Freedom 1	2	3	4	5	6	8	10	12	15	20	24	30
1	4050	5000	5400	5620	5760	5860	5980	6060	6110	6160	6210	6235	6260
2	98.5	99.0	99.2	99.2	99.3	99.3	99.4	99.4	99.4	99.4	99.4	99.5	99.5
3	34.1	30.8	29.5	28.7	28.2	27.9	27.5	27.3	27.1	26.9	26.7	26.6	26.5
4	21.2	18.0	16.7	16.0	15.5	15.2	14.8	14.5	14.4	14.2	14.0	13.9	13.8
5	16.3	13.3	12 1.	11.4	11.0	10.7	10.3	10.1	9.89	9.72	9.55	9.47	9.38
6	13.7	10.9	9.78	9.15	8.75	8.47	8.10	7.87	7.72	7.56	7.40	7.31	7.23
7	12.2	9.55	8.45	7.85	7.46	7.19	6.84	6.62	6.47	6.31	6.16	6.07	5.99
8	11.3	8.65	7.59	7.01	6.63	6.37	6.03	5.81	5.67	5.52	5.36	5.28	5.20
9	10.6	8.02	6.99	6.42	6.06	5.80	5.47	5.26	5.11	4.96	4.81	4.73	4.65
10	10.0	7.56	6.55	5.99	5.64	5.39	5.06	4.85	4.71	4.56	4.41	4.33	4.25
11	9.65	7.21	6.22	5.67	5.32	5.07	4.74	4.54	4.40	4.25	4.10	4.02	3.94
12	9.33	6.93	5.95	5.41	5.06	4.82	4.50	4.30	4.16	4.01	3.86	3.78	3.70
13	9.07	6.70	5.74	5.21	4.86	4.62	4.30	4.10	3.96	3.82	3.66	3.59	3.51
14	8.86	6.51	5.56	5.04	4.69	4.46	4.14	3.94	3.80	3.66	3.51	3.43	3.35
15	8.68	6.36	5.42	4.89	4.56	4.32	4.00	3.80	3.67	3.52	3.37	3.29	3.21
16	8.53	6.23	5.29	4.77	4.44	4.20	3.89	3.69	3.55	3.41	3.26	3.18	3.10
17	8.40	6.11	5.18	4.67	4.34	4.10	3.79	3.59	3.46	3.31	3.16	3.08	3.00
18	8.29	6.01	5.09	4.58	4.25	4.01	3.71	3.51	3.37	3.23	3.08	3.00	2.92
19	8.18	5.93	5.01	4.50	4.17	3.94	3.63	3.43	3.30	3.15	3.00	2.92	2.84
20	8.10	5.85	4.94	4.43	4.10	3.87	3.56	3.37	3.23	3.09	2.94	2.86	2.78
21	8.02	5.78	4.87	4.37	4.04	3.81	3.51	3.31	3.17	3.03	2.88	2.80	2.72
22	7.95	5.72	4.82	4.31	3.99	3.76	3.45	3.26	3.12	2.98	2.83	2.75	2.67
23	7.88	5.66	4.76	4.26	3.94	3.71	3.41	3.21	3.07	2.93	2.78	2.70	2.62
24	7.82	5.61	4.72	4.22	3.90	3.67	3.36	3.17	3.03	2.89	2.74	2.66	2.58
25	7.77	5.57	4.68	4.18	3.86	3.63	3.32	3.13	2.99	2.85	2.70	2.62	2.54
30	7.56	5.39	4.51	4.02	3.70	3.47	3.17	2.98	2.84	2.70	2.55	2.47	2.39
40	7.31	5.18	4.31	3.83	3.51	3.29	2.99	2.80	2.66	2.52	2.37	2.29	2.20
60	7.08	4.98	4.13	3.65	3.34	3.12	2.82	2.63	2.50	2.35	2.20	2.12	2.03

Table V Distribution of the Standardized Range $W = R/\sigma$ (assuming a normal population)

	Sample size										
	2	3	4	5	6	7	8	9	10	12	15
$E(W)$	1.128	1.693	2.059	2.326	2.534	2.704	2.847	2.970	3.078	3.258	3.472
σ_W	.853	.888	.880	.864	.848	.833	.820	.808	.797	.778	.755
	Percentiles:										
.005	.01	.13	.34	.55	.75	.92	1.08	1.21	1.33	1.55	1.80
.01	.02	.19	.43	.66	.87	1.05	1.20	1.34	1.47	1.68	1.93
.025	.04	.30	.59	.85	1.06	1.25	1.41	1.55	1.67	1.88	2.14
.05	.09	.43	.76	1.03	1.25	1.44	1.60	1.74	1.86	2.07	2.32
.10	.18	.62	.98	1.26	1.49	1.68	1.83	1.97	2.09	2.30	2.54
.20	.36	.90	1.29	1.57	1.80	1.99	2.14	2.28	2.39	2.59	2.83
.30	.55	1.14	1.53	1.82	2.04	2.22	2.38	2.51	2.62	2.82	3.04
.40	.74	1.36	1.76	2.04	2.26	2.44	2.59	2.71	2.83	3.01	3.23
.50	.95	1.59	1.98	2.26	2.47	2.65	2.79	2.92	3.02	3.21	3.42
.60	1.20	1.83	2.21	2.48	2.69	2.86	3.00	3.12	3.23	3.41	3.62
.70	1.47	2.09	2.47	2.73	2.94	3.10	3.24	3.35	3.46	3.63	3.83
.80	1.81	2.42	2.78	3.04	3.23	3.39	3.52	3.63	3.73	3.90	4.09
.90	2.33	2.90	3.24	3.48	3.66	3.81	3.93	4.04	4.13	4.29	4.47
.95	2.77	3.31	3.63	3.86	4.03	4.17	4.29	4.39	4.47	4.62	4.80
.975	3.17	3.68	3.98	4.20	4.36	4.49	4.61	4.70	4.79	4.92	5.09
.99	3.64	4.12	4.40	4.60	4.76	4.88	4.99	5.08	5.16	5.29	5.45
.995	3.97	4.42	4.69	4.89	5.03	5.15	5.26	5.34	5.42	5.54	5.70

Table VI Acceptance Limits for the Kolmogorov–Smirnov Test of Goodness of Fit

Sample size (n)	Significance level				
	.20	.15	.10	.05	.01
1	.900	.925	.950	.975	.995
2	.684	.726	.776	.842	.929
3	.565	.597	.642	.708	.829
4	.494	.525	.564	.624	.734
5	.446	.474	.510	.563	.669
6	.410	.436	.470	.521	.618
7	.381	.405	.438	.486	.577
8	.358	.381	.411	.457	.543
9	.339	.360	.388	.432	.514
10	.322	.342	.368	.409	.486
11	.307	.326	.352	.391	.468
12	.295	.313	.338	.375	.450
13	.284	.302	.325	.361	.433
14	.274	.292	.314	.349	.418
15	.266	.283	.304	.338	.404
16	.258	.274	.295	.328	.391
17	.250	.266	.286	.318	.380
18	.244	.259	.278	.309	.370
19	.237	.252	.272	.301	.361
20	.231	.246	.264	.294	.352
25	.21	.22	.24	.264	.32
30	.19	.20	.22	.242	.29
35	.18	.19	.21	.23	.27
40				.21	.25
50				.19	.23
60				.17	.21
70				.16	.19
80				.15	.18
90				.14	
100				.14	
Asymptotic Formula:	$\frac{1.07}{\sqrt{n}}$	$\frac{1.14}{\sqrt{n}}$	$\frac{1.22}{\sqrt{n}}$	$\frac{1.36}{\sqrt{n}}$	$\frac{1.63}{\sqrt{n}}$

Note: Reject the hypothetical distribution $F(x)$ if $D_n = \max|F_n(x) - F(x)|$ exceeds the tabulated value.

Table VII Cumulative Probabilities for the Wilcoxon Signed-Rank Statistic

c \ n	3	4	5	6	7	8	9	10	11	12	13	14	15	n / c
0	.125	.062	.031	.016	.008									0
1	.250	.125	.062	.031	.016	.008								1
2		.188	.094	.047	.023	.012								2
3			.156	.078	.039	.020	.010							3
4				.109	.055	.027	.014	.007						4
5				.156	.078	.039	.020	.010						5
6					.109	.055	.027	.014						6
7					.148	.074	.037	.019	.009					7
8					.188	.098	.049	.024	.012					8
9						.125	.064	.032	.016					9
10						.156	.082	.042	.021	.010				10
11							.102	.053	.027	.013				11
12							.125	.065	.034	.017	.009			12
13							.150	.080	.042	.021	.011			13
14								.097	.051	.026	.013			14
15									.062	.032	.016	.008		15
16									.074	.039	.020	.010		16
17									.087	.046	.024	.012		17
18										.055	.029	.015		18
19										.065	.034	.018	.009	19
20										.076	.040	.021	.011	20
21										.088	.047	.025	.013	21
22											.055	.029	.015	22
23											.064	.034	.018	23
24											.073	.039	.021	24
25											.084	.045	.024	25
26											.095	.052	.028	26
27												.059	.032	27
28												.068	.036	28
29												.077	.042	29
30												.086	.047	30
31												.097	.053	31
32													.060	32
33													.068	33
34													.076	34
35													.084	35
36													.094	36

1. Table entries are $P(R_+ \leq c) = P(R_+ \geq n(n + 1)/2 - c)$. For example, a two-sided test at $\alpha \leq .10$ would reject the null hypothesis if $R_+ \leq 8$ or $R_+ \geq 37$, when $n = 9$, since

$$P(R_+ \leq 8) = P(R_+ \geq 9 \cdot 10/2 - 8) = .049.$$

2. For $n > 15$, use a normal approximation (Section 11.5).

Table VIII Cumulative Probabilities for the Rank-Sum Statistic

c \ m, n	3, 3	3, 4	3, 5	3, 6	3, 7	3, 8	3, 9	3, 10
6	.050	.029	.018	.012	.008	.006		
7	.100	.057	.036	.024	.017	.012	.009	.007
8			.071	.048	.033	.024	.018	.014
9				.083	.058	.042	.032	.024
10					.092	.067	.050	.034
11						.097	.073	.056
12								.080
M	21	24	27	30	33	36	39	42

c \ m, n	4, 4	4, 5	4, 6	4, 7	4, 8	4, 9	4, 10
10	.014	.008					
11	.029	.016	.010	.006			
12	.057	.032	.019	.012	.008		
13	.100	.056	.033	.021	.014	.010	.007
14		.095	.057	.036	.024	.017	.012
15			.086	.055	.036	.025	.018
16				.082	.055	.038	.027
17					.077	.053	.038
18						.074	.053
19						.099	.071
20							.094
M	36	40	44	48	52	56	60

1. m is the size of the smaller sample.
2. Entry opposite M is the sum of the max and min rank sums.
3. Entry opposite c is the cumulative probability in a tail:

$$P(R \leq c) = P(R \geq M - c),$$

where R is the rank-sum for the smaller sample. For example, if $m = 4$, $n = 8$, a two-tailed test at $\alpha \leq .05$ rejects the null hypothesis if $R \leq 14$ or $R \geq 52 - 14$ (the exact level is $.024 + .024$).

4. For $n, m > 10$, use a normal approximation (Section 12.3).

Table VIII *(continued)*

c \ m, n	5, 5	5, 6	5, 7	5, 8	5, 9	5, 10
15	.004					
16	.008					
17	.016	.009				
18	.028	.015	.009			
19	.048	.026	.015	.009		
20	.075	.041	.024	.015	.009	
21		.063	.037	.023	.014	.010
22		.089	.053	.033	.021	.014
23			.074	.047	.030	.020
24				.064	.041	.028
25				.085	.056	.038
26					.073	.050
27					.095	.065
28						.082
M	55	60	65	70	75	80

c \ m, n	6, 6	6, 7	6, 8	6, 9	6, 10
24	.008				
25	.013	.007			
26	.021	.011			
27	.032	.017	.010		
28	.047	.026	.015	.009	
29	.066	.037	.021	.013	.008
30	.090	.051	.030	.018	.011
31		.069	.041	.025	.016
32		.090	.054	.033	.021
33			.071	.044	.028
34			.091	.057	.036
35				.072	.047
36				.091	.059
37					.074
38					.090
M	78	84	90	96	102

c \ m, n	7, 7	7, 8	7, 9	7, 10
34	.009			
35	.013	.007		
36	.019	.010		
37	.027	.014	.008	
38	.036	.020	.011	
39	.049	.027	.016	.009
40	.064	.036	.021	.012
41	.082	.047	.027	.017
42		.060	.036	.022
43		.076	.045	.028
44		.095	.057	.035
45			.071	.044
46			.087	.054
47				.067
48				.081
49				.097
M	105	112	119	126

Table VIII *(continued)*

c \ m, n	m=8, n=8	m=8, n=9	m=8, n=10
46	.010		
47	.014	.008	
48	.019	.010	
49	.025	.014	.008
50	.032	.018	.010
51	.041	.023	.013
52	.052	.030	.017
53	.065	.037	.022
54	.080	.046	.027
55	.097	.057	.034
56		.069	.042
57		.084	.051
58			.061
59			.073
60			.086
M	136	144	152

c \ m, n	m=9, n=9	m=9, n=10
59	.009	
60	.012	
61	.016	.009
62	.020	.011
63	.025	.014
64	.031	.017
65	.039	.022
66	.047	.027
67	.057	.033
68	.068	.039
69	.081	.047
70	.095	.056
71		.067
72		.078
73		.091
M	171	180

c \ m, n	m=10, n=10
74	.009
75	.012
76	.014
77	.018
78	.022
79	.026
80	.032
81	.038
82	.045
83	.053
84	.062
85	.072
86	.083
87	.095
M	210

Table IX Binomial Coefficients $\binom{n}{k}$

n \ k	2	3	4	5	6	7	8	9	10
2	1								
3	3	1							
4	6	4	1						
5	10	10	5	1					
6	15	20	15	6	1				
7	21	35	35	21	7	1			
8	28	56	70	56	28	8	1		
9	36	84	126	126	84	36	9	1	
10	45	120	210	252	210	120	45	10	1
11	55	165	330	462	462	330	165	55	11
12	66	220	495	792	924	792	495	220	66
13	78	286	715	1287	1716	1716	1287	715	286
14	91	364	1001	2002	3003	3432	3003	2002	1001
15	105	455	1365	3003	5005	6435	6435	5005	3003
16	120	560	1820	4368	8008	11440	12870	11440	8008
17	136	680	2380	6188	12376	19448	24310	24310	19448
18	153	816	3060	8568	18564	31824	43758	48620	43758
19	171	969	3876	11628	27132	50388	75582	92378	92378
20	190	1140	4845	15504	38760	77520	125970	167960	184756

Table X Binomial Probabilities: $\binom{n}{k} p^k (1-p)^{n-k}$

n	k \ p	.01	.05	.10	.15	.20	.25	.30	1/3	.35	.40	.45	.50
5	0	.9510	.7738	.5905	.4437	.3277	.2373	.1681	.1317	.1160	.0778	.0503	.0312
	1	.0480	.2036	.3280	.3915	.4096	.3955	.3601	.3292	.3124	.2592	.2059	.1562
	2	.0010	.0214	.0729	.1382	.2048	.2637	.3087	.3292	.3364	.3456	.3369	.3125
	3	.0000	.0011	.0081	.0244	.0512	.0879	.1323	.1646	.1811	.2304	.2757	.3125
	4	.0000	.0000	.0004	.0022	.0064	.0146	.0283	.0412	.0488	.0768	.1128	.1562
	5	.0000	.0000	.0000	.0001	.0003	.0010	.0024	.0041	.0053	.0102	.0185	.0312
6	0	.9415	.7351	.5314	.3771	.2621	.1780	.1176	.0878	.0754	.0467	.0277	.0156
	1	.0571	.2321	.3543	.3993	.3932	.3560	.3025	.2634	.2437	.1866	.1359	.0938
	2	.0014	.0305	.0984	.1762	.2458	.2966	.3241	.3293	.3280	.3110	.2780	.2344
	3	.0000	.0021	.0146	.0415	.0819	.1318	.1852	.2195	.2355	.2765	.3032	.3125
	4	.0000	.0001	.0012	.0055	.0154	.0330	.0595	.0823	.0951	.1382	.1861	.2344
	5	.0000	.0000	.0001	.0004	.0015	.0044	.0102	.0165	.0205	.0369	.0609	.0938
	6	.0000	.0000	.0000	.0000	.0001	.0002	.0007	.0014	.0018	.0041	.0083	.0156
7	0	.9321	.6983	.4783	.3206	.2097	.1335	.0824	.0585	.0490	.0280	.0152	.0078
	1	.0659	.2573	.3720	.3960	.3670	.3115	.2471	.2049	.1848	.1306	.0872	.0547
	2	.0020	.0406	.1240	.2097	.2753	.3115	.3177	.3073	.2985	.2613	.2140	.1641
	3	.0000	.0036	.0230	.0617	.1147	.1730	.2269	.2561	.2679	.2903	.2918	.2734
	4	.0000	.0002	.0026	.0109	.0287	.0577	.0972	.1280	.1442	.1935	.2388	.2734
	5	.0000	.0000	.0002	.0012	.0043	.0115	.0250	.0384	.0466	.0774	.1172	.1641
	6	.0000	.0000	.0000	.0001	.0004	.0013	.0036	.0064	.0084	.0172	.0320	.0547
	7	.0000	.0000	.0000	.0000	.0000	.0001	.0002	.0005	.0006	.0016	.0037	.0078
8	0	.9227	.6634	.4305	.2725	.1678	.1001	.0576	.0390	.0319	.0168	.0084	.0039
	1	.0746	.2793	.3826	.3847	.3355	.2670	.1977	.1561	.1373	.0896	.0548	.0312
	2	.0026	.0515	.1488	.2376	.2936	.3115	.2965	.2731	.2587	.2090	.1569	.1094
	3	.0001	.0054	.0331	.0839	.1468	.2076	.2541	.2731	.2786	.2787	.2568	.2187
	4	.0000	.0004	.0046	.0185	.0459	.0865	.1361	.1707	.1875	.2322	.2627	.2734
	5	.0000	.0000	.0004	.0026	.0092	.0231	.0467	.0683	.0808	.1239	.1719	.2187
	6	.0000	.0000	.0000	.0002	.0011	.0038	.0100	.0171	.0217	.0413	.0703	.1094
	7	.0000	.0000	.0000	.0000	.0001	.0004	.0012	.0024	.0033	.0079	.0164	.0312
	8	.0000	.0000	.0000	.0000	.0000	.0000	.0001	.0002	.0002	.0007	.0017	.0039
9	0	.9135	.6302	.3874	.2316	.1342	.0751	.0404	.0260	.0207	.0101	.0046	.0020
	1	.0830	.2985	.3874	.3679	.3020	.2253	.1556	.1171	.1004	.0605	.0339	.0176
	2	.0034	.0629	.1722	.2597	.3020	.3003	.2668	.2341	.2162	.1612	.1110	.0703
	3	.0001	.0077	.0446	.1069	.1762	.2336	.2668	.2731	.2716	.2508	.2119	.1641
	4	.0000	.0006	.0074	.0283	.0661	.1168	.1715	.2048	.2194	.2508	.2600	.2461
	5	.0000	.0000	.0008	.0050	.0165	.0389	.0735	.1024	.1181	.1672	.2128	.2461
	6	.0000	.0000	.0001	.0006	.0028	.0087	.0210	.0341	.0424	.0743	.1160	.1641
	7	.0000	.0000	.0000	.0000	.0003	.0012	.0039	.0073	.0098	.0212	.0407	.0703
	8	.0000	.0000	.0000	.0000	.0000	.0001	.0004	.0009	.0013	.0035	.0083	.0176
	9	.0000	.0000	.0000	.0000	.0000	.0000	.0000	.0001	.0001	.0003	.0008	.0020
10	0	.9044	.5987	.3487	.1969	.1074	.0563	.0282	.0173	.0135	.0060	.0025	.0010
	1	.0914	.3151	.3874	.3474	.2684	.1877	.1211	.0867	.0725	.0403	.0207	.0098
	2	.0042	.0746	.1937	.2759	.3020	.2816	.2335	.1951	.1757	.1209	.0763	.0439
	3	.0001	.0105	.0574	.1298	.2013	.2503	.2668	.2601	.2522	.2150	.1665	.1172
	4	.0000	.0010	.0112	.0401	.0881	.1460	.2001	.2276	.2377	.2508	.2384	.2051
	5	.0000	.0001	.0015	.0085	.0264	.0584	.1029	.1366	.1536	.2007	.2340	.2461
	6	.0000	.0000	.0001	.0012	.0055	.0162	.0368	.0569	.0689	.1115	.1596	.2051
	7	.0000	.0000	.0000	.0001	.0008	.0031	.0090	.0163	.0212	.0425	.0746	.1172
	8	.0000	.0000	.0000	.0000	.0001	.0004	.0014	.0030	.0043	.0106	.0229	.0439
	9	.0000	.0000	.0000	.0000	.0000	.0000	.0001	.0003	.0005	.0016	.0042	.0098
	10	.0000	.0000	.0000	.0000	.0000	.0000	.0000	.0000	.0000	.0001	.0003	.0010

1. For $p > .5$, reverse roles of p and q. Thus if $p = .7$ and $n = 5$:

$$f(0; .7) = \binom{5}{0}(.7)^0(.3)^5 = f(5; .3) = .0024.$$

2. For $n > 10$, see approximations in Sections 7.1 and 8.5.

Table XI Natural Logarithms

N	0	1	2	3	4	5	6	7	8	9
0.0		5.395	6.088	6.493	6.781	7.004	7.187	7.341	7.474	7.592
0.1	7.697	7.793	7.880	7.960	8.034	8.103	8.167	8.228	8.285	8.339
0.2	8.391	8.439	8.486	8.530	8.573	8.614	8.653	8.691	8.727	8.762
0.3	8.796	8.829	8.861	8.891	8.921	8.950	8.978	9.006	9.032	9.058
0.4	9.084	9.108	9.132	9.156	9.179	9.201	9.223	9.245	9.266	9.287
0.5	9.307	9.327	9.346	9.365	9.384	9.402	9.420	9.438	9.455	9.472
0.6	9.489	9.506	9.522	9.538	9.554	9.569	9.584	9.600	9.614	9.629
0.7	9.643	9.658	9.671	9.685	9.699	9.712	9.726	9.739	9.752	9.764
0.8	9.777	9.789	9.802	9.814	9.826	9.837	9.849	9.861	9.872	9.883
0.9	9.895	9.906	9.917	9.927	9.938	9.949	9.959	9.970	9.980	9.990
1.0	0.00000	0995	1980	2956	3922	4879	5827	6766	7696	8618
1.1	9531	*0436	*1333	*2222	*3103	*3976	*4842	*5700	*6511	*7395
1.2	0.1 8232	9062	9885	*0701	*1511	*2314	*3111	*3902	*4686	*5464
1.3	0.2 6236	7003	7763	8518	9267	*0010	*0748	*1481	*2208	*2930
1.4	0.3 3647	4359	5066	5767	6464	7156	7844	8526	9204	9878
1.5	0.4 0547	1211	1871	2527	3178	3825	4469	5108	5742	6373
1.6	7000	7623	8243	8858	9470	*0078	*0682	*1282	*1879	*2473
1.7	0.5 3063	3649	4232	4812	5389	5962	6531	7098	7661	8222
1.8	8779	9333	9884	*0432	*0977	*1519	*2058	*2594	*3127	*3658
1.9	0.6 4185	4710	5233	5752	6269	6783	7294	7803	8310	8813
2.0	9315	9813	*0310	*0804	*1295	*1784	*2271	*2755	*3237	*3716
2.1	0.7 4194	4669	5142	5612	6081	6547	7011	7473	7932	8390
2.2	8846	9299	9751	*0200	*0648	*1093	*1536	*1978	*2418	*2855
2.3	0.8 3291	3725	4157	4587	5015	5442	5866	6289	6710	7129
2.4	7547	7963	8377	8789	9200	9609	*0016	*0422	*0826	*1228
2.5	0.9 1629	2028	2426	2822	3216	3609	4001	4391	4779	5166
2.6	5551	5935	6317	6698	7078	7456	7833	8208	8582	8954
2.7	9325	9695	*0063	*0430	*0796	*1160	*1523	*1885	*2245	*2604
2.8	1.0 2962	3318	3674	4028	4380	4732	5082	5431	5779	6126
2.9	6471	6815	7158	7500	7841	8181	8519	8856	9192	9527
3.0	9861	*0194	*0526	*0856	*1186	*1514	*1841	*2168	*2493	*2817
3.1	1.1 3140	3462	3783	4103	4422	4740	5057	5373	5688	6002
3.2	6315	6627	6938	7248	7557	7865	8173	8479	8784	9089
3.3	9392	9695	9996	*0297	*0597	*0896	*1194	*1491	*1788	*2083
3.4	1.2 2378	2671	2964	3256	3547	3837	4127	4415	4703	4990
3.5	5276	5562	5846	6130	6413	6695	6976	7257	7536	7815
3.6	8093	8371	8647	8923	9198	9473	9746	*0019	*0291	*0563
3.7	1.3 0833	1103	1372	1641	1909	2176	2442	2708	2972	3237
3.8	3500	3763	4025	4286	4547	4807	5067	5325	5584	5841
3.9	6098	6354	6609	6864	7118	7372	7624	7877	8128	8379
4.0	8629	8879	9128	9377	9624	9872	*0118	*0364	*0610	*0854
N	**0**	**1**	**2**	**3**	**4**	**5**	**6**	**7**	**8**	**9**

Take tabular value—10 (for N = 0.1 to 0.9)

Table XI Natural Logarithms *(continued)*

N	0	1	2	3	4	5	6	7	8	9
4.0	8629	8879	9128	9377	9624	9872	*0118	*0364	*0610	*0854
4.1	1.4 1099	1342	1585	1828	2070	2311	2552	2792	3031	3270
4.2	3508	3746	3984	4220	4456	4692	4927	5161	5395	5629
4.3	5862	6094	6326	6557	6787	7018	7247	7476	7705	7933
4.4	8160	8387	8614	8840	9065	9290	9515	9739	9962	*0185
4.5	1.5 0408	0630	0851	1072	1293	1513	1732	1951	2170	2388
4.6	2606	2823	3039	3256	3471	3687	3902	4116	4330	4543
4.7	4756	4969	5181	5393	5604	5814	6025	6235	6444	6653
4.8	6862	7070	7277	7485	7691	7898	8104	8309	8515	8719
4.9	8924	9127	9331	9534	9737	9939	*0141	*0342	*0543	*0744
5.0	1.6 0944	1144	1343	1542	1741	1939	2137	2334	2531	2728
5.1	2924	3120	3315	3511	3705	3900	4094	4287	4481	4673
5.2	4866	5058	5250	5441	5632	5823	6013	6203	6393	6582
5.3	6771	6959	7147	7335	7523	7710	7896	8083	8269	8455
5.4	8640	8825	9010	9194	9378	9562	9745	9928	*0111	*0293
5.5	1.7 0475	0656	0838	1019	1199	1380	1560	1740	1919	2098
5.6	2277	2455	2633	2811	2988	3166	3342	3519	3695	3871
5.7	4047	4222	4397	4572	4746	4920	5094	5267	5440	5613
5.8	5786	5958	6130	6302	6473	6644	6815	6985	7156	7326
5.9	7495	7665	7834	8002	8171	8339	8507	8675	8842	9009
6.0	9176	9342	9509	9675	9840	*0006	*0171	*0336	*0500	*0665
6.1	1.8 0829	0993	1156	1319	1482	1645	1808	1970	2132	2294
6.2	2455	2616	2777	2938	3098	3258	3418	3578	3737	3896
6.3	4055	4214	4372	4530	4688	4845	5003	5160	5317	5473
6.4	5630	5786	5942	6097	6253	6408	6563	6718	6872	7026
6.5	7180	7334	7487	7641	7794	7947	8099	8251	8403	8555
6.6	8707	8858	9010	9160	9311	9462	9612	9762	9912	*0061
6.7	1.9 0211	0360	0509	0658	0806	0954	1102	1250	1398	1545
6.8	1692	1839	1986	2132	2279	2425	2571	2716	2862	3007
6.9	3152	3297	3442	3586	3730	3874	4018	4162	4305	4448
7.0	4591	4734	4876	5019	5161	5303	5445	5586	5727	5869
7.1	6009	6150	6291	6431	6571	6711	6851	6991	7130	7269
7.2	7408	7547	7685	7824	7962	8100	8238	8376	8513	8650
7.3	8787	8924	9061	9198	9334	9470	9606	9742	9877	*0013
7.4	2.0 0148	0283	0418	0553	0687	0821	0956	1089	1223	1357
7.5	1490	1624	1757	1890	2022	2155	2287	2419	2551	2683
7.6	2815	2946	3078	3209	3340	3471	3601	3732	3862	3992
7.7	4122	4252	4381	4511	4640	4769	4898	5027	5156	5284
7.8	5412	5540	5668	5796	5924	6051	6179	6306	6433	6560
7.9	6686	6813	6939	7065	7191	7317	7443	7568	7694	7819
8.0	7944	8069	8194	8318	8443	8567	8691	8815	8939	9063
N	**0**	**1**	**2**	**3**	**4**	**5**	**6**	**7**	**8**	**9**

Table XI Natural Logarithms *(continued)*

N	0	1	2	3	4	5	6	7	8	9
8.0	7944	8069	8194	8318	8443	8567	8691	8815	8939	9063
8.1	9186	9310	9433	9556	9679	9802	9924	*0047	*0169	*0291
8.2	2.1 0413	0535	0657	0779	0900	1021	1142	1263	1384	1505
8.3	1626	1746	1866	1986	2106	2226	2346	2465	2585	2704
8.4	2823	2942	3061	3180	3298	3417	3535	3653	3771	3889
8.5	4007	4124	4242	4359	4476	4593	4710	4827	4943	5060
8.6	5176	5292	5409	5524	5640	5756	5871	5987	6102	6217
8.7	6332	6447	6562	6677	6791	6905	7020	7134	7248	7361
8.8	7475	7589	7702	7816	7929	8042	8155	8267	8380	8493
8.9	8605	8717	8830	8942	9054	9165	9277	9389	9500	9611
9.0	9722	9834	9944	*0055	*0166	*0276	*0387	*0497	*0607	*0717
9.1	2.2 0827	0937	1047	1157	1266	1375	1485	1594	1703	1812
9.2	1920	2029	2138	2246	2354	2462	2570	2678	2786	2894
9.3	3001	3109	3216	3324	3431	3538	3645	3751	3858	3965
9.4	4071	4177	4284	4390	4496	4601	4707	4813	4918	5024
9.5	5129	5234	5339	5444	5549	5654	5759	5863	5968	6072
9.6	6176	6280	6384	6488	6592	6696	6799	6903	7006	7109
9.7	7213	7316	7419	7521	7624	7727	7829	7932	8034	8136
9.8	8238	8340	8442	8544	8646	8747	8849	8950	9051	9152
9.9	9253	9354	9455	9556	9657	9757	9858	9958	*0058	*0158
10.0	2.3 0259	0358	0458	0558	0658	0757	0857	0956	1055	1154
N	0	1	2	3	4	5	6	7	8	9

Table XII Cumulative Poisson Probabilities $\sum_{0}^{c} \frac{m^k}{k!} e^{-m}$

m (expected value)

c	.02	.04	.06	.08	.10	.15	.20	.25	.30	.35	.40
0	.980	.961	.942	.923	.905	.861	.819	.779	.741	.705	.670
1	1.000	.999	.998	.997	.995	.990	.982	.974	.963	.951	.938
2		1.000	1.000	1.000	1.000	.999	.999	.998	.996	.994	.992
3						1.000	1.000	1.000	1.000	1.000	.999
4											1.000

c	.45	.50	.55	.60	.65	.70	.75	.80	.85	.90	.95
0	.638	.607	.577	.549	.522	.497	.472	.449	.427	.407	.387
1	.925	.910	.894	.878	.861	.844	.827	.809	.791	.772	.754
2	.989	.986	.982	.977	.972	.966	.959	.953	.945	.937	.929
3	.999	.998	.998	.997	.996	.994	.993	.991	.989	.987	.984
4	1.000	1.000	1.000	1.000	.999	.999	.999	.999	.998	.998	.997
5					1.000	1.000	1.000	1.000	1.000	1.000	1.000

c	1.0	1.1	1.2	1.3	1.4	1.5	1.6	1.7	1.8	1.9	2.0
0	.368	.333	.301	.273	.247	.223	.202	.183	.165	.150	.135
1	.736	.699	.663	.627	.592	.558	.525	.493	.463	.434	.406
2	.920	.900	.879	.857	.833	.809	.783	.757	.731	.704	.677
3	.981	.974	.966	.957	.946	.934	.921	.907	.891	.875	.857
4	.996	.996	.992	.989	.986	.981	.976	.970	.964	.956	.947
5	.999	.999	.998	.998	.997	.996	.994	.992	.990	.987	.983
6	1.000	1.000	1.000	1.000	.999	.999	.999	.998	.997	.997	.995
7					1.000	1.000	1.000	1.000	.999	.999	.999
8									1.000	1.000	1.000

c	2.2	2.4	2.6	2.8	3.0	3.2	3.4	3.6	3.8	4.0	4.2
0	.111	.091	.074	.061	.050	.041	.033	.027	.022	.018	.015
1	.355	.308	.267	.231	.199	.171	.147	.126	.107	.092	.078
2	.623	.570	.518	.469	.423	.380	.340	.303	.269	.238	.210
3	.819	.779	.736	.692	.647	.603	.558	.515	.473	.433	.395
4	.928	.904	.877	.848	.815	.781	.744	.706	.668	.629	.590
5	.975	.964	.951	.935	.916	.895	.871	.844	.816	.785	.753
6	.993	.988	.983	.976	.966	.955	.942	.927	.909	.889	.867
7	.998	.997	.995	.992	.988	.983	.977	.969	.960	.949	.936
8	1.000	.999	.999	.998	.996	.994	.992	.988	.984	.979	.972
9		1.000	1.000	.999	.999	.998	.997	.996	.994	.992	.989
10				1.000	1.000	1.000	.999	.999	.998	.997	.996
11							1.000	1.000	.999	.999	.999
12									1.000	1.000	1.000

Table XII Cumulative Poisson Probabilities *(continued)*

c	4.4	4.6	4.8	5.0	5.2	5.4	5.6	5.8	6.0	6.2	6.4
0	.012	.010	.008	.007	.006	.005	.004	.003	.002	.002	.002
1	.066	.056	.048	.040	.034	.029	.024	.021	.017	.015	.012
2	.185	.163	.143	.125	.109	.095	.082	.072	.062	.054	.046
3	.359	.326	.294	.265	.238	.213	.191	.170	.151	.134	.119
4	.551	.513	.476	.440	.406	.373	.342	.313	.285	.259	.235
5	.720	.686	.651	.616	.581	.546	.512	.478	.446	.414	.384
6	.844	.818	.791	.762	.732	.702	.670	.638	.606	.574	.542
7	.921	.905	.887	.867	.845	.822	.797	.771	.744	.716	.687
8	.964	.955	.944	.932	.918	.903	.886	.867	.847	.826	.803
9	.985	.980	.975	.968	.960	.951	.941	.929	.916	.902	.886
10	.994	.992	.990	.986	.982	.977	.972	.965	.957	.949	.939
11	.998	.997	.996	.995	.993	.990	.988	.984	.980	.975	.969
12	.999	.999	.999	.998	.997	.996	.995	.993	.991	.989	.986
13	1.000	1.000	1.000	.999	.999	.999	.998	.997	.996	.995	.994
14				1.000	1.000	1.000	.999	.999	.999	.998	.997
15							1.000	1.000	.999	.999	.999
16									1.000	1.000	1.000

c	6.6	6.8	7.0	7.2	7.4	7.6	7.8	8.0	8.5	9.0	9.5
0	.001	.001	.001	.001	.001	.001	.000	.000	.000	.000	.000
1	.010	.009	.007	.006	.005	.004	.004	.003	.002	.001	.001
2	.040	.034	.030	.025	.022	.019	.016	.014	.009	.006	.004
3	.105	.093	.082	.072	.063	.055	.048	.042	.030	.021	.015
4	.213	.192	.173	.156	.140	.125	.112	.100	.074	.055	.040
5	.355	.327	.301	.276	.253	.231	.210	.191	.150	.116	.089
6	.511	.480	.450	.420	.392	.365	.338	.313	.256	.207	.165
7	.658	.628	.599	.569	.539	.510	.481	.453	.386	.324	.269
8	.780	.755	.729	.703	.676	.648	.620	.593	.523	.456	.392
9	.869	.850	.830	.810	.788	.765	.741	.717	.653	.587	.522
10	.927	.915	.901	.887	.871	.854	.835	.816	.763	.706	.645
11	.963	.955	.947	.937	.926	.915	.902	.888	.849	.803	.752
12	.982	.978	.973	.967	.961	.954	.945	.936	.909	.876	.836
13	.992	.990	.987	.984	.980	.976	.971	.966	.949	.926	.898
14	.997	.996	.994	.993	.991	.989	.986	.983	.973	.959	.940
15	.999	.998	.998	.997	.996	.995	.993	.992	.986	.978	.967
16	.999	.999	.999	.999	.998	.998	.997	.996	.993	.989	.982
17	1.000	1.000	1.000	.999	.999	.999	.999	.998	.997	.995	.991
18				1.000	1.000	1.000	1.000	.999	.999	.998	.996
19								1.000	.999	.999	.998
20									1.000	1.000	.999
21											1.000

Table XII Cumulative Poisson Probabilities *(continued)*

c	10.0	10.5	11.0	11.5	12.0	12.5	13.0	13.5	14.0	14.5	15.0
2	.003	.002	.001	.001	.001	.000					
3	.010	.007	.005	.003	.002	.002	.001	.001	.000		
4	.029	.021	.015	.011	.008	.005	.004	.003	.002	.001	.001
5	.067	.050	.038	.028	.020	.015	.011	.008	.006	.004	.003
6	.130	.102	.079	.060	.046	.035	.026	.019	.014	.010	.008
7	.220	.179	.143	.114	.090	.070	.054	.041	.032	.024	.018
8	.333	.279	.232	.191	.155	.125	.100	.079	.062	.048	.037
9	.458	.397	.341	.289	.242	.201	.166	.135	.109	.088	.070
10	.583	.521	.460	.402	.347	.297	.252	.211	.176	.145	.118
11	.697	.629	.579	.520	.462	.406	.353	.304	.260	.220	.185
12	.792	.742	.689	.633	.576	.519	.463	.409	.358	.311	.268
13	.864	.825	.781	.733	.682	.628	.573	.518	.464	.413	.363
14	.917	.888	.854	.815	.772	.725	.675	.623	.570	.518	.466
15	.951	.932	.907	.878	.844	.806	.764	.718	.669	.619	.568
16	.973	.960	.944	.924	.899	.869	.835	.798	.756	.711	.664
17	.986	.978	.968	.954	.937	.916	.890	.861	.827	.790	.749
18	.993	.988	.982	.974	.963	.948	.930	.908	.883	.853	.819
19	.997	.994	.991	.986	.979	.969	.957	.942	.923	.901	.875
20	.998	.997	.995	.992	.988	.983	.975	.965	.952	.936	.917
21	.999	.999	.998	.996	.994	.991	.986	.980	.971	.960	.947
22	1.000	.999	.999	.998	.997	.995	.992	.989	.983	.976	.967
23		1.000	1.000	.999	.999	.998	.996	.994	.991	.986	.981
24				1.000	.999	.999	.998	.997	.995	.992	.989
25					1.000	.999	.999	.998	.997	.996	.994
26						1.000	1.000	.999	.999	.998	.997
27								1.000	.999	.999	.998
28									1.000	.999	.999
29										1.000	1.000

Table XIII Random Digits

35997	30761	97081	09501	68887	32876	01705	34260	95065	45528
88241	30402	12318	52430	40139	96986	84900	72408	42027	31676
54382	73370	26184	14024	57444	57660	52173	30274	93448	63273
77681	74946	02099	69091	19372	66961	14595	58642	75760	52253
53148	26074	52293	65359	63971	04833	86492	01227	54505	19515
89889	46933	13364	33883	83389	36952	52505	67513	40071	31001
03105	87912	29610	75108	37363	28479	43546	89992	19550	54863
82633	19209	21548	35022	21960	57961	11815	95867	00559	26428
69386	57453	70147	73538	49562	46806	64550	36653	25718	68792
31113	07607	48037	71020	22666	65957	11141	39227	07990	19849
65972	74528	40888	55386	95918	92088	91125	53648	66122	00138
79933	71058	34826	97725	69513	22915	18246	52244	91161	40861
40374	13239	56162	04703	95851	22824	41271	28202	62852	84238
46625	20031	08524	20077	65817	21174	29279	57712	22401	67500
30980	74485	26480	21343	30031	61921	35744	57308	71196	01865
49234	62616	54021	29008	83672	85839	96836	45077	80900	66906
63526	93824	71820	11033	20183	85704	04683	63512	39144	56880
64424	95979	17709	94849	31771	05737	84286	16757	46256	24478
73180	59978	08254	78963	95437	86351	33824	32540	18357	02668
99260	21284	81351	70961	10255	06911	47394	72408	23827	59865
96395	30665	43699	03593	29165	23388	26628	92402	16731	86740
29493	09069	78653	90094	42735	33682	95041	89887	92369	57949
81585	50593	14698	04737	72551	57271	59433	00156	33966	58773
59108	49578	18100	59836	73221	21110	01650	11058	47770	66141
84576	58388	40915	94507	32209	17272	65674	95552	25685	05345
36995	36302	07971	67001	62062	75939	36005	26739	56484	46885
66348	87666	78055	44485	82955	85936	09219	01847	92687	72579
45457	78252	98239	40000	75563	92408	17175	78845	32638	26959
35406	59553	57852	07506	00009	93172	77713	93880	40981	27924
09678	24538	52426	84852	83781	23712	82490	77890	22482	66668
55850	25644	44972	62275	78089	28894	98685	32998	98766	89119
34355	75127	69797	71419	62067	57990	96514	50603	79807	26135
29207	43632	32905	38513	18924	88872	20758	70232	60425	01116
24077	21369	93541	75329	78656	44251	42014	98154	42552	14575
30765	00348	01134	71581	68420	78141	21105	63305	09718	54851
65867	08595	47390	39182	51174	41478	64433	59628	31945	87322
78667	95282	05622	26224	19972	97269	98376	14779	51138	49658
45345	04972	52794	15737	00496	48939	63485	42780	16061	59631
37171	13483	56058	51093	62290	88227	17400	88433	67363	89507
26482	85964	71336	67799	28342	37747	61722	27180	78755	18603
42953	06606	23875	56766	01932	36113	62807	84012	21103	09685
69662	76755	13701	95168	13169	44726	15284	16702	89617	54397
52052	12835	37741	86434	22400	37947	95763	86337	35189	22756
47473	16618	42479	47405	14055	64262	66670	89692	54032	94591
44149	29854	76691	33263	62048	25116	88598	16119	62116	54517
31883	86707	18895	81790	71294	02684	15292	48107	14341	91416
75609	92564	39987	02283	89970	95855	80970	05432	89860	90293
99851	94648	05598	32171	28793	92305	64244	08277	93391	96717
34464	29838	10664	28050	60122	77934	10758	84922	92220	45071
97697	36368	17792	84792	76594	67319	51886	05665	45201	11348

Table XIII Random Digits *(continued)*

85604	54908	38992	77961	59337	75563	59755	31862	45033	91503
80508	76285	17630	09429	30293	16391	87516	20628	53159	80261
12043	94593	02328	43332	83707	12201	23088	39829	76777	55495
41717	72807	33686	73225	30173	05410	91541	45387	48084	21855
54866	57899	13389	68475	77825	01301	74831	15970	68803	14519
12030	92278	86864	04430	50868	04949	08820	98949	33713	87279
71744	72285	82724	45846	69682	89838	70910	26386	16527	21698
07607	46148	29548	08230	93459	69788	43771	50812	60337	40035
25584	34039	92437	61873	07874	43107	56212	48897	48008	83125
64572	02625	39993	32573	88828	19036	19394	51921	68629	84838
22089	96239	65157	03977	92561	41314	80082	60159	74429	34535
58590	90320	67095	28958	62803	05097	08269	63296	92249	80332
21640	45655	94143	89051	22782	29086	38014	11641	54398	85092
81936	35183	97146	90677	41012	62425	19569	40059	32565	23037
34506	67652	56534	21287	58697	36165	43304	52134	22272	75345
64575	80559	38389	21713	36749	30055	39889	04287	21294	77790
28400	71414	73453	62631	85191	18446	81309	33305	67816	56922
00797	10584	63075	31922	48847	34738	32528	01884	71241	34618
39544	26038	86456	29624	76562	21853	31395	81509	72150	35599
94881	65970	22406	21125	41074	63283	61007	22211	21082	73175
74166	39761	35695	43436	38419	00937	68925	63631	90667	15306
81282	51396	92605	34582	75716	00563	85458	98519	95291	28719
27598	45667	77617	09543	59492	80179	95095	05757	87364	82473
97034	28089	39063	71062	11200	42848	74844	60021	42546	97331
80634	67739	96041	57015	92100	97403	19269	98036	10903	80518
77896	10740	83670	67658	58485	87918	55526	68299	59299	82190
01283	11951	80432	60004	67206	72482	81741	23414	99285	55717
05949	28563	05514	32008	51159	61218	10172	01302	87465	81485
89689	04867	34303	15893	84495	38167	35321	04292	14843	83461
90508	90873	64852	32793	64464	75215	70933	70719	27386	37924
15783	01737	35510	21366	54979	31337	43469	98473	90915	50441
20779	84571	58354	34236	53161	58091	40332	96450	44850	64913
59022	81361	34099	09325	62447	16298	65423	38905	44505	65521
41161	69968	56531	92298	22547	40605	07198	34011	47837	04156
05101	61976	75740	71764	22726	58391	25497	48681	64279	30092
10938	69708	44018	30646	74933	47964	77638	60987	47121	36748
83088	93442	10017	02240	39942	21743	14976	98483	68418	75537
00579	56704	47455	03455	01725	08959	87671	76914	80139	35079
97868	40700	80345	96368	40054	89076	10675	12199	24137	66889
09711	15318	19689	34252	70632	75170	04407	86853	59651	68194
31789	88212	17274	83864	47527	40799	75637	23308	00004	73403
34532	89800	71622	31468	86314	77136	16149	38935	40487	30025
26057	79898	15817	95290	49993	97314	54353	96343	41199	59863
40196	52895	35391	07148	53757	27107	25989	04982	86187	05431
39542	20073	46940	03292	81653	20670	35522	77340	85154	22598
75945	04357	35302	90752	37844	24999	34389	37872	20274	50215
86180	59251	09172	36760	49093	94483	73259	09651	26947	99041
45163	28966	15577	69733	43612	49510	18029	56461	80917	25340
10920	28307	25105	45935	44072	51297	97347	03892	70827	63643
33755	95467	98153	49384	39956	57617	40684	77645	86474	66051

Table XIII Random Digits *(continued)*

84808	43031	48554	82458	87384	19590	06969	39413	00490	46371
75732	11167	81847	67665	62859	65585	36680	73201	53940	43287
84914	95208	39319	84116	87796	57141	94519	52247	11165	60096
72667	56841	62432	60656	43342	26220	92608	18328	32980	81617
18390	21474	84512	96440	04766	69049	65751	30281	78239	81369
13318	22031	13835	50508	26929	01665	45360	41295	08740	18617
09728	68854	89430	54618	63571	75724	73614	53595	95706	14704
54217	62829	07425	70136	97000	60189	60879	22515	38314	61529
58570	90187	91255	39523	48443	09426	38733	85223	47797	33018
71672	67977	36018	27377	55411	55640	38621	11145	30444	45297
64917	66341	80757	32247	25363	66505	23082	62579	94977	63445
12223	95956	79445	45645	73969	87528	03058	16936	68101	52699
68708	42538	38378	38118	20182	85867	84491	15485	60390	20316
73460	37140	72734	04144	82561	82752	93494	48507	16629	60373
99321	33832	12440	64249	04280	16799	70404	84303	26350	10273
75463	86244	38187	89003	31003	88562	91817	02069	19204	91970
66008	82094	02096	24799	66195	53896	44461	74904	74873	50640
29442	21353	62576	88604	72886	97588	75540	49080	78483	63536
12406	24728	27360	62888	49129	85255	13757	97367	26061	83547
48286	77396	81989	73584	42864	39132	37792	83695	40548	56832
53083	50542	03848	64942	43743	41942	33798	03064	39132	36488
96892	57009	63690	95371	19347	33715	50962	76426	92746	56077
43995	12197	80445	98603	92036	59567	86852	88842	61005	27222
67740	75694	32936	34038	33741	74345	56314	09761	02411	51896
47466	23182	23954	15369	93500	26580	70913	90378	02911	27425
66710	47453	17772	44528	01950	21711	41791	68131	02385	23176
18391	00912	94640	43488	47020	47584	01447	84400	13559	51632
52156	36861	31229	85868	83275	45152	75322	91962	44405	87033
64173	14597	38466	92943	71605	79843	46965	34527	34469	77054
83632	28935	44536	97770	62994	94353	72468	24548	52293	27024
85602	75788	84883	35231	88331	85273	23739	31370	69888	09007
56610	18964	42826	00447	30303	43632	09969	28540	66455	83999
33788	71359	41447	92560	18632	65375	09251	64684	64257	97168
79704	02936	79534	54777	87688	92474	91935	97536	50014	58777
82515	28962	53740	23583	68920	91591	78464	70903	96795	16822
80223	77642	29131	11656	29357	98885	72115	40082	96237	75506
72109	59902	96152	35299	77254	44999	44741	15814	97245	53367
76656	74256	93022	52309	07931	32243	52029	17912	58734	73931
65523	52849	82421	34211	81116	09902	76312	00428	21891	91809
51468	68349	43905	69053	33060	49048	86650	40404	78447	83966
73723	32825	63192	78713	69518	39899	82987	92956	71545	59099
68505	03166	68506	77924	61139	63491	34523	71448	66685	28487
40692	67295	17059	18489	01754	37909	21806	96590	37544	19776
28489	98954	88189	65791	46488	44483	65443	54561	55880	32310
29409	11662	72479	73065	71675	06015	12443	97592	66569	46773
15755	71956	22084	04784	30292	82839	92860	92070	37167	29411
17512	04000	22248	52834	06094	92698	76625	84248	79468	54038
96487	71990	14488	46343	05298	67548	89166	27843	83610	37814
83939	20296	35464	30465	03821	98114	34571	99957	15494	04518
98370	54035	97385	08583	15678	92224	44465	91994	15765	61535

Table XIII Random Digits *(continued)*

93880	30067	11137	06696	77494	27526	88922	57619	02337	86662
76692	57656	99298	30095	02558	82919	25651	28120	23522	66263
04424	76978	71903	75842	31784	80399	93582	73670	02739	94643
88242	36928	66833	09970	17683	06931	09337	01020	31477	68849
21698	00855	90324	64854	64085	99148	84241	99699	85404	28371
78731	32497	56331	38406	98132	99669	25402	47197	09682	33835
82808	34677	71302	75892	49462	10275	73918	67115	20224	44925
99669	97878	35554	87900	04135	40700	02766	17443	09538	98772
08085	66902	45228	63702	97678	84155	53760	98409	65789	90750
49333	93906	43205	93404	16765	15665	32425	09293	00513	06712
19727	21123	06113	87759	71520	00040	93197	87436	09712	99234
08322	30093	19533	71269	82284	56203	89683	72041	33476	71856
06408	41290	36686	99287	18048	11168	90761	39183	93279	37994
18310	33376	12655	07615	59982	87924	93692	61118	36910	49622
97474	78227	28139	89395	60111	20995	11236	57144	90898	34917
39982	67482	78091	79694	00970	13710	64101	16277	07816	13879
28006	54888	78643	68059	67424	26880	63383	26194	97576	84261
45645	49941	04369	70130	94352	40112	43857	23164	47664	99511
13472	43394	03599	73615	49058	58048	80515	06741	56118	75052
27850	25453	21032	29743	48509	21267	03681	74642	67082	11904
59014	90744	42987	69613	20332	30392	11715	95052	58113	16536
55480	31996	85226	22210	84892	19453	80510	56212	45910	84855
24339	43283	59724	62822	15548	56141	89220	05213	36311	46352
34541	15246	09754	49137	68131	56389	64410	71732	19148	65009
49561	38668	87621	74272	67268	82594	08537	48996	83825	74170
99125	39759	25049	11813	99569	73368	50004	07974	73607	13360
04875	42765	61690	54915	31801	79108	18289	81161	97928	58360
50607	19272	53424	58923	54316	89854	71677	66815	19877	47235
30290	03075	11667	16122	32999	77755	12604	19115	97502	27969
28274	48942	27579	91896	80234	39543	95392	79920	83016	96471
16959	76737	16654	13542	20728	13928	30394	43895	21995	85942
04994	99280	94784	55171	14880	40420	15667	84423	62403	77012
25594	15045	07410	61877	55872	77197	82477	33640	06882	89657
63364	34022	88784	94445	84722	77505	64618	61370	18801	16361
08694	33467	29511	18888	57516	63467	16959	21585	35113	25620
50046	65487	22656	74729	84950	60194	80964	92915	53704	66919
28960	79829	02580	24360	40273	75814	04835	00662	29784	88808
22308	89960	58428	34381	49539	89565	73880	27768	79531	29346
98592	23403	19929	67535	74484	80083	45577	62977	13548	04909
18040	39633	74903	07420	15147	31795	84481	90974	24042	24544
04314	06864	55314	44064	68632	71109	01162	53475	24321	76906
66875	54661	48942	26027	65858	26851	72889	09511	10760	49833
29794	47770	62871	15559	49146	59030	34857	62411	60031	67962
26351	53194	42764	10192	05613	76994	67324	14248	35070	48409
45360	18780	69171	57162	59969	15610	44283	93966	03249	10961
25773	46930	91698	54007	16852	63493	16731	11774	63152	22423
34331	59199	03614	22220	64023	26306	78073	25201	02566	32066
65774	44518	93113	80244	86535	97521	27092	20736	24130	72258
40982	17521	22630	61308	90794	17845	30663	83489	35955	15812
84187	91945	63475	71413	87856	95561	88319	69211	11872	46843

Table XIII Random Digits *(continued)*

30188	13165	42327	89186	27240	51475	19084	01658	95939	42242
29260	36702	97498	27870	21999	45430	15279	42815	09683	06634
12698	73654	63795	85377	78064	31331	67956	01890	66299	97281
87590	68966	05510	18719	76529	32568	79512	50737	24240	88013
33776	17733	77028	94252	63704	47794	32551	56121	54926	44099
66923	65455	21726	24004	35958	70388	47181	80114	44393	95822
22851	61064	49863	53279	32119	86375	12438	54748	30937	04016
73937	66302	22601	01597	35595	40183	93217	35330	08129	23157
06605	94854	80177	80420	91128	58026	97434	02305	67726	92404
67392	32869	27712	63769	95480	05856	89340	07983	40790	50733
94418	38671	24761	20757	67165	07112	89044	62020	86202	76771
36819	32887	71331	09233	34971	34624	33038	88939	17552	75324
87495	03564	94267	34851	88708	13548	03087	76093	22550	62537
72669	49656	06160	31219	55804	23433	64733	10714	03945	00774
61154	27475	38261	43140	86358	05317	79974	15852	44198	74379
31335	88879	34354	44536	07427	42242	68317	73923	87079	36150
04118	73475	00063	96418	23398	84148	19541	92994	83983	94374
46613	25413	08897	80130	47931	86219	54789	13477	88560	01041
40174	20364	33677	97450	88056	69695	25333	13845	02196	66952
28556	34497	06416	64599	36939	58812	49009	25225	03467	14737
13689	88776	86085	70559	61091	68788	81146	39182	86314	49760
44883	51127	41120	88062	94853	70794	41750	04166	77851	68804
89660	40387	01296	42873	73371	16992	26361	60660	06412	26810
22852	54621	07075	98314	28673	87411	64601	26239	87265	18640
69917	37727	32781	75126	85640	94493	75273	02898	51556	21788
60703	62586	56277	21287	60160	79217	30686	37135	11855	25546
39734	88161	06705	20574	29691	07649	19880	82844	31621	26134
69718	22606	20449	88216	55589	66005	46014	82664	91379	35866
01820	01898	51631	17022	49883	48785	28978	81393	98702	88148
51365	67294	31747	05780	67140	71584	08691	01685	50799	09931
39816	24496	92068	58664	67457	41030	27589	21936	79622	98704
99927	40974	21355	69282	64955	49478	84395	11261	52108	87075
60526	61540	31224	64740	70860	37315	97594	82585	13615	04376
55793	36106	13752	45954	64082	05354	61770	22923	91118	30852
06943	40899	06839	09155	23301	17333	15068	66413	41941	41413
68225	21378	91780	13103	22094	81029	93778	62740	78006	85206
66278	42054	12481	13818	99451	60559	40825	66105	33483	35180
79953	47432	72632	26078	50973	04703	80743	11052	86725	01724
20001	17872	32032	46190	29714	02116	25718	98186	24250	08942
64452	97761	18509	94831	21073	97805	66900	92196	56401	96024
52589	31610	90208	38242	52248	33515	62763	82129	50971	96281
94602	48232	95201	41543	49516	21054	74436	25550	67091	98491
63386	13036	13586	98065	05321	29855	62456	57802	25697	73456
86307	35369	11978	61855	69235	07772	68694	86795	91955	94214
72414	77593	88984	79264	70313	26421	39524	48498	42915	49827
54708	70188	25904	50140	78021	39060	56463	25953	41446	70694
64650	02236	63322	45683	24874	35386	96878	40419	77392	28566
51006	55169	94738	01522	86380	81992	84746	31348	97524	18687
32302	49942	52830	13982	30058	81341	26841	32310	90199	38088
53090	47393	72579	11405	27519	00263	00375	15250	51472	07703

Table XIII Random Digits *(continued)*

29257	36060	81080	67493	23666	22251	17616	60716	77125	18653
04426	95304	83272	18379	46498	60045	80649	35179	03185	57068
72622	55513	82844	85553	16852	57931	84063	57516	46529	47030
63755	08166	33097	46244	16769	48531	56618	90035	88363	04097
23931	55916	48477	33067	76572	84835	96208	68558	23560	89245
96443	13697	61186	63971	20547	14846	77137	62636	88927	34322
33595	83707	92545	83866	06895	28019	08547	04275	79277	28833
87400	72301	05172	25637	13665	86725	45970	42670	35291	22685
10716	35521	73850	99275	97475	11064	93492	05362	57562	99582
81465	80905	77978	42899	65518	48688	96755	83554	76916	15224
77706	53575	16463	00350	44697	94868	22697	33740	60701	04034
75622	35864	56564	40277	66044	78417	52968	52982	82340	92970
18418	16826	26355	51841	01235	15986	65898	74181	51391	11313
02205	51273	87582	80276	88583	30633	50721	65017	48735	04476
08345	03180	15659	86285	09579	07969	17850	88197	14309	25013
43734	89733	87697	28098	70926	22790	79293	01093	72673	60257
89981	80872	87829	56857	78208	85949	60249	30159	45499	29735
20660	70109	01273	42633	77445	02439	03144	12100	12971	59574
48210	18773	36169	60470	37941	68015	76627	38973	81699	34262
65433	21735	36444	40569	32023	60078	31045	99679	76253	81056
09638	39323	55045	58988	30967	16461	57491	58625	14327	19825
89418	30419	67738	14877	17948	30083	54764	62024	68310	21207
23083	08913	59531	11403	96757	23468	57382	48057	70725	59933
30343	01516	78510	15138	49538	58588	06080	23844	44412	87882
18284	64484	57502	80120	95894	34977	74098	73551	81743	58364
47943	40974	82162	34485	44930	27794	66180	76055	69035	91186
55480	50090	91328	09979	23824	71199	65640	69121	12071	46595
29905	46328	06091	12419	35540	09817	63102	80393	76848	09779
82268	73981	37205	65249	07751	92537	92368	09251	29892	53919
42183	74789	78645	68686	12399	27592	36651	05904	45764	11336
82178	50120	06396	62927	27126	40350	87365	28918	39266	22549
49836	15424	10577	89926	89407	37622	38430	48356	93062	86591
96881	32775	85851	60822	08530	83330	99819	24740	34199	90176
20414	02812	65040	27108	34291	90243	18207	85800	53786	28723
62174	64315	82589	31761	26476	95984	54304	13061	77554	92988
48808	13755	89592	43057	35053	58853	20086	12130	90105	26139
93162	69261	87781	82538	75030	08496	25862	41090	06267	07824
81009	59154	63054	77841	35348	61706	27750	03274	64051	62435
11796	19222	73310	68903	54526	56338	54017	47320	67925	40813
48728	67741	07718	37592	95126	68160	55907	46168	06840	96853
73347	09803	63349	39332	23487	69045	51759	99114	86318	60963
34704	15020	45067	29373	24970	94427	23023	88559	65920	94063
24147	47285	01070	86152	20380	13113	60138	69612	38184	41601
38113	53850	17580	01850	33698	60239	84065	82019	52464	11886
01435	59660	45587	96635	57407	97292	15584	05498	54391	33149
76230	93764	38424	43810	11252	02843	64856	71551	66186	36921
76698	49005	35125	73104	93675	83189	38168	69103	75653	64780
21081	41735	47975	16487	37540	89936	16311	78716	51188	87349
25671	32206	93142	93930	57886	14113	18177	97474	32669	21337
03245	76312	63143	57914	10925	61894	91043	64156	61020	55339

Table XIII Random Digits *(continued)*

22719	92549	10907	35994	63461	83659	24494	53825	97047	79069
17618	88357	52487	79816	74600	50436	88823	19806	33960	30928
25267	35973	80231	60039	50253	63457	97444	13799	35853	03149
88594	69428	66934	27705	51262	63941	77660	66418	84755	29197
60482	33679	03078	08047	39891	34068	81957	02985	83113	36981
30753	19458	02849	30366	83892	80912	91335	41703	79401	97251
60551	24788	35764	57453	06341	10178	91896	70819	46440	98356
35612	09972	98891	92625	70599	95484	34858	13499	28966	88287
43713	18448	45922	55179	18442	31186	91047	37949	76542	79361
73998	97374	66685	06639	34590	17935	79544	15475	74765	11199
14971	68806	49122	16124	61905	22047	17229	46703	39727	16753
78976	48382	25242	97656	51686	15537	73857	35398	91783	92825
37868	82946	73732	63230	85306	56988	15570	98029	42208	00190
01666	48114	95183	02628	05355	97627	74554	91267	31240	34723
56638	70054	19427	24811	37164	71641	50515	88231	99539	75745
43973	07496	17405	08966	65989	68017	56975	94080	93689	98889
05540	72301	36504	00187	90375	22891	22205	27777	84803	39220
95141	07885	94399	41145	50210	92423	13303	09621	94153	18691
75954	68499	42308	38387	52163	64563	02843	45577	93125	25294
97905	05301	98496	20682	68082	68537	70220	78282	02396	10002
23458	57782	67537	38813	00377	93873	97813	10039	25457	28716
03954	14799	63187	46191	22805	50502	08810	19572	48024	58206
52251	06804	85959	20974	73104	15009	25486	09306	24721	04187
62361	59105	39338	59358	69193	15586	57695	89518	59788	04215
54954	90337	99346	60442	90933	58323	83183	90041	44236	90815
70773	03331	84228	01405	61494	72064	24713	39851	01431	60841
68702	08331	08923	83173	67081	87472	47980	08802	95495	78745
39599	33465	96705	41458	34670	55385	25484	71068	15155	85371
54958	34935	16858	16523	54262	63310	50348	53457	39440	80411
98124	08864	36485	78766	52802	56315	43523	06513	50899	86432
43099	88373	80091	35058	35755	47556	98602	71744	70442	92312
88667	44515	80435	17140	32588	98708	93010	98590	23656	85664
87009	95736	76930	71090	27143	95229	24799	02313	17436	20273
70581	40618	16631	54178	44737	02544	81368	08078	46740	52583
03723	25551	03816	97612	99833	06779	47619	12901	60179	23780
49943	30139	07932	29267	01934	19584	13356	35803	90284	97565
71559	30728	83499	65977	37442	72526	53123	99948	59762	19952
75500	16143	79028	81790	57747	87972	54981	10079	17490	15215
59894	59543	13668	27197	51979	38403	23989	38549	82968	53300
29757	26942	08736	15184	73650	51130	59167	89866	06030	88929
87650	08162	90596	70312	84462	07653	80962	96692	07030	62470
84094	70059	86833	23531	31749	23930	04763	89322	67576	38627
92101	17194	06003	99847	12781	38729	88072	92589	61828	36504
26641	99088	65294	37138	75881	12627	19461	69536	64419	82106
04920	91233	46959	14735	15153	28306	76351	28109	86078	45234
25417	97570	91045	09929	75140	43926	90282	99088	93605	03547
98874	96989	84371	87624	74090	71983	62424	62130	44470	74725
82127	82000	84618	58572	56716	79862	59896	50702	31938	18336
26311	59516	98602	47197	31139	27631	64619	01504	77617	30219
76176	03499	17999	84361	63898	97861	63620	23931	87903	91566

Table XIII Random Digits *(continued)*

44553	29642	20317	69470	57789	27631	68040	73201	51302	66497
01914	36106	71351	69176	53353	57353	42430	68050	47862	61922
00768	37958	69915	17709	31629	49587	07136	42959	56207	03625
29742	67676	62608	54215	97167	07008	77130	15806	53081	14297
07721	20143	56131	56112	23451	48773	38121	74419	11696	42614
99158	07133	04325	43936	83619	77182	55459	28808	38034	01054
97168	13859	78155	55361	04871	78433	58538	78437	14058	79510
07508	63835	83056	74942	70117	91928	10383	93793	31015	60839
68400	66460	67212	28690	66913	90798	71714	07698	31581	31086
88512	62908	65455	64015	00821	23970	58118	93174	02201	16771
94549	31145	62897	91582	94064	14687	47570	83714	45928	32685
02307	86181	44897	60884	68072	77693	83413	61680	55872	12111
28922	89390	66771	39185	04266	55216	91537	36500	48154	04517
73898	85742	97914	74170	10383	16366	37404	73282	20524	85004
66220	81596	18533	84825	43509	16009	00830	13177	54961	31140
64452	91627	21897	31830	62051	00760	43702	22305	79009	15065
26748	19441	87908	06086	62879	99865	50739	98540	54002	98337
61328	52330	17850	53204	29955	48425	84694	11280	70661	27303
89134	85791	73207	93578	62563	37205	97667	61453	01067	31982
91365	23327	81658	56441	01480	09677	86053	11505	30898	82143
54576	02572	60501	98257	40475	81401	31624	27951	60172	21382
39870	60476	02934	39857	06430	59325	84345	62302	98616	13452
82288	29758	35692	21268	35101	77554	35201	22795	84532	29927
57404	93848	87288	30246	34990	50575	49485	60474	17377	46550
22043	17104	49653	79082	45099	24889	04829	49097	58065	23492
61981	00340	43594	22386	41782	94104	08867	68590	61716	36120
96056	16227	74598	28155	23304	66923	07918	15303	44988	79076
64013	74715	31525	62676	75435	93055	37086	52737	89455	83016
59515	37354	55422	79471	23150	79170	74043	49340	61320	50390
38534	33169	40448	21683	82153	23411	53057	26069	86906	49708
41422	50502	40570	59748	59499	70322	62416	71408	06429	70123
38633	80107	10241	30880	13914	09228	68929	06438	17749	81149
48214	75994	31689	25257	28641	14854	72571	78189	35508	26381
54799	37862	06714	55885	07481	16966	04797	57846	69080	49631
25848	27142	63477	33416	60961	19781	65457	23981	90348	24499
27576	47298	47163	69614	29372	24859	62090	81667	50635	08295
52970	93916	81350	81057	16962	56039	27739	59574	79617	45698
69516	87573	13313	69388	32020	66294	99126	50474	04258	03084
94504	41733	55936	77595	55959	90727	61367	83645	80997	62103
67935	14568	27992	09784	81917	79303	08616	83509	64932	34764
63345	09579	40232	51061	09455	36491	04810	06040	78959	41435
87119	21605	86917	97715	91250	79587	80967	39872	52512	78444
02612	97319	10487	68923	58607	38261	67119	36351	48521	69965
69860	16526	41420	01514	46902	03399	12286	52467	80387	10561
27669	67730	53932	38578	25746	00025	98917	18790	51091	24920
59705	91472	01302	33123	35274	88433	55491	27609	02824	05245
36508	74042	44014	36243	12724	06092	23742	90436	33419	12301
13612	24554	73326	61445	77198	43360	62006	31038	54756	88137
82893	11961	19656	71181	63201	44946	14169	72755	47883	24119
97914	61228	42903	71187	54964	14945	20809	33937	13257	66387

Table XIV Gaussian Deviates

.117	.136-	.820	1.213-	.131	.738-	.918	1.002	.846-	.288
.519	.787-	1.128-	1.100	1.609	.797	.382	1.157-	1.320-	2.056-
.876-	.832-	.788-	1.490	.923-	.710-	2.149-	1.967-	.088	1.158
.311	.494	.357	.025	.016-	.448	.733	.199-	.440	.609
1.041-	.627	.957-	.777	.304	.581-	1.495	1.564-	1.471-	1.097-
.239	.061	1.091	.060-	.521	.777-	.461	.919	.091-	1.412
.151-	.664	.596	.370	.346-	.526-	1.557-	.180-	.323-	.918
.962	.502-	.967-	.859	.916	1.525-	.064	1.023	.001	1.577-
1.573	1.912-	1.010-	1.780	.771-	2.390	.188-	.593-	.608-	.561-
.742-	.137	.563	.887	.740-	1.410-	.818-	.545-	1.130	.741-
.143-	1.299-	1.869-	.191	.789-	.296-	2.232-	.268	1.582-	.389
1.433-	1.169	.733-	1.176	.582-	1.060	.447	.305	2.418-	1.209-
1.946-	1.045	1.705-	1.544-	1.701	.972	.346	.341-	1.240-	.194-
.885-	.247	1.230-	1.461-	.175	2.072	1.174	.223-	1.106-	.028
.046-	.513	.201-	.740-	.727	.668	.433-	.991-	.174-	1.421
.683-	.161-	.964	1.182-	.485	.901	1.321	.803	.727-	.569-
.749-	.029-	1.150-	.122	.016-	.690-	1.261	1.884	.758	.035-
.995	.542	.448	.796	.616	.261	1.072	1.153-	1.866-	1.029-
.274	.188-	.846-	1.557	.554	.514	.723	.322-	.805-	.178
1.120	.396-	2.110	1.469-	.589-	.779	.338	.093-	1.629	.134
.668-	.678-	.406	.092	.944	.728-	.358-	1.206-	.783-	.510
1.583	.730-	.911-	.126	1.864	.296-	.980-	1.022-	.315	.274
1.050	1.162	1.236	2.039-	1.299-	.722-	.630-	1.359	.511	.448
.477	.433-	.110	.182-	.363-	.716	1.355-	1.579	.574-	.043
1.538-	.137	.382-	.578	1.053	.489	1.552	1.520	.391	1.026-
.314-	.889-	.913-	.417	.537	.426-	.100-	1.467	.483	.627-
.730	.946-	.231-	.671-	.798-	1.330	1.006-	.123-	.442	1.513
.276	.473-	.477	1.076	.316	.600-	.146-	.090	.608-	1.198-
.638-	1.270-	.447-	1.101-	1.107-	1.433-	.349	.546	.283-	.887
.497	.829-	.745	.469	1.975	.130	.367	.202	.433-	.630
.769-	.866-	1.034-	1.615-	.120	.493	.103	.639-	1.732	1.066
1.384-	.453	.586	1.549-	.421-	.815	1.319-	.805-	.009-	.100-
.784	1.980	1.265-	.239	1.189	.382-	.047	.582-	.806	1.336-
.035-	.514-	.087-	.202-	.925	.047-	.926-	1.157-	.498	1.066-
.678	.917	.376	1.282	1.176-	.622	2.123	.646	.730-	.026
.179	.841	.298-	2.437-	.740-	.039-	.226	.247	1.614-	.492
.111	.044-	.209	.527	.598	.206-	1.042-	.012-	.757	.840
1.006	.919-	.956	.808	1.793	.079-	1.953	1.494-	.559	1.290
.307-	1.174-	.858-	.039	1.505-	.037	.107-	.120	.557	1.809
2.467-	.273	.899-	.691-	1.092-	1.374-	1.238	2.046	.879	.296
.275	1.313-	.331-	.305	.404	.399-	.591	.280	1.802-	1.207
.514-	.713-	.501	1.214	.001	.360	.124-	1.373	1.857	1.135-
.982	.139-	1.113	.433-	.761-	.182	.405-	.714	.616-	1.402-
.071-	.115-	.344-	.429	.316	.667-	1.676	.155-	1.085	1.780-
1.975-	1.416-	1.367	.592-	.480	.406	.701	1.077	1.475-	1.024
.027	1.446-	.464-	1.180-	1.223	1.116-	1.017-	1.051	.051	.853-
.016	1.118-	1.228-	1.382	.502-	.494	.612-	2.755	.809-	1.216-
.584-	1.410-	.551-	.602-	.381-	.078-	1.310-	1.198	1.359	.115
.669	.611-	.452-	.302	1.026-	.331-	1.047-	.618	.931	.218-
.070	1.598-	.506-	.812-	1.203	2.110-	.049	.059	1.890	.421

Reprinted with permission from *A Million Random Digits with 100,000 Normal Deviates*, Rand Corporation, Santa Monica, Calif.

.236-	.062	.602-	.744-	.752-	.431-	1.222	.638	.813	.448
1.265	.851	.332	.991-	.122-	2.061-	.558	.792-	.108	.586
1.604	1.566	1.035	1.028-	.135-	1.377-	.831	.533	1.284-	2.014-
.176-	.166	.477	1.112	1.909-	.313-	.064-	1.288	1.371-	.934-
.030	.223	.833	2.071-	.862	.765-	.314	.291-	.788-	.107-
.450	.507	.393	1.246	.342	.006	.001	.016-	.379	.471-
1.126-	.682	.126-	.826	1.619	.400	.130-	.030-	.205	.643
.676	1.066-	1.239-	1.513	.169	.003-	.465	1.170-	.630	.772
2.638-	1.318	1.416-	2.152	.290	1.794-	.934	.728	1.453	.631-
1.398-	1.549-	.875-	.138-	1.429-	.773-	1.202-	.043-	.183	.485
1.797-	.428-	1.671-	.600	1.362	.845-	.257	.253-	.645-	.650
.045	.202-	.565	1.136-	1.891	1.175	2.118-	.760	.053	.435-
.669	.703	.514-	.315	.680-	.602-	1.342	1.481	.140-	.261
.921-	1.361	1.105	.173	1.088-	1.374-	.056	.512	.302	1.332
2.255-	1.821-	.559-	2.111-	1.170-	1.377	.331	1.331-	.053-	.460
.540-	1.989-	.140	.668-	.315-	1.549	1.346	.398	.415-	.621
1.127	.143	1.443	.443	.901	.624	1.167-	.817-	.033-	.592-
.686-	.669	2.295-	.098-	.459-	.530-	.766	1.598-	.168	.321
.092-	.648-	1.763-	.095-	.203-	.191-	1.587	1.068-	.440-	1.415-
.384-	1.139	.926	1.489	.452-	.372-	1.178-	.106	.022	.942
.049	.631	.147-	.550-	.972	.236	1.221	.799-	1.422-	.498-
.395	.419	.392-	.781-	2.430	.914	1.331-	.155-	.936	.149
1.548	.048-	.599	.633-	.377-	.414-	.296-	1.342-	.518	.127-
.450-	1.073-	.731-	.282-	.324-	1.885-	.355-	.574	.669-	1.236-
.250-	.055-	.034	1.880	.448-	.859-	.689	.945-	1.018	.628
.388	.269-	.109-	1.205	2.347-	.563	1.908	1.342-	.381-	.221-
1.342	1.000	1.107	.365	1.167	.619-	.097	1.331-	.005	.261-
1.229	.388	.563	.350	.615-	.007	1.022-	2.142	.473-	1.530
.430	.137-	.068-	1.122-	1.139	1.782	.601-	1.219	.397-	.441-
1.453-	1.175-	.157-	1.517-	1.211	.849	.154	.958-	.357-	.705
1.088-	1.266	.993	.564-	1.276	.746	1.203	.406-	1.350-	.812
.162	1.048	.305-	.971-	1.256-	.612-	.407	.652-	1.024-	.130-
.700	.106-	.948	1.708-	2.893	1.484-	.414-	.955	1.816-	1.355
.786	1.432	.576-	1.082	.627-	2.111-	.967	.680-	.742	.122-
.741	.853	2.088	1.435	.162	.717-	1.405-	.430-	.397	.719
.112-	.956-	.125-	1.288	1.020	.759	1.669-	.300-	.001-	.119-
1.980	1.872-	.264-	.374-	1.229	.854	1.816-	1.163	.707	1.082-
1.732	.087-	.196-	.962	.427	.985	.371-	.661-	.868-	1.426-
.388-	.215-	1.618	.994	1.492	1.302-	1.850-	1.705-	1.429	1.095-
.085-	1.196-	.221-	1.676-	.377-	1.179-	1.514	.187	.280	.165
2.173-	1.466	.369	.663-	2.001-	.301-	1.028-	.491	1.531	.300
.146-	.600-	.566	.875	.229	.021	1.143-	.734-	.842	1.193-
1.309	2.183-	2.904	.115	.777	.722-	.502	.489-	2.010-	.152
.651	.285	.698-	.297-	.534	.526-	1.357	1.319-	1.319-	.943-
1.422	.414	1.588	.630-	.344-	.758-	1.187-	2.100	.724	1.258-
.092	.580	.769-	.492	1.360-	1.395-	.008	.255-	1.387	.268
1.017-	.343-	.728	.093-	1.916-	.518	.070-	.157	.485	.841-
.076	1.086	.486	.009-	2.231-	.344	.196-	.361-	.270-	.014
.575-	.921	1.823-	.674	1.461	.493	.739	1.128-	1.451-	.387
1.150	.678	.377-	.319-	.312	2.244	.607-	1.382	1.731-	.721

.436	.161	.111	.400	.382-	.054-	.190-	2.198-	.849	.312-
.569-	3.086-	.238	1.328-	.325-	.301-	.011-	.481	1.783-	.453
.064	.418-	.497	.153	.579	.808-	1.536-	.098	.477	.610-
1.229-	.029	.569-	.968-	.280	1.627	.304-	.028	1.574	.922
.329-	.729	.163-	2.665	.893-	.356-	.924	.468-	1.019	.116
.259-	.698	.205-	1.109-	.763-	.865	1.160-	.404	.352-	1.657
.241	.805	.785	.372	1.125-	.694	.737-	.118-	.037	.966
.215-	.065	.226-	.430-	.746	.856	.558	.709	.322-	.661
1.026-	.148	2.133	.560-	.149	1.250	.479-	.431	.573-	2.529-
.101	.003	.847-	1.945-	1.372	.712	.018-	.326-	1.259	.329
.235-	.046	1.344	1.264	1.029-	.001	1.041	1.305	1.615	.121
.787-	1.455	.459-	.588	.586	1.376-	1.754-	.304-	.969	.695
1.225	2.064-	.818	1.021-	.500-	2.217	.246	.585-	.247-	.026-
.976-	.035	.449-	.377-	.175-	1.120	.477	.387-	1.605	1.255-
.550	.220	1.141-	.111	.060	.359-	.306	.298	2.503-	.881-
1.497	.767	.461	.108-	1.065-	.347-	.355-	1.453	.830-	.599
.076	.038	.471-	.144-	.269-	.125	1.038	1.730-	1.431-	.176-
.909-	1.364	.656	.552	.226	.678-	.207-	.945-	1.252-	1.769-
1.932	1.977-	.324-	.663	.473-	.625-	1.124-	.382-	.100	1.532
.111	.848-	1.086	1.387	.433	1.384	1.225-	.085	.979	.738-
.862-	2.809	.946	.222	1.003-	.612	1.215	1.962-	.733-	.727-
1.661-	1.773	.796-	.855	.149-	1.003-	.798-	.856	.681	.037-
1.774	.119	.650	.394	1.265	.720-	.156-	.021-	.413-	.495-
.318	.572-	1.911	1.792-	1.892	.941	1.590	1.962-	.052-	1.268
.306-	1.616	.570-	1.332-	1.392-	.074	1.228-	.500	.064	.316-
2.623-	1.571-	.491-	.334	.253	.129	.623-	1.510-	.602	.746
.513-	.728	.020-	1.014-	.866	1.923	.652	.110-	1.927-	1.339
.678-	.181	.708-	.857	.274	1.103-	.225-	.585-	.519	.117
.299-	.976	.080	1.298	.252	1.033	1.418	1.656-	.320-	.202
.167-	.722	1.230-	1.323-	1.422	.678	1.490-	.386	.668	1.326
.844-	1.034-	.272-	1.623	.221-	.613-	.091	.929-	.165	.110-
.700-	.294	.946	.053-	.097	.882	1.559-	.549	.254	.461
1.019-	.169-	1.416	.924-	1.040-	.547	1.248-	1.257	.944	2.032-
.328	.275-	1.860	.902-	.010	.699	.383	1.265-	.635	2.122-
1.115	.072	.233-	.443-	1.054	.900	.441	.649-	1.412	1.217-
1.019-	.319	.643	.161	.671	.142-	.306-	.127	1.119	.418-
1.659-	.200-	.114-	.132	3.063-	.022	.101	1.637	.095	.497
.936	.486	.785	.556	2.411	.108-	.729	.186-	1.350	.408
1.848	1.191	1.828-	1.404-	1.551	.380-	1.475-	.034-	.138-	.942
.657-	1.762	.932-	.239-	.338-	1.060-	.490-	1.742-	.678-	.600
.566-	.566	1.743	1.271	1.119-	1.433	.178-	.890	.033	.156
.186-	.672-	1.068-	.822	1.091	1.064-	.580	1.462-	.749	.474-
1.089	.788-	.804	.457-	.721-	.820-	.108-	.655-	.202-	.058-
1.698-	1.121	.838-	.108-	.950-	1.075	.332	1.407	.505	1.856
1.056	.880	1.277-	1.259-	.974	.690	1.216-	1.279-	.145-	2.119
1.204-	.485	.581-	.289-	.204	.067-	2.140	1.155	1.003	1.765
.337	.845	1.538	.035	1.531-	1.695-	1.294	1.006	.802	.286
.987-	.242-	.174-	.684	1.892	.570-	.249	.972	1.929	.704
.570-	1.166	.525-	.661-	1.467	.359	1.778	1.424-	.947	.497-
.924	1.444-	.111	.613-	.177	1.496-	.330-	1.220	1.037	.885-

Table XIV Gaussian Deviates

1.718-	.751	.671-	.239	1.236	1.147	1.184	.754	.059	.216-
2.665	.832	.484	1.104	.445	.751-	.914	2.605	.974	.727-
.834	.930	.082-	.088	.236	1.098-	.108-	.798-	.743-	.651-
.314-	.479-	.343	.497-	.256	1.060-	.476	.740	1.457-	1.238-
.612-	.425-	.947	1.460-	.478	.550-	.172-	.891-	.841	.555
1.048-	.236	.034	.103-	.600-	.143	.403	1.383-	.326	1.442
2.169-	.281	.146	.978	.201	2.322	1.530	.804	.635	.490
.052	1.064-	.344	.707	1.333	.409	.634-	.544-	.094	.337
.428-	1.326	.295	.731	2.302-	.889	.582-	.443-	.331-	.300
1.138-	.414-	.799	1.192-	.205-	.170	.407	.042-	.481	1.360-
.657-	1.783	1.359-	.079	1.315-	.916-	.650	.617	1.146-	1.750
.659	.685-	1.376	.412	.887	.783-	.025-	.082	.011-	1.412-
1.104	.834	.434-	1.550	.037	1.939	.342-	.202	1.425-	.629-
.532-	1.440-	1.770-	.090	.037-	1.233	1.990	2.311-	.708-	.522
.081-	.326	.686	.129	1.015-	.211-	.686	.920-	2.617-	.709
.201	.026-	.266	.897	1.350	.072-	.344	.644-	.034-	.790-
1.160-	1.327-	.284-	.355-	.355	.521	.163-	.266-	.219	.816-
.179-	1.349	1.336-	.413-	.332-	1.630	.457	.778	.588-	1.319-
1.463	.165-	.394-	1.188	.054-	1.534	.361-	1.630	1.150-	1.211
.371-	1.573	.030-	.312	2.206-	.466-	.937	.459-	.572-	.433-
1.007	1.351	2.325-	.101-	.059	.130-	.481-	2.668-	.428	.960
.321-	.849-	1.182	.572-	1.865-	1.453-	1.540	.580	.896-	1.111-
.144-	.327	.894	.297-	.916-	2.071	.185-	1.026-	.242	1.582
.039-	.763	.645-	1.549-	1.285-	.169-	1.233-	.793-	.993-	.297
.832	1.808	1.309	1.260-	.838	1.417-	.103	.284-	1.263-	1.310
.117	1.383	.912	1.156-	1.128	.356-	2.127-	1.555-	.092-	.771-
2.782	.276-	.117	.116-	.552-	1.072	.522	1.220	1.011	.778
2.273-	1.251-	.484-	.871	1.040	.682-	1.289	.421	.377-	1.657-
.816	.523	1.241-	.569-	.005	1.079	.504	.831-	1.470-	.556-
.870	.348	.211	1.164-	.759-	.498-	2.102-	1.627-	1.100	.548-
1.373-	.416-	.685-	.067	.105	1.032-	.018	2.642	.230	.207
.689-	.026-	.295-	1.109	.954	.604-	.060	.180-	.707-	.306
.860	.330-	.310-	.297-	.144	1.057-	.148-	.999	.987-	.178
1.663	.137	.092	.578-	.310	.936-	.555-	1.327-	.277-	.413
.007	.611	1.209-	1.723-	.969-	1.754-	2.253	.807-	1.456	1.042
.493	1.538	.532	.770-	1.547	1.549	.861-	1.332	1.026-	2.595-
.344-	1.702	2.089-	.352	.831	.133	.623	.092-	.734-	.285
.204-	1.316-	1.561	.501	.105	2.354	.970-	1.223-	.201-	.513-
.633	.965	.162-	.429-	.047	.757-	.528	.633-	1.250-	.447-
.792-	1.492	.074-	.194-	.678-	.158-	.081-	1.296	.051	2.030-
1.163-	1.081	1.294	.186	.454	1.875-	2.825-	.352	.945-	.225
.504-	.110-	.676-	.863-	.854-	.770	.652-	.386	.022-	.637
.967	2.124-	.755-	.490-	.733	.694-	1.059	.561	.094	1.341-
.797-	.526	1.448-	.194-	.607-	.544-	2.190-	1.539	.925-	.707
.586	.364	.407	.089	.530-	.546-	1.257	.137	.330	1.279
.120	1.429-	1.390-	1.401	1.266-	.999	.916	.193	.760	.107-
.055	.192-	.568	.464	.711-	.720-	1.906-	.936-	.106	.087-
1.130	.710-	1.135	.840-	.306-	.724-	.988-	.134-	.517-	1.409-
.614-	.235	.583-	2.139-	.696-	.240-	1.310	1.192	.639	.659-
.894-	1.518	1.991-	1.929-	.017-	.344	.139-	1.585-	.784-	.880-

Answers

Chapter 1

1. (a) $\{1, 2, 3, \ldots\}$
(b) $\{0, 1, 2, \ldots, 30\}$
(c) {Democrat, Republican, Independent}
(d) Interval of real numbers (e.g., $0 < x < 1000$)
(e) Points in the plane of the target
(f) $\{t: 0 \le t < \infty\}$

2. (a) 16
(b) $\{hH\}, \{hT, tH, hH\}, \{hT, hH\}$
(c) $\{hT, hH\}, \{hH\}, \{tT\}$

3. (a) $\{1, 2\}$
(b) $\{1, 2, 4, 6\}$
(c) $\{3, 4, 5\}$

5. $P(E + F) = P(E) + P(F) - P(EF)$

7. (a) 3/4 (b) 5/13 (c) 5/26 (d) 2/13

9. (a) .3 (b) .25 (c) .19 (d) .6

11. Yes (2/3 of the time the unopened box would have the prize).

Chapter 2

1. (a) 5040 (b) 24 (c) 24 (d) 3 (e) 6

2. (a) 6^4 (b) $(6)_4$ (c) 500

3. (a) 1024
(b) Probability that a combination that has occurred will occur on the next trial is about 1.18×10^{21} (sextillion = 10^{21}).

4. $(364)_5/365^5 \doteq .96$ (but in a room of 23 persons, the odds are about even—probability = .49).

5. 3/8

7. 36

9. (a) 30/91
(b) 230/273
(c) 2/35

11. 3/10

13. (a) 2×11^{14} (b) No (c) 11/12

15. (a) 56 (b) 105 (c) 2520

Chapter 3

1. (a)

0	1	2	3
1/8	3/8	3/8	1/8

(b) 3/2

2. 1/4

3. (a)

0	1	2
7/15	7/15	1/15

(b) 3/5

4. (a) 10/13; 290/169 (b) 10

5. (a)

0	1	2	4
9/24	8/24	6/24	1/24

(b) 1

7. (a) $p(x) = 1/4, x = 1, 2, 3, 4$ (*b*) 5/2, 5/4

9. (a) $9000 (b) Hard to say, even though the expectations are equal (see Problem 10).

(c) 12/5 (d) 3/5 (e)

0	1	2
7/15	7/15	1/15

11. Professional golfer: 3.95, .64; club member: 5.53, 1.37.

13. (a)

	0	1	2	3
0	0	1/30	3/30	1/30
1	1/30	8/30	6/30	0
2	3/30	6/30	0	0
3	1/30	0	0	0

Chapter 4

1. The 13 hearts, probability 1/13 each

2. 1/10
3. (a) 25/36 (b) 24/36
4. .7
5. 49/2048
7. No
9. .71
11. (a) No (b) 1/3 (c) 8/3
13. (a) .20 (b) .60 (c) 3/4 (d) 5/8 (e) No (probabilities do not factor).
15. $(.9)^8$

Chapter 5

1. (a) 9 (b) .23 (c) .30, (d) 3/64
2. 20, $\sqrt{10}$
3. 50, 5 (seems weak)
4. (a)

x	0	1	2	3	4	5
$p(x)$	1/32	5/32	10/32	10/32	5/32	1/32

(b) 1/2
5. 0.34
7. (a) 6/7 (b) $p(0) = 2/7, p(1) = 4/7, p(2) = 1/7$
9. 35/128
11. (a) 4 (b) 1.95 (c) 24
13. (a) $2q/p^2$ (b) q/p^2 (c) rq/p^2

Chapter 6

3. (a) (1/4 (b) $(x + 1)/2$ for $-1 < x < 1$ (c) Let $Y = 2U - 1$.
5. (a) 3/4 (b) 4/3, 2/9 (c) $x^2/4, 0 < x < 2$ (d) $y/4, 0 < y < 4$
7. .224
9. (a) $.1/\pi = .032$ (b) 0 (c) Integral is divergent. (d) $\frac{1}{\pi}\left(\frac{\pi}{2} + \text{Arctan } x\right)$
11. About .195
13. a—E, b—A, c—C, d—D, e—F, f—B

Chapter 7

1. (a) .303 (b) .09 (c) .082
2. .135
3. (a) .223 (b) 10
4. (a) $\left[\binom{950}{20} + \binom{950}{19}\binom{50}{1}\right]\Big/\binom{1000}{20}$ (b) $20(.05)(.95)^{19} + (.95)^{20}$ (c) .736
5. .594
7. (a) .96 (b) $.97 = \dfrac{1 - 6e^{-5}}{1 - e^{-5}}$ (c) .85
9. (a) $6e^2$ (b) $36e^{-6}$

11. (a) $30e^{-30t}(t > 0)$ (b) 2 minutes, 20 minutes
13. $1/(k\lambda)$

Chapter 8

1. (a) .0228 (b) .1587 (c) 712.7
2. .9772
3. (a) 433.4 (b) 450 ± 34, (1/2)
5. 144.7
7. .31
9. .0005
11. .003
13. .0399 vs. .0398
15. (a) Normal (b) Poisson (c) Direct (d) Not clear
17. .0297
19. .0005

Chapter 9

1. $\bar{X} = 67.28$; median = 67; midrange = 64.5; range = 45; $S = 13.52$.
2. $\bar{X} = 22.12$; $S = 5.54$
3.

$\bar{X}$	S
243.75	46.62
244.83	46.57

4. Median = 245.5; range = 241
5. 20.45
7. −7.22, 3.89
9. (a) Binomial ($n = 25, p = 1/2$) (b) Normal ($\mu = 12.5, \sigma^2 = 6.25$)
11. (a) .0002 (b) .92 (c) 49.535 (d) .465
13. .0456

Chapter 10

1. (a) 4.60 (b) 3.25 (c) .19 (d) .16
2. $n = 2020$
3. 625
4. 31.5 per cent
5. B
7. (a) 190 (b) No (the true value would usually never be discovered).
9. $236.67 < \mu < 254.71$. $233.84 < \mu < 257.54$
11. $34.0 < \mu < 42.0$
13. $.43 < p < .57$ both ways
15. 95 per cent: $40.69 < \sigma < 53.90$
17. (a) Sample is from wrong population. (b) 2 (c) No

Chapter 11

1. $Z = -1.67 < -1.645$ (barely significant at .05 level)
2. (a) $Z = -2.3 < -1.64$, reject at $\alpha = .05$. (b) Yes
3. (a) $Z = -2 < -1.96$, reject H_0 at $\alpha = .05$. (b) .023 (c) .84 (.036)
4. (a) Yes (it *is* better). (b) The evidence suggests it is not ($Z = 2$).
5. (a) $Z = 2.24$, reject H_0 at $\alpha = .05$. (b) $|T| = 1.67 < 2.78$, accept H_0 at $\alpha = .05$.
7. (a) 10 ± 7.97 (b) Reject H_0.
9. (a) $R_+ = 50$ ($R_- = 5$), reject H_0 at $\alpha = .05$. (b) 11 per cent
11. $\pi(\mu) = \Phi\left(\frac{5.08 - \mu}{.01}\right)$
13. $n = 66$, $|\bar{X} - 2| > .0121$
15. (d) .0163

Chapter 12

1. $Z' = 2.15 < 2.33$, not significant at .01
2. $Z' = 1.92 > 1.645$, reject no difference at .05. (Variances are quite different.)
3. $Z' = .95$, do not reject H_0.
4. (a) $|T| = 1.50 < 1.76$, accept H_0. (b) $R_c = 55$, accept H_0 ($52 < 55 < 84$).
5. (a) $R_B = 45$, reject H_0. (b) $|T| = 2.78 > 1.76$, reject H_0.
7. (a) 6 (b) 18 (c) .056 (d) .056 (e) It would be larger.
9. $T = 4.66$ (5 df), reject H_0.
11. (a) $T = 1.97 > 1.90$, reject equality of means at $\alpha = .05$.
    (b) $R_- = 6$, $P = .055$, accept equality of means at $\alpha = .05$.

Chapter 13

1. .91
2. .072
3. (a) .4 (b) .4
5. (a)

r	3	4	5
$f(r)$	1/15	4/15	10/15

   (b) 4/5
7. (a) 25 (b) 11,323; 0.988

Chapter 14

1. $\alpha = 0$, $\beta \leq (1/2)^{10}$, for this test: Stop and reject "n is prime" if statement is ever found true; accept if statement is found false at all 10 trials.
2. Continue sampling if $n/2 - 3.285 < \Sigma X_i < n/2 + 2.15$.
3. Continue sampling if ΣX_i is between $n/2 \pm 3.89$.
5. 12.6, under H_0; 18.6, under H_1

Chapter 15

1. $\chi^2 = .47$ (almost too good to be true)
2. (a) Reject H_0. (b) Accept $p = 1/2$ (has no bearing on binomiality).

3. $\chi^2 = 7.73 > 5.99$, reject H_0 at $\alpha = .05$.
4. $Z = 2.8$, reject at $\alpha = .01$.
5. (a) Reject H_0. (b) Accept H_0.
7. $Z = 1.66$ (not very significant)
9. Reject H_0 ($\chi^2 = 20.5 > 7.81$).
11. Not at $\alpha = .05$ ($\chi^2 = 7.34 < 9.49$)

Chapter 16

1. (a) 78 (b) 6
2. (a) $y = 1$ (b) $y = 1$ (c) 0 (d) r measures *linear* correlation.
3. (b) .783 (c) $y = .58x + 31.6$
5. $.122 + .0105x$
7. $r = .67$
9. (a) .36 (b) Predict $Y = 3.15$ (r.m.s. error .29).

Chapter 17

1. (a) $F = 1.8 < 5.14 = F_{.95}(2, 6)$ (b) 0; no ($F = \infty$)
2. $F = 2.25 = (1.50)^2$
3. $F = 17.045 > 5.78 = F_{.99}(2, 21)$, reject H_0.
5. $F = 6.54 > 5.14$, reject H_0.

Chapter 18

1. $\sum_0^2 \binom{50}{k}\binom{450}{75-k} \Big/ \binom{500}{75} \doteq .02$

2. (a)

M	0	1	2	3	4	5	6	7	8
56(OC), (1)	56	35	20	10	4	1	0	0	0
56(OC), (2)	56	56	50	40	28	16	6	0	0

(b) 5/56, 45/224 (c) 3/8, 1/14; 0, 1/2

3. $\left(1 - \frac{M}{8}\right)\left(1 - \frac{M}{7}\right)\left(1 - \frac{M}{6}\right)$, as compared with $\left(1 - \frac{M}{8}\right)^3$

5. (a) $\bar{X}$: 15 ± 3.71, R: 9.98 ± 8.62 (*b*) .0143 (*c*) .0284
7. (a) 3.70 (b) LCL = 3.47, UCL = 3.93 (c) LCL = 0, UCL = .719

Appendix A

1. $\frac{1}{1 - t/\lambda}, \frac{1}{\lambda}, \frac{2}{\lambda^2}$
2. $e^{t^2/2}$; 0, 1, 0, 3
3. $(1/6)(t + t^2 + t^3 + t^4 + t^5 + t^6)$; 21/6, 70/6
5. $e^{nt^2/2}$; 0, n
7. $(1 - t/\lambda)^{-n}$, n/λ

Index